IDEAS IN CONTEXT 52

MODELS AS MEDIATORS

Models as Mediators discusses the ways in which models function in modern science, particularly in the fields of physics and economics. Models play a variety of roles in the sciences: they are used in the development, exploration and application of theories and in measurement methods. They also provide instruments for using scientific concepts and principles to intervene in the world. The editors provide a framework which covers the construction and function of scientific models, and explore the ways in which they enable us to learn about both theories and the world. The contributors to the volume offer their own individual theoretical perspectives to cover a wide range of examples of modelling from physics, economics and chemistry. These papers provide ideal case-study material for understanding both the concepts and typical elements of modelling, using analytical approaches from the philosophy and history of science.

MARY S. MORGAN is Reader in the History of Economics at the London School of Economics and Political Science, and Professor in the History and Philosophy of Economics at the University of Amsterdam. Her first book, *The History of Econometric Ideas* (1990), has been followed by investigations into the methodology of econometrics and the method of economic modelling as well as writings on the history of twentieth-century economics.

MARGARET MORRISON is Professor of Philosophy at the University of Toronto. She writes widely in the field of history and philosophy of science and is author of *Unifying Scientific Theories* (forthcoming).

T0283701

IDEAS IN CONTEXT

Edited by QUENTIN SKINNER *(General Editor)*
LORRAINE DASTON, DOROTHY ROSS
and JAMES TULLY

The books in this series will discuss the emergence of intellectual traditions and of related new disciplines. The procedures, aims and vocabularies that were generated will be set in the context of the alternatives available within the contemporary frameworks of ideas and institutions. Through detailed studies of the evolution of such traditions, and their modification by different audiences, it is hoped that a new picture will form of the development of ideas in their concrete contexts. By this means, artificial distinctions between the history of philosophy, of the various sciences, of society and politics, and of literature may be seen to dissolve.

The series is published with the support of the Exxon Foundation.

A list of books in the series will be found at the end of the volume.

MODELS AS MEDIATORS

Perspectives on Natural and Social Science

EDITED BY

MARY S. MORGAN and MARGARET MORRISON

CAMBRIDGE
UNIVERSITY PRESS

CAMBRIDGE UNIVERSITY PRESS
Cambridge, New York, Melbourne, Madrid, Cape Town, Singapore, São Paulo

Cambridge University Press
The Edinburgh Building, Cambridge CB2 8RU, UK

Published in the United States of America by Cambridge University Press, New York

www.cambridge.org
Information on this title: www.cambridge.org/9780521650977

© Cambridge University Press 1999

This publication is in copyright. Subject to statutory exception
and to the provisions of relevant collective licensing agreements,
no reproduction of any part may take place without the written
permission of Cambridge University Press.

First published 1999

A catalogue record for this publication is available from the British Library

Library of Congress Cataloguing in Publication data
Models as mediators / edited by Mary S. Morgan and Margaret Morrison.
p.   cm. – (Ideas in context)
Includes index.
ISBN 0 521 65097 6 (hb). – ISBN 0 521 65571 4 (pb)
1. Science–Mathematical models.   2. Physics–Mathematical models.
I. Morgan, Mary S.   II. Morrison, Margaret.   III. Series.
Q158.5   .M63   1999
510′.1′5118–dc21   98-41630   CIP

ISBN 978-0-521-65097-7 hardback
ISBN 978-0-521-65571-2 paperback

Transferred to digital printing 2008

Contents

Contents

Contributors

ADRIENNE VAN DEN BOGAARD completed her thesis "Configuring the Economy: the Emergence of a Modelling Practice in The Netherlands, 1920–1955" at the University of Amsterdam and now holds a post–doctoral research fellowship at the Center for Science, Technology and Society at the University of Twente to work on a history of twentieth century technology.

MARCEL BOUMANS studied engineering and philosophy of science and now teaches at the Faculty of Economics and Econometrics at the University of Amsterdam. Author of *A Case of Limited Physics Transfer: Jan Tinbergen's Resources for Re-shaping Economics* (1992), his current work focuses on philosophical questions about modelling and measurement, and the relations between the two, in twentieth century economics.

NANCY CARTWRIGHT is Professor in both the Department of Philosophy, Logic and Scientific Method at the London School of Economics and Department of Philosophy at University of California San Diego. Author of *How the Laws of Physics Lie* (1983) and *Nature's Capacities and Their Measurement* (1989), her philosophical analyses of physics and economics are again combined in *The Dappled World: A Study of the Boundaries of Science* (forthcoming).

URSULA KLEIN is author of *Verbindung und Affinität. Die Grundlegung der Neuzeitlichen Chemie an der Wende vom 17. zum 18. Jahrundert* (1994). She is Research Director at the Max-Planck Institute for the History of Science in Berlin, with a research group doing historical and philosophical research into chemistry.

STEPHAN HARTMANN is Assistant Professor in the Department of Philosophy at the University of Konstanz following graduate studies in both physics and philosophy. He has written on the methodological and foundational problems of modern physics and published papers in particle physics and quantum optics.

R.I.G. HUGHES is Professor at the Department of Philosophy at the University of South Carolina. He is author of *The Structure and Interpretation of Quantum Mechanics* (1989) and publishes widely in the philosophy and history of physics.

MARY S. MORGAN teaches at both the London School of Economics and the University of Amsterdam. Author of *The History of Econometric Ideas* (1990), she continues to publish on econometrics along with historical studies on American economics, and on the history and methodology of twentieth century economics.

MARGARET MORRISON is Professor of Philosophy at the University of Toronto and writes widely in the fields of history and philosophy of science. She is author of *Unifying Scientific Theories* (forthcoming).

GEERT REUTEN is Associate Professor of Economics at the University of Amsterdam. Author of *Value-Form and the State* (1989, jointly with Michael Williams), and many papers in economics, he also specialises in the history of economics and its methodologies.

MAURICIO SUÁREZ teaches at the University of Bristol in the Department of Philosophy where his interests lie in the history and philosophy of science, particularly in relation to physics. His current research focuses on philosophical problems of the quantum theory of measurement and more widely on questions about scientific modelling.

Preface

This project has a long and varied history, both intellectually and geographically. Our interest in models arose in the context of independent work done by Morrison in the history and philosophy of physics and by Morgan in the historical and philosophical aspects of econometrics. Our first conversation about models occurred at the History of Science meeting in Seattle in Fall 1990 and lively discussions comparing modelling in our two fields continued through the early 1990s. Thereafter, from 1993–1996, through the generous support of Nancy Cartwright and the Centre for the Philosophy of the Natural and the Social Sciences (CPNSS) at the London School of Economics, we organised a working research group focusing on issues relevant to modelling in physics and economics. The goal of the project was not to find or construct general principles that these two disciplines shared; instead, our initial focus was empirical – to examine the ways in which modelling works in each field and then move on to investigate how the presuppositions behind those practices influenced the way working scientists looked on models as a source of knowledge. The project at the LSE was directed by Cartwright, Morgan and Morrison and consisted of a group of graduate students who attended regularly, as well as a constant stream of European and American visitors and faculty from various LSE departments such as economics, statistics and philosophy. The CPNSS not only provided support for graduate students working on the project but enabled us to organise seminars and workshop meetings on the topic of modelling.

While the LSE group work was primarily focused on physics, another group, concentrating mainly on economics, was organised by Morgan at the University of Amsterdam from 1993 onwards. Though the meetings of this group were less eclectic in subject coverage, their membership and approach spanned from the history and philosophy of economics over to social studies of science. The group involved, as core members,

Geert Reuten and Marcel Boumans (from the Faculty of Economics and Econometrics) and Adrienne van den Bogaard (from Science Dynamics). The research work of the group was supported by the Tinbergen Institute and by the Faculty of Economics and Econometrics at the University of Amsterdam. Joint meetings (both formal and informal) between members of the two groups took place on a regular basis thereby enhancing the cooperative nature of the project and extending our 'data base' of studies on various aspects of modelling.

The year 1995–96 was a critical year for the project in two respects. First, Morgan and Morrison were fellows at the Wissenschaftskolleg zu Berlin. It was there that we felt our ideas finally began to take shape during a period of intense reading and discussion. Much to our delight a number of the other fellows were also interested in modelling and we benefited enormously from conversations about philosophical and practical issues concerning the construction and function of models in fields as diverse as biology, architecture and civic planning. Our time at the Wissenschaftskolleg was invaluable for solidifying the project, giving us the time to think through together the issues of why and how models can act as mediators in science. We are extremely grateful to the director Professor Wolf Lepenies and the entire staff of the Kolleg for their generous support of our work.

The year was also crucial for the genesis of this volume. During the year we organised two small-scale workshop meetings on modelling. The first took place in Amsterdam through the generous financial support of the Royal Dutch Academy of Sciences (KNAW) and the Tinbergen Institute. This three-day meeting involved the main researchers from the LSE and the University of Amsterdam together with some other international speakers, all expert on the topic of models in science. The present volume was conceived in the enthusiastic discussions amongst members of the two research groups that followed that workshop. The year culminated with a second small two-day conference at the Wissenschaftskolleg in Berlin, supported by the Otto and Martha Fischbeck Stiftung, where once again LSE, Amsterdam and now Berlin interests were represented. These two workshops together saw early presentations of most of the essays contained in this volume.

There are many scholars who made a significant contribution to the project but whose work is not represented in the volume. Three of the LSE research students come immediately to mind – Marco Del Seta, Towfic Shomar and George Zouros. Mike Williams (from De Montfort University) contributed to our Amsterdam discussions. Special symposia resulting from our work on modelling have been presented at the Joint British and North American History of Science Society Meeting in Edinburgh in July 1996; the Philosophy of Science Meeting in Cleveland in October 1996 and the History of Science Society Meeting in San Diego in November 1997. Many papers have been given at specialist (subject-based) conferences and individual seminars and lectures. We would like to thank the audiences for their enthusiasm, questions and criticisms. We also thank many unnamed individuals who have taken the trouble to comment on individual papers, Cambridge University Press's anonymous referees who helped us improve the structure of the volume and our editors at Cambridge University Press: Patrick McCartan, Richard Fisher and Vicky Cuthill as well as Adam Swallow who saw the book through production. Our thanks go also to Linda Sampson and Kate Workman at the LSE, the secretariat at the University of Amsterdam, and Elfie Bonke at the Tinbergen Institute for their help with the research projects and with the volume. Margaret Morrison would like to thank the Social Sciences and Humanities Research Council of Canada for its continued and generous support of her research as well as the Philosophy Department at the University of Toronto who allowed her periods of time away to pursue the research. Mary Morgan would like to thank the Tinbergen Institute, the Department of Economic History at the London School of Economics, the LSE Staff Research Fund, and the British Academy for their support in terms of research time and expenses. Finally we thank Professor George Fisher for permission to use a figure from his grandfather, Irving Fisher's, work on the front cover of this book.

MARGARET MORRISON AND MARY S. MORGAN

CHAPTER I

Introduction

Margaret Morrison and Mary S. Morgan

Typically, the purpose of an introduction for an edited volume is to give the reader some idea of the main themes that will be explored in the various papers. We have chosen, instead, to take up that task in chapter 2. Here we want to simply provide a brief overview of the literature on models in the philosophy of science and economics, and to provide the audience with a sense of how issues relevant to modelling have been treated in that literature. By specifying a context and point of departure it becomes easier to see how our approach differs, both in its goals and methods, from its predecessors.

The use of models in scientific practice has a rich and varied history with their advantages and disadvantages discussed by philosophically minded scientists such as James Clerk Maxwell and his contemporaries Lord Kelvin and Sir George Francis FitzGerald. In fact, it was the use of mechanical models by British field theorists that became the focus of severe criticism by the French scientist and philosopher Pierre Duhem (1954). In Duhem's view models served only to confuse things, a theory was properly presented when cast in an orderly and logical manner using algebraic form. By contrast, mechanical models introduced disorder, allowing for diverse representations of the same phenomena. This emphasis on logical structure as a way of clarifying the nature of theories was also echoed in the early twentieth century by proponents of logical empiricism. This is not to suggest that their project was the same as Duhem's; we draw the comparison only as a way of highlighting the importance of logical form in philosophical appraisals of theories. The emphasis on logic is also significant because it was in this context that models came to be seen as an essential part of theory structure in twentieth-century philosophy of science.

It is perhaps not surprising that much of the early literature on theory structure and models in philosophy of science takes physics as its starting point. Physical theories are not only highly mathematical but they are certainly more easily cast into an axiomatic form than theories in other sciences. According to the logical empiricist account of theories, sometimes referred to as the received view or the syntactic view, the proper characterisation of a scientific theory consists of an axiomatisation in first-order logic. The axioms were formulations of laws that specified relationships between theoretical terms such as electron, charge, etc. The language of the theory was divided into two parts, the observation terms that described observable macroscopic objects or processes and theoretical terms whose meaning was given in terms of their observational consequences. In other words, the meaning of 'electron' could be explicated by the observational terms 'track in a cloud chamber'. Any theoretical terms for which there were no corresponding observational consequences were considered meaningless. The theoretical terms were identified with their observational counterparts by means of correspondence rules, rules that specified admissible experimental procedures for applying theories to phenomena. For example, mass could be defined as the result of performing certain kinds of measurements. One can see then why this account of theory structure was termed the syntactic view; the theory itself was explicated in terms of its logical form with the meanings or semantics given by an additional set of definitions, the correspondence rules. That is to say, although the theory consisted of a set of sentences expressed in a particular language, the axioms were syntactically describable. Hence, without correspondence rules one could think of the theory itself as uninterpreted.

An obvious difficulty with this method was that one could usually specify more than one procedure or operation for attributing meaning to a theoretical term. Moreover, in some cases the meanings could not be fully captured by correspondence rules; hence the rules were considered only partial interpretations for these terms.[1] A possible solution to these problems was to provide a semantics for a theory (T) by specifying a model (M) for the theory, that is, an interpretation on which all the axioms of the theory are true. As noted above, this notion of a model comes from the field of mathematical logic and, some argue, has little to do with the way working scientists use models. Recall, however, that the goal of the logical empiricist programme was a clarification of the nature of theories; and to the extent that that remains a project worthy of pursuit, one might want to retain the emphasis on logic as a means to that end.

---

[1] A number of other problems, such as how to define dispositional theoretical terms, also plagued this approach. For an extensive account of the growth, problems with, and decline of the received view, see Suppe (1977).

But the significance of the move to models as a way of characterising theories involves replacing the syntactic formulation of the theory with the theory's models. Instead of formalising the theory in first-order logic, one defines the intended class of models for a particular theory. This view still allows for axiomatisation provided one can state a set of axioms such that the models of these axioms are exactly the models in the defined class. One could still formulate the axioms in a first-order language (predicate calculus) in the manner of the syntactic view; the difference however is that it is the models (rather than correspondence rules) that provide the interpretation for the axioms (or theory). Presenting a theory by identifying a class of structures as its models means that the language in which the theory is expressed is no longer of primary concern. One can describe the models in a variety of different languages, none of which is the basic or unique expression of the theory. This approach became known as the semantic view of theories (see Suppes (1961) and (1967); Suppe (1977); van Fraassen (1980) and Giere (1988)) where 'semantic' refers to the fact that the model provides a realisation in which the theory is satisfied. That is, the notion of a model is defined in terms of truth. In other words the claims made by the theory are true in the model and in order for M to be a model this condition must hold.

But what exactly are these models on the semantic view? According to Alfred Tarski (1936), a famous twentieth-century logician, a model is a non-linguistic entity. It could, for example, be a set theoretical entity consisting of an ordered tuple of objects, relations and operations on these objects (see Suppes (1961)). On this account we can define a model for the axioms of classical particle mechanics as an ordered quintuple containing the following primitives $\mathscr{P} = \langle P, T, s, m, f \rangle$ where P is a set of particles, T is an interval or real numbers corresponding to elapsed times, s is a position function defined on the Cartesian product of the set of particles and the time interval, m is a mass function and f is a force function defined on the Cartesian product of the set of particles, the time interval and the positive integers (the latter enter as a way of naming the forces). Suppes claims that this set theoretical model can be related to what we normally take to be a physical model by simply interpreting the set of particles to be, for instance, the set of planetary bodies. The idea is that the abstract set-theoretical model will contain a basic set consisting of objects ordinarily thought to constitute a physical model. The advantage of the logicians' sense of model is that it supposedly renders a more precise and clear account of theory structure, experimental design and data analysis (see Suppes (1962)).

Other proponents of the semantic view including van Fraassen and Giere have slightly different formulations yet both subscribe to the idea that models are non-linguistic entities. Van Fraassen's version incorporates the notion of a state space. If we think of a system consisting of physical entities developing in time, each of which has a space of possible states, then we can define a model as representing one of these possibilities. The models of the system will be united by a common state space with each model having a domain of objects plus a history function that assigns to each object a trajectory in that space. A physical theory will have a number of state spaces each of which contains a cluster of models. For example, the laws of motion in classical particle mechanics are laws of succession. These laws select the physically possible trajectories in the state space; in other words only the trajectories in the state space that satisfy the equations describing the laws of motion will be physically possible. Each of these physical possibilities is represented by a model.[2] We assess a theory as being empirically adequate if the empirical structures in the world (those that are actual and observable) can be embedded in some model of the theory, where the relationship between the model and a real system is one of isomorphism.

Giere's account also emphasises the non-linguistic character of models but construes them in slightly less abstract terms. On his account, the idealised systems described in mechanics texts, like the simple harmonic oscillator, is a model. As such the model perfectly satisfies the equations of motion for the oscillator in the way that the logicians' model satisfies the axioms of a theory. Models come in varying degrees of abstraction, for example, the simple harmonic oscillator has only a linear restoring force while the damped oscillator incorporates both a restoring and a damping force. These models function as representations in 'one of the more general senses now current in cognitive psychology' (Giere 1988, 80). The relationship between the model and real systems is fleshed out in terms of similarity relations expressed by theoretical hypotheses of the form: 'model M is similar to system S in certain respects and degrees'. On this view a theory is not a well-defined entity since there are no necessary nor sufficient conditions determining which models or hypotheses belong to a particular theory. For example, the models for classical mechanics do not comprise a well-defined group because there are no specific conditions for what constitutes an admissible force function. Instead we classify the models on the basis of their

---

[2] Suppe (1977) has also developed an account of the semantic view that is similar to van Fraassen's.

family resemblance to models already in the theory: a judgement made in a pragmatic way by the scientists using the models.

Two of the things that distance Giere from van Fraassen and Suppes respectively are (1) his reluctance to accept isomorphism as the way to characterise the relation between the model and a real system, and (2) his criticism of the axiomatic approach to theory structure. Not only does Giere deny that most theories have the kind of tightly knit structure that allows models to be generated in an axiomatic way, but he also maintains that the axiomatic account fails even to capture the correct structure of classical mechanics. General laws of physics like Newton's laws and the Schrodinger equation are not descriptions of real systems but rather part of the characterisation of models, which can in turn represent different kinds of real systems. But a law such as $F = ma$ does not by itself define a model of anything; in addition we need specific force functions, boundary conditions, approximations etc. Only when these conditions are added can a model be compared with a real system.

We can see then how Giere's account of the semantic view focuses on what many would call 'physical models' as opposed to the more abstract presentation characteristic of the set theoretic approach. But this desire to link philosophical accounts of models with more straightforward scientific usage is not new; it can be traced to the work of N. R. Campbell (1920) but was perhaps most widely discussed by Mary Hesse (1966).[3] The physical model is taken to represent, in some way, the behaviour and structure of a physical system; that is, the model is structurally similar to what it models. If we think of the Bohr atom as modelled by a system of billiard balls moving in orbits around one ball, with some balls jumping into different orbits at different times, then as Hesse puts it, we can think of the relation between the model and the real system as displaying different kinds of analogies. There is a positive analogy where the atom is known to be analogous to the system of billiard balls, a negative analogy where they are disanalogous and neutral where the similarity relation is not known. The kinds of models that fulfil this characterisation can be scale models like a model airplane or a mathematical model of a theory's formalism. An example of the latter is the use of the Langevin equations to model quantum statistical relations in the behaviour of certain kinds of laser phenomena. In this case we model the Schrodinger equation in a specific kind

---

[3] There are also other noteworthy accounts of models such as those of Max Black (1962) and R. B. Braithwaite (1953, 1954).

of way depending on the type of phenomena we are interested in. The point is, these physical models can be constructed in a variety of ways; some may be visualisable, either in terms of their mathematical structure or by virtue of their descriptive detail. In all cases they are thought to be integral components of theories; they suggest hypotheses, aid in the construction of theories and are a source of both explanatory and predictive power.

The tradition of philosophical commentary on models in economic science is relatively more recent, for despite isolated examples in previous centuries, economic modelling emerged in the 1930s and only became a standard method in the post-1950 period. In practical terms, economists recognise two domains of modelling: one associated with building mathematical models and the activity of theorising; the other concerned with statistical modelling and empirical work.

Given that mathematical economists tend to portray their modelling activity within the domain of economic theory, it is perhaps no surprise that philosophical commentaries about mathematical models in economics have followed the traditional thinking about models described above. For example, Koopmans' (1957) account can be associated with the axiomatic tradition, while Hausman's (1992) position is in many ways close to Giere's semantic account, and McCloskey's (1990) view of models as metaphors can surely be related to Hesse's analogical account. Of these, both Koopmans and Hausman suggest that models have a particular role to play in economic science. Koopmans sees economics beginning from abstract theory (as for example the formulation of consumer choice theory within a utility maximising framework) and 'progressing' through a sequence of ever more realistic mathematical models;[4] whereas for Hausman, models are a tool to help form and explore theoretical concepts.

In contrast to these mathematical concerns, discussions about empirical models in economics have drawn on the foundations of statistics and probability theory. The most important treatment in this tradition is the classic thesis by Haavelmo (1944) in which econometric models are defined in terms of the probability approach, and their function is to act as the bridge between economic theories and empirical economics. Given that economists typically face a situation where data are not generated from controlled experiments, Haavelmo proposed using models

---

[4] Recent literature on idealisation by philosophers of economics has also supposed that models might be thought of as the key device by which abstract theories are applied to real systems and the real world simplified for theoretical description. See Hamminga and De Marchi (1994).

in econometrics as the best means to formulate and to solve a series of correspondence problems between the domains of mathematical theory and statistical data.

The account of Gibbard and Varian (1978) also sees models as bridging a gap, but this time between mathematical theory and the evidence obtained from casual observation of the economy. They view models as caricatures of real systems, in as much as the descriptions provided by mathematical models in economics often do not seek to approximate, but rather to distort the features of the real world (as for example in the case of the overlapping generations model). Whereas approximation models aim to capture the main characteristics of the problem being considered and omit minor details, caricature models take one (or perhaps more) of those main characteristics and distorts that feature into an extreme case. They claim that this distortion, though clearly false as a description, may illuminate certain relevant aspects of the world. Thus even small mathematical models which are manifestly unrealistic can help us to understand the world. Although they present their account within the tradition of logical positivism described above, it is better viewed as a practise-based account of economic modelling in the more modern philosophy of science tradition seen in the work of Cartwright (1983), Hacking (1983) and others. Their treatments, emphasising the physical characteristics of models (in the sense noted above), attempt to address questions concerning the interplay among theories, models, mathematical structures and aspects of creative imagination that has come to constitute the practice we call modelling.

Despite this rather rich heritage there remains a significant lacuna in the understanding of exactly how models in fact function to give us information about the world. The semantic view claims that models, rather than theory, occupy centre stage, yet most if not all of the models discussed within that framework fall under the category 'models of theory' or 'theoretical models' as in Giere's harmonic oscillator or Hausman's account of the overlapping generations model. Even data models are seen to be determined, in part, by theories of data analysis (as in Haavelmo's account) in the same way that models of an experiment are linked to theories of experimental design. In that sense, literature on scientific practice still characterises the model as a subsidiary to some background theory that is explicated or applied via the model. Other examples of the tendency to downplay models in favour of theory include the more mundane references to models as tentative hypotheses; we have all heard the phrase 'it's just a model at this stage', implying that the hypothesis has not yet acquired the level of consensus reserved for theory. The result is that we have very little sense of what a model is in itself and how it is able to function in an autonomous way.

Yet clearly, autonomy is an important feature of models; they provide the foundation for a variety of decision making across contexts as diverse as economics, technological design and architecture. Viewing models strictly in terms of their relationship to theory draws our attention away from the processes of constructing models and manipulating them, both of which are crucial in gaining information about the world, theories and the model itself. However, in addition to emphasising the autonomy of models as entities distinct from theory we must also be mindful of the ways that models and theory do interact. It is the attempt to understand the dynamics of modelling and its impact on the broader context of scientific practice that motivates much of the work presented in this volume. In our next chapter, we provide a general framework for understanding how models can act as mediators and illustrate the elements of our framework by drawing on the contributions to this volume and on many other examples of modelling. Our goal is to clarify at least some of the ways in which models can act as autonomous mediators in the sciences and to uncover the means by which they function as a source of knowledge.

REFERENCES

Black, Max (1962). *Models and Metaphors*. Ithaca, NY: Cornell University Press.
Braithwaite, R. B. (1953). *Scientific Explanation*. New York: Harper Torchbooks.
    (1954). 'The Nature of Theoretical Concepts and the Role of Models in an Advanced Science', *Revue International de Philosophie*, 8, 114–31.
Campbell, N. R. (1920). *Physics: The Elements*. Cambridge: Cambridge University Press.
Cartwright, N. (1983). *How the Laws of Physics Lie*. Oxford: Clarendon Press.
Duhem, P. (1954). *The Aim and Structure of Physical Theory*. Princeton: Princeton University Press.
Fraassen, B. van (1980). *The Scientific Image*. Oxford: Clarendon Press.
Gibbard A. and H. R. Varian (1978). 'Economic Models', *The Journal of Philosophy*, 75, 664–77.
Giere, R. N. (1988). *Explaining Science: A Cognitive Approach*. Chicago: University of Chicago Press.
Haavelmo, T. M. (1944). 'The Probability Approach in Econometrics', *Econometrica*, 12 (Supplement), i–viii, 1–118.
Hacking, I. (1983). *Representing and Intervening*. Cambridge: Cambridge University Press.

Hamminga, B. and De Marchi (1994). *Idealization in Economics*. Amsterdam: Rodopi.

Hausman, D. (1992). *The Inexact and Separate Science of Economics*. Cambridge: Cambridge University Press.

Hesse, M. (1966). *Models and Analogies in Science*. Notre Dame, IN: University of Indiana Press.

Koopmans, T. J. (1957). *Three Essays on the State of Economic Science*. New York: McGraw Hill.

McCloskey, D. N. (1990). *If You're So Smart*. Chicago: University of Chicago Press.

Suppe, F. (1977). *The Structure of Scientific Theories*. Chicago: University of Illinois Press.

Suppes, P. (1961). 'A Comparison of the Meaning and Use of Models in the Mathematical and Empirical Sciences', pp. 163–77 in H. Freudenthal (ed.), *The Concept and Role of the Model in Mathematics and Natural and Social Sciences*. Dordrecht: Reidel.

(1962). 'Models of Data', pp. 252–61 in E. Nagel, P. Suppes and A. Tarski (eds.), *Logic Methodology and Philosophy of Science: Proceedings of the 1960 International Congress*. Stanford, CA: Stanford University Press.

(1967). 'What is a Scientific Theory?' pp. 55–67 in S. Morgenbesser (ed.), *Philosophy of Science Today*. New York: Basic Books.

Tarski, A. (1936). 'Der Wahrheitsbegriff in den formalisierten Sprachen', *Studia Philosophica*, 1, 261–405. Trans. in (1956). *Logic, Semantics, Metamathematics* Oxford: Clarendon Press.

CHAPTER 2

Models as mediating instruments

Margaret Morrison and Mary S. Morgan

Models are one of the critical instruments of modern science. We know that models function in a variety of different ways within the sciences to help us to learn not only about theories but also about the world. So far, however, there seems to be no systematic account of *how* they operate in both of these domains. The semantic view as discussed in the previous chapter does provide some analysis of the relationship between models and theories and the importance of models in scientific practice; but, we feel there is much more to be said concerning the dynamics involved in model construction, function and use. One of the points we want to stress is that when one looks at examples of the different ways that models function, we see that they occupy an autonomous role in scientific work. In this chapter we want to outline, using examples from both the chapters in this volume and elsewhere, an account of models as *autonomous agents*, and to show how they function as *instruments* of investigation. We believe there is a significant connection between the autonomy of models and their ability to function as instruments. It is precisely because models are partially independent of both theories and the world that they have this autonomous component and so can be used as instruments of exploration in both domains.

In order to make good our claim, we need to raise and answer a number of questions about models. We outline the important questions here before going on to provide detailed answers. These questions cover four basic elements in our account of models, namely how they are constructed, how they function, what they represent and how we learn from them.

CONSTRUCTION   What gives models their autonomy? Part of the answer lies in their construction. It is common to think that models can be derived entirely from theory or from data. However, if we look closely at the way models are constructed we can begin to see the sources of their independence. It is because they are neither one thing nor the other, neither just theory nor data, but typically involve some of both (and often additional 'outside' elements), that they can mediate between theory and the world. In addressing these issues we need to isolate the nature of this partial independence and determine why it is more useful than full independence or full dependence.

FUNCTIONING   What does it mean for a model to function autonomously? Here we explore the various tasks for which models can be used. We claim that what it means for a model to function autonomously is to function like a tool or instrument. Instruments come in a variety of forms and fulfil many different functions. By its nature, an instrument or tool is independent of the thing it operates on, but it connects with it in some way. Although a hammer is separate from both the nail and the wall, it is designed to fulfil the task of connecting the nail to the wall. So too with models. They function as tools or instruments and are independent of, but mediate between things; and like tools, can often be used for many different tasks.

REPRESENTING   Why can we learn about the world and about theories from using models as instruments? To answer this we need to know what a model consists of. More specifically, we must distinguish between instruments which can be used in a purely instrumental way to effect something and instruments which can also be used as investigative devices for learning something. We do not learn much from the hammer. But other sorts of tools (perhaps just more sophisticated ones) can help us learn things. The thermometer is an instrument of investigation: it is physically independent of a saucepan of jam, but it can be placed into the boiling jam to tell us its temperature. Scientific models work like these kinds of investigative instruments – but how? The critical difference between a simple tool, and a tool of investigation is that the latter involves some form of representation: models typically represent either some aspect of the world, or some aspect of our theories about the world, or both at once. Hence the model's representative power allows it to function not just instrumentally, but to teach us something about the thing it represents.

LEARNING   Although we have isolated representation as the mechanism that enables us to learn from models we still need to know *how* this learning takes place and we need to know what else is involved in a model functioning as a mediating instrument. Part of the answer comes from seeing how models are used in scientific practice. We do not learn much from looking at a model – we learn more from building the model and from manipulating it. Just as one needs to use or observe the use of a hammer in order to really understand its function, similarly, models have to be used before they will give up their secrets. In this sense, they have the quality of a technology – the power of the model only becomes apparent in the context of its use. Models function not just as a means of intervention, but also as a means of representation. It is when we manipulate the model that these combined features enable us to learn how and why our interventions work.

Our goal then is to flesh out these categories by showing how the different essays in the volume can teach us something about each of the categories. Although we want to argue for some general claims about models – their autonomy and role as mediating instruments, we do not see ourselves as providing a 'theory' of models. The latter would provide well-defined criteria for identifying something as a model and differentiating models from theories. In some cases the distinction between models and theories is relatively straightforward; theories consist of general principles that govern the behaviour of large groups of phenomena; models are usually more circumscribed and very often several models will be required to apply these general principles to a number of different cases. But, before one can even begin to identify criteria for determining what comprises a model we need much more information about their place in practice. The framework we have provided will, we hope, help to yield that information.

2.1 CONSTRUCTION

2.1.1 Independence in construction

When we look for accounts of how to construct models in scientific texts we find very little on offer. There appear to be no general rules for model construction in the way that we can find detailed guidance on principles of experimental design or on methods of measurement. Some might argue that it is because modelling is a tacit skill, and has to be learnt not taught. Model building surely does involve a large amount of craft skill, but then so does designing experiments and any other part of scientific practice. This omission in scientific texts may also point to the creative element involved in model building, it is, some argue, not only a craft but also an art, and thus not susceptible to rules. We find a similar lack of advice available in philosophy of science texts. We are given definitions of models, but remarkably few accounts of how they are constructed. Two accounts which do pay attention to construction, and to which we refer in this part of our discussion, are the account of models as analogies by Mary Hesse (1966) and the simulacrum account of models by Nancy Cartwright (1983).

Given the lack of generally agreed upon rules for model building, let us begin with the accounts that emerge from this volume of essays. We have an explicit account of model construction by Marcel Boumans who argues that models are built by a process of choosing and integrating a set of items which are considered relevant for a particular task. In order to build a mathematical model of the business cycle, the economists that he studied typically began by bringing together some bits of theories, some bits of empirical evidence, a mathematical formalism and a metaphor which guided the way the model was conceived and put together. These disparate elements were integrated into a formal (mathematically expressed) system taken to provide the key relationships between a number of variables. The integration required not only the translation of the disparate elements into something of the same form (bits of mathematics), but also that they be fitted together in such a way that they could provide a solution equation which represents the path of the business cycle.

Boumans' account appears to be consistent with Cartwright's simulacrum account, although in her description, models involve a rather more straightforward marriage of theory and phenomena. She suggests that models are made by fitting together prepared descriptions from the empirical domain with a mathematical representation coming from the theory (Cartwright 1983). In Boumans' description of the messy, but probably normal, scientific work of model building, we find not only the presence of elements other than theory and phenomena, but also the more significant claim that theory does not even determine the model form. Hence, in his cases, the method of model construction is carried out in a way which is to a large extent independent of theory. A similar situation arises in Mauricio Suárez's discussion of the London brothers' model of superconductivity. They were able to construct an equation for the superconducting current that accounted for an effect that could not be accommodated in the existing theory. Most importantly, the London equation was not derived from electromagnetic theory, nor was it arrived at by simply adjusting parameters in the theory governing superconductors. Instead, the new equation emerged as a result of a completely new conceptualisation of superconductivity that was supplied by the model. So, not only was the model constructed without the aid of theory, but it became the impetus for a new theoretical understanding of the phenomena.

The lesson we want to draw from these accounts is that models, by virtue of their construction, embody an element of independence from both theory and data (or phenomena): it is because they are made up from a *mixture* of elements, including those from outside the original domain of investigation, that they maintain this partially independent status.

But such partial independence arises even in models which largely depend on and are derived from bits of theories – those with almost no empirical elements built in. In Stephan Hartmann's example of the MIT-Bag Model of quark confinement, the choice of bits which went into the model is motivated in part by a story of how quarks can exist in nature. The story begins from the empirical end: that free quarks were not observed experimentally. This led physicists to hypothesise that quarks were confined – but how, for confinement does not follow from (cannot be derived from) anything in the theory of quantum chromodynamics that supposedly governs the behaviour of quarks. Instead, various models, such as the MIT-Bag Model, were proposed to account for confinement. When we look at the way these models are constructed, it appears that the stories not only help to legitimise the model after its construction, but also play a role in both selecting and putting together the bits of physical theories involved. Modelling confinement in terms of the bag required modelling what happened inside the bag, outside the bag and, eventually, on the surface of the bag itself.

At first sight, the pendulum model used for measuring the gravitational force, described in Margaret Morrison's account, also seems to have been entirely derived from theory without other elements involved. It differs importantly from Hartmann's case because there is a very close relationship between one specific theory and the model. But there is also a strong empirical element. We want to use the pendulum to measure gravitational force and in that sense the process starts not with a theory, but with a real pendulum. But, we also need a highly detailed theoretical account of how it works in all its physical respects. Newtonian mechanics provides all the necessary pieces for describing the pendulum's motion but the laws of the theory cannot be applied directly to the object. The laws describe various kinds of motion in idealised circumstances, but we still need something separate that allows us to apply these laws to concrete objects. The model of the pendulum plays this role; it provides a more or less idealised context where theory is applied.[1] From an initially idealised model we can then build in the appropriate corrections so that the model becomes an increasingly realistic representation of the real pendulum.

It is equally the case that models which look at first sight to be constructed purely from data often involve several other elements. Adrienne van den Bogaard makes a compelling case for regarding the business barometer as a model, and it is easy to see that such a 'barometer' could not be constructed without imposing a particular structure onto the raw data. Cycles are not just there to be seen, even if the data are mapped into a simple graph with no other adjustments to them. Just as the bag story told MIT physicists what bits were needed and how to fit them together, so a particular conception of economic life (that it consists of certain overlapping, but different time-length, cycles of activity) was required to isolate, capture and then combine the cyclical elements necessary for the barometer model. The business barometer had to be constructed out of concepts and data, just as a real barometer requires some theories to interpret its operations, some hardware, and calibration from past measuring devices.

We claim that these examples are not the exception but the rule. In other words, as Marcel Boumans suggests, models are typically constructed by fitting together a set of bits which come from disparate sources. The examples of modelling we mentioned involve elements of theories and empirical evidence, as well as stories and objects which could form the basis for modelling decisions. Even in cases where it initially seemed that the models were derived purely from theory or were simply data models, it became clear that there were other elements involved in the models' construction. It is the presence of these other elements in their construction that establish scientific models as separate from, and partially independent of, both theory and data.

But even without the process of integrating disparate elements, models typically still display a degree of independence. For example, in cases where models supposedly remain true to their theories (and/or to the world) we often see a violation of basic theoretical assumptions.

---

[1] Although the model pendulum is an example of a notional object, there are many cases where real objects have functioned as models. Everyone is familiar with the object models of atoms, balls connected with rods, which are widely used in chemistry, appearing both in the school room and in the hands of Nobel prize winners. Most people are also familiar with the early object models of the planetary system. Such object models have played an important role in scientific research from an early time, and were particularly important to nineteenth-century physicists as we shall see later.

Geert Reuten's account of Marx's Schema of Reproduction, found in volume II of *Capital*, shows how various modelling decisions created a structure which was partly independent of the general requirements laid down in Marx's verbal theories. On the one hand, Marx had to deliberately set aside key elements of his theory (the crisis, or cycle, element) in order to fix the model to demonstrate the transition process from one stable growth path to another. On the other hand, it seems that Marx became a prisoner to certain mathematical conditions implied by his early cases which he then carried through in constructing later versions of the model. Even Margaret Morrison's example of the pendulum model, one which is supposed to be derived entirely from theory and to accurately represent the real pendulum, turns out to rely on a series of modelling decisions which simplify both the mathematics and the physics of the pendulum.

In other words, theory does not provide us with an algorithm from which the model is constructed and by which all modelling decisions are determined. As a matter of practice, modelling always involves certain simplifications and approximations which have to be decided independently of the theoretical requirements[2] or of data conditions.

Another way of characterising the construction of models is through the use of analogies. For example, in the work of Mary Hesse (1966), we find a creative role for neutral analogical features in the construction of models. We can easily reinterpret her account by viewing the neutral features as the means by which something independent and separate is introduced into the model, something which was not derived from our existing knowledge or theoretical structure. This account too needs extending, for in practice it is not only the neutral features, but also the negative features which come in from outside. Mary Morgan (1997a, and this volume) provides two examples from the work of Irving Fisher in which these negative analogical features play a role in the construction of models. In one of the cases, she describes the use of the mechanical balance as an analogical model for the equation of exchange between money and goods in the economy. The balance provides not only neutral features, but also negative features, which are incorporated into the economic model, providing it with independent elements which certainly do not appear in the original equation of exchange. Her second example, a model of how

---

[2] For a discussion of when modelling decisions are independent of theory, see M. Suárez's paper in this volume.

economic laws interact with the institutional arrangements for money in the economy, involves a set of 'vessels' supposed to contain the world stores of gold and silver bullion. These vessels constitute negative analogical features, being neither part of the monetary theories of the time nor the available knowledge in the economic world. Both models depend on the addition of these negative analogical features and enabled Fisher to develop theoretical results and explain empirical findings for the monetary system.

2.1.2 Independence and the power to mediate

There is no *logical* reason why models should be constructed to have these qualities of partial independence. But, in practice they are. And, if models are to play an autonomous role allowing them to mediate between our theories and the world, and allowing us to learn about one or the other, they *require* such partial independence. It has been conventional for philosophers of science to characterise scientific methodology in terms of theories and data. Full dependence of a theory on data (and vice versa) is regarded as unhelpful, for how can we legitimately use our data to test our theory if it is not independent? This is the basis of the requirement for independence of observation from theory. In practice however, theory ladenness of observation is allowed provided that the observations are at least neutral with respect to the theory under test.

We can easily extend this argument about theories and data to apply to models: we can only expect to use models to learn about our theories or our world if there is at least partial independence of the model from both. But models must also connect in some way with the theory or the data from the world otherwise we can say nothing about those domains. The situation seems not unlike the case of correlations. You learn little from a perfect correlation between two things, for the two sets of data must share the same variations. Similarly, you learn little from a correlation of zero, for the two data sets share nothing in common. But any correlation between these two end-values tell you both the degree of association and provides the starting point for learning more.

The crucial feature of partial independence is that models are *not* situated in the middle of an hierarchical structure between theory and the world. Because models typically include other elements, and model building proceeds in part independently of theory and data, we construe models as being outside the theory–world axis. It is this feature which enables them to mediate effectively between the two.

Before we can understand how it is that models help us to learn new things via this mediating role, we need to understand how it is that models function autonomously and more about how they are connected with theories and the world.

2.2 FUNCTION

Because model construction proceeds in part independently of theory and data, models can have a life of their own and occupy a unique place in the production of scientific knowledge. Part of what it means to situate models in this way involves giving an account of what they do – how it is that they can function autonomously and what advantages that autonomy provides in investigating both theories and the world. One of our principle claims is that the autonomy of models allows us to characterise them as instruments. And, just as there are many different kinds of instruments, models can function as instruments in a variety of ways.

2.2.1 Models in theory construction and exploration

One of the most obvious uses of models is to aid in theory construction.[3] Just as we use tools as instruments to build things, we use models as instruments to build theory. This point is nicely illustrated in Ursula Klein's discussion of how chemical formulas, functioning as models or paper tools, altered theory construction in organic chemistry. She shows how in 1835 Dumas used his formula equation to introduce the notion of substitution, something he would later develop into a new theory about the unitary structure of organic compounds. This notion of substitution is an example of the construction of a chemical conception that was constrained by formulas and formula equations. Acting as

---

[3] This of course raises the sometimes problematic issue of distinguishing between a model and a theory; at what point does the model become subsumed by, or attain the status of, a theory. The rough and ready distinction followed by scientists is usually to reserve the word model for an account of a process that is less certain or incomplete in important respects. Then as the model is able to account for more phenomena and has survived extensive testing it evolves into a theory. A good example is the 'standard model' in elementary particle physics. It accounts for particle interactions and provides extremely accurate predictions for phenomena governed by the weak, strong and electromagnetic forces. Many physicists think of the standard model as a theory; even though it has several free parameters its remarkable success has alleviated doubts about its fundamental assumptions.

models, these chemical formulas were not only the referents of the new conception but also the tools for producing it. Through these models the conception of a substitution linked, for the first time, the theory of proportion to the notions of compound and reaction. We see then how the formulas (models) served as the basis for developing the concept of a substitution which in turn enabled nineteenth-century chemists to provide a theoretical representation for empirical knowledge of organic transformations.

What we want to draw attention to however is a much wider characterisation of the function of models in relation to theory. Models are often used as instruments for exploring or experimenting on a theory that is already in place. There are several ways in which this can occur; for instance, we can use a model to correct a theory. Sir George Francis FitzGerald, a nineteenth-century British physicist, built mechanical models of the aether out of pulleys and rubber bands and used these models to correct Maxwell's electromagnetic theory. The models were thought to represent particular mechanical processes that must occur in the aether in order for a field theoretic account of electrodynamics to be possible. When processes in the model were not found in the theory, the latter was used as the basis of correction for the former.

A slightly different use is found in Geert Reuten's analysis of how Marx used his model to explore certain characteristics of his theory of the capitalist economy. In particular, Marx's modelling enabled him to see which requirements for balanced growth in the economy had to hold and which (such as price changes) could be safely neglected. Marx then developed a sequence of such models to investigate the characteristics required for successful transition from simple reproduction (no growth) to expanded reproduction (growth). In doing so he revealed the now well-known 'knife-edge' feature of the growth path inherent in such models.

But we also need models as instruments for exploring processes for which our theories do not give good accounts. Stephan Hartmann's discussion of the MIT-Bag Model shows how the model provided an explanation of how quark confinement might be physically realised. Confinement seemed to be a necessary hypothesis given experimental results yet theory was unable to explain how it was possible.

In other cases, models are used to explore the implications of theories in concrete situations. This is one way to understand the role of the twentieth-century conception of 'rational economic man'. This idealised and highly simplified characterisation of real economic behaviour has been widely used in economists' microeconomic theories as a tool to explore the theoretical implications of the most single-minded econom-ising behaviour (see Morgan 1997b). More recently this 'model man' has been used as a device for benchmarking the results from experimental economics. This led to an explosion of theories accounting for the diver-gence between the observed behaviour of real people in experimental situations and that predicted from the theory of such a model man in the same situation.

Yet another way of using models as instruments focuses not on explor-ing how theories work in specific contexts but rather on applying theo-ries that are otherwise inapplicable. Nancy Cartwright's contribution to the volume provides an extended discussion of how interpretative models are used in the application of abstract concepts like force func-tions and the quantum Hamiltonian. She shows how the successful use of theory depends on being able to apply these abstract notions not to just any situation but only to those that can be made to fit the model. This fit is *carried out* via the bridge principles of the theory, they tell us what concrete form abstract concepts can take; but these concepts can only be applied when their interpretative models fit. It is in this sense that the models are crucial for applying theory – they limit the domain of abstract concepts. Her discussion of superconductivity illustrates the cooperative effort among models, fundamental theory, empirical knowl-edge and an element of guesswork.

In other cases, we can find a model functioning directly as an instru-ment for experiment. Such usage was prominent in nineteenth-century physics and chemistry. The mechanical aether models of Lord Kelvin and FitzGerald that we mentioned above were seen as replacements for actual experiments on the aether. The models provided a mechanical structure that embodied certain kinds of mechanical properties, connec-tions and processes that were supposedly necessary for the propagation of electromagnetic waves. The successful manipulation of the models was seen as equivalent to experimental evidence for the existence of these properties in the aether. That is, manipulating the model was tan-tamount to manipulating the aether and, in that sense, the model func-tioned as both the instrument and object of experimentation.

Similarly, Ursula Klein shows us how chemical formulas were applied to represent and structure experiments – experiments that were paradigmatic in the emerging sub-discipline of organic chemistry. Using the formulas, Dumas could calculate how much chlorine was needed for the production of chloral and how much hydrochloric acid was simultaneously produced. Due to these calculational powers, the formulas became surrogates for the concrete measurement of substances involved in chemical transformations. They functioned as models capable of singling out pathways of reactions in new situations. Because the formulas could link symbols with numbers it was possible to balance the ingredients and products of a chemical transformation – a crucial feature of their role as instruments for experiments.

2.2.2 Models and measurement

An important, but overlooked function of models is the various but specific ways in which they relate to measurement.[4] Not only are models instruments that can both structure and display measuring practices but the models themselves can function directly as measuring instruments. What is involved in structuring or displaying a measurement and how does the model function as an instrument to perform such a task? Mary Morgan's analysis of Irving Fisher's work on models illustrates just how this works. The mechanical balance, as used by merchants for weighing and measuring exchange values of goods, provided Fisher with an illustration of the equation of exchange for the whole economy. What is interesting about Fisher's model is that he did not actually use the balance model directly as a measuring instrument, but he did use it as an instrument to display measurements that he had made and calibrated. He then used this calibrated display to draw inferences about the relative changes that had taken place in the trade and money supply in the American economy over the previous eighteen years. In a more subtle way, he also used the model of the mechanical balance to help him conceptualise certain thorny measurement problems in index number theory.

An example where it is considerably more difficult to disentangle the measurement functions from model development is the case of national income accounts and macroeconometric models discussed by Adrienne van den Bogaard. She shows how intimately the two were connected. The model was constructed from theories which involved a certain aggregate conception of the economy. This required the reconception of economic measurements away from business-cycle data and toward national income measures, thereby providing the

---

[4] We do not include here examples of using models as calculation devices – these are discussed in section 2.3.2, when we consider simulations.

model with its empirical base. At the same time, the particular kinds of measurements which were taken imposed certain constraints on the way the model was built and used: for example, the accounting nature of national income data requires certain identities to hold in the model. Models could fulfil their primary measurement task – measuring the main *relationships* in the economy from the measurements on the *individual variables* – only because the model and the measurements had already been structured into a mutually compatible form.

As we mentioned above, models themselves can also function directly as measuring instruments. A good example of this is the Leontief input–output model. Based on the Marxian reproduction model (discussed by Reuten), the Leontief model can be used to measure the technical coefficients of conversion from inputs to outputs in the economy. This Leontief matrix provides a measurement device to get at the empirical structure of the economy, and can be applied either at a very fine-grained or a very coarse-grained level, depending on the number of sectors represented within the model. Another good example is provided in Margaret Morrison's discussion of the pendulum referred to above. It is possible using a plane pendulum to measure local gravitational acceleration to four significant figures of accuracy. This is done by beginning with an idealised pendulum model and adding corrections for the different forces acting on various parts of the real pendulum. Once all the corrections have been added, the pendulum model has become a reasonably good approximation to the real system. And, although the sophistication of the apparatus (the pendulum itself) is what determines the *precision* of the measurement, it is the analysis and addition of all the correction factors necessary for the model that determines the *accuracy* of the measurement of the gravitational acceleration. What this means is that the model functions as the source for the numerical calculation of G; hence, although we use the real pendulum to perform the measurement, that process is only possible given the corrections performed on the model. In that sense the model functions as the instrument that in turn enables us to use the pendulum to measure G.

Models can also serve as measuring instruments in cases where the model has less structure than either the pendulum or the input–output cases. One example is the use of multivariate structural time-series models in statistical economics. These are the direct descendants of the business barometer models discussed above and share their general assumption that certain economic time series consist of trends and cycles, but they do not specify the time length of these components in advance. When these models are run on a computer, they generate relatively precise measurements of whatever trend and cyclical components are present in the data and provide an analysis of the interrelationships between them.

2.2.3 Models for design and intervention

The final classification of models as instruments includes those that are used for design and the production of various technologies. The interesting feature of these kinds of models is that they are by no means limited to the sorts of scale models that we usually associate with design. That is, the power of the model as a design instrument comes not from the fact that it is a replica (in certain respects) of the object to be built; instead the capacity of mathematical/theoretical models to function as design instruments stems from the fact that they provide the kind of information that allows us to intervene in the world.

A paradigm case of this is the use of various kinds of optics models in areas that range from lens design to building lasers. Models from geometrical optics that involve no assumptions about the physical nature of light are used to calculate the path of a ray so that a lens can be produced that is free from aberration. A number of different kinds of geometrical models are available depending on the types of rays, image distance and focal lengths that need to be considered. However, technology that relies on light wave propagation requires models from physical optics and when we move to shorter wave lengths, where photon energies are large compared with the sensitivity of the equipment, we need to use models from quantum optics. For example, the design of lasers sometimes depends on quantum models and sometimes on a combination of quantum and classical. The interesting point is that theory plays a somewhat passive role; it is the model that serves as an independent guideline for dealing with different kinds of technological problems (see Morrison 1998).

A similar situation occurs in nuclear physics. Here there are several different models of nuclear structure, each of which describes the nucleus in a way different from and incompatible with the others. The liquid drop model is useful in the production of nuclear fission while the optical model serves as the basis for high energy scattering experiments. Although we know that each individual model fails to incorporate significant features of the nucleus, for example, the liquid drop ignores quantum statistics and treats the nucleus classically while others ignore different quantum mechanical properties, they nevertheless are able to map onto technologies in a way that makes them successful, independent sources of knowledge.

In economics, we can point to the way that central banks use economic models to provide a technology of intervention to control money and price movements in the economy. There is no one model that governs all situations – each bank develops a model appropriate for its own economy. This modelling activity usually involves tracking the growth in various economic entities and monitoring various relationships between them. More recently monetary condition indicators (MCIs) have been developed; these indicators are derived from models and function as measurement tools. With the help of their model(s) and MCIs, the central bank decides when and how much to intervene in the money market in order to prevent inflation. The model provides the technology of intervention by prompting the timing, and perhaps indicating the amount of intervention needed. Sometimes the model-based intervention is triggered almost automatically, sometimes a large amount of judgement is involved. (Of course some central banks are more successful than others at using this technology!) The more complex case of macroeconometric modelling and its use as a technology of intervention is discussed below (in section 2.4 on learning).

As we stressed above, part of the reason models can function as instruments is their partial independence from both theory and data. Yet, as we have seen in this section, models fulfil a wide range of functions in building, exploring and applying theories; in various measurement activities; and in the design and production of technologies for intervention in the world. These examples demonstrate the variety of ways in which models mediate between theories and the world by utilising their points of intersection with both domains. Indeed, these intersections are especially evident in cases like the optical models and nuclear models in physics and the monetary and macroeconomic models in economics. Although they draw on particular aspects of high level theory, they are by no means wholly dependent on theory for either their formulation or decisions to use a particular model in a specific context.

We want to caution, however, that our view of models as instruments is not one that entails a classical instrumentalist interpretation of models. To advocate instrumentalism would be to undermine the various ways in which models do teach us about both theories and the world by providing concrete information about real physical and economic systems. They can do this because, in addition to playing the role of instruments, they fulfil a representative function, the nature of which is sometimes not obvious from the structure of the model itself.

2.3 REPRESENTATION

The first question we need to ask is how an instrument can represent. We can think of a thermometer representing in a way that includes not simply the measurement of temperature but the representation of the rise and fall in temperature through the rise and fall of the mercury in the column. Although the thermometer is not a model, the model as an instrument can also incorporate a representational capacity. Again, this arises because of the model's relation to theory or through its relation to the world or to both.

2.3.1 Representing the world, representing theory

Above we saw the importance of maintaining a partial independence of the model from both theory and the world; but, just as partial independence is required to achieve a level of autonomy so too a *relation* to at least one domain is necessary for the model to have any representative function whatsoever. In some cases the model may, in the first instance, bear its closest or strongest relation to theory. For example, in Morrison's case the model of a pendulum functions specifically as a model of a theory – Newtonian mechanics – that describes a certain kind of motion. In other words, the pendulum model is an instance of harmonic motion. Recall that we need the model because Newton's force laws alone do not give us an adequate description of how a physical pendulum (an object in the world) behaves. The pendulum model represents certain kinds of motion that are both described by the theory and produced by the real pendulum. To that extent, it is also a model of the physical object. Fisher's mechanical balance model (discussed by Morgan) provided a representation of the theory of the monetary system. This model enabled him to explore theoretical aspects of the dynamic adjustment processes in the monetary economy and the phenomena of the business cycle in a way that the existing theoretical representation (the equation of exchange) did not allow.

Alternatively, the model-world representation may be the more prominent one. The early statistical business barometers, constructed to represent (in graphic form) the path of real-world economic activity through time, were used to help determine the empirical relationships between various elements in the economy and to forecast the turning points in that particular economy's cycle. In contrasting cases, such model-world representations may be used to explore theory by extending its basic structure or developing a new theoretical framework. Such was the case with the nineteenth-century mechanical aether models of Kelvin and FitzGerald discussed above. Recall that their function was to represent dynamical relations that occurred in the aether, and based on the workings of the model FitzGerald was able to make corrections to Maxwell's field equations. In the previous section we saw how manipulating these models had the status of experiment. This was possible only because the model itself was taken as a *representation* of the aether.

The more interesting examples are where the practice of model building provides representations of both theory and the world, enabling us to see the tremendous power that models can have as representative instruments. Margaret Morrison's discussion of Prandtl's hydrodynamic model of the boundary layer is a case in point. At the end of the nineteenth century the theory of fluid flow was in marked conflict with experiment; no account could be given of why the very small frictional forces present in the flow of water and air around a body created a no-slip condition at the solid boundary. What Prandtl did was build a small water tunnel that could replicate fluid flows past different kinds of bodies. In a manner similar to a wind tunnel, this mechanical model supplied a *representation* of different kinds of flows in different regions of the fluid, thereby allowing one to understand the nature of the conflict with experiment. That is, the water tunnel furnished a visualisation of different areas in the fluid, those close to the body and those more remote. The understanding of the various flow patterns produced by the tunnel then provided the elements necessary to construct a mathematical model that could represent certain kinds of theoretical structures applicable to the fluid.

But, the idea that a model can represent a theoretical structure is one that needs clarification. In the hydrodynamics case the two theories used to describe fluids, the classical theory and the Navier-Stokes equations were inapplicable to real fluid flow. The former could not account for frictional forces and the latter was mathematically intractable. The mathematical model, developed on the basis of the phenomena observed in the water tunnel, allowed Prandtl to apply theory in a specific way. The tunnel enabled him to see that, in certain areas of fluid flow, frictional forces were not important, thereby allowing the use of classical hydrodynamics. And, in areas where frictional forces were present the mathematical model provided a number of approximations to the Navier-Stokes equations that could apply in the boundary layer. The fluid flow was divided conceptually into two regions, one of which treated the fluid as ideal while the other required taking account of the boundary layer close to a solid body. The mathematical model of a fluid with a boundary layer functioned as a representation of both classical theory and the Navier-Stokes equations because each played a role in describing the fluid, yet neither was capable of such description taken on its own. In that sense the model was a representation of certain aspects of theoretical structure in addition to representing the actual phenomena involved in fluid flow past a solid body. In the first instance, however, the model-world representation was established by the water tunnel and it was this that formed the foundation for the model-theory representation as exemplified by the mathematical account of fluid flow.

Another case where the model bears a relation to both theory and the world is Fisher's hydraulic model of the monetary system discussed by Mary Morgan. The representative power of the model stems from both domains, with the structure of the model (its elements, their shapes and their relationships) coming from theory while the model could be manipulated to demonstrate certain empirical phenomena in the world. Because the model represented both certain well-accepted theories (e.g. the quantity theory of money) and could be shown to represent certain well-known empirical phenomena (e.g. Gresham's law that 'bad money drives out good'), the model could be used to explore both the contested theory and problematic phenomena of bimetallism.

As we can see from the examples above, the idea of representation used here is not the traditional one common in the philosophy of science; in other words, we have not used the notion of 'representing' to apply only to cases where there exists a kind of mirroring of a phenomenon, system or theory by a model.[5] Instead, a representation is seen as a kind of rendering – a partial representation that either abstracts from, or translates into another form, the real nature of the system or theory, or one that is capable of embodying only a portion of a system.

Morrison's example of the pendulum is about as close to the notion of 'mirroring' that we get. The more corrections that are added to the pendulum model the closer it approximates the real object and gives us accurate measurements. Many, perhaps most cases, are not like this. Even cases where we begin with data (rather than theory) do not produce reflecting models. For example, the business barometers of van den Bogaard's

---

[5] See R. I. G. Hughes (1997) for a discussion of the notion of representation.

chapter are thought to reflect rather closely the time path of the economy. But they are by no means simple mirrors. Such a model involves both the abstraction of certain elements from a large body of data provided by the economy and their transformation and recombination to make a simple time-series graphic representation which forms the barometer.

Often, models are partial renderings and in such cases, we cannot always add corrections to a stable structure to increase the accuracy of the representation. For example, models of the nucleus are able to represent only a small part of its behaviour and sometimes represent nuclear structure in ways that we know are not accurate (e.g. by ignoring certain quantum mechanical properties). In this case, the addition of parameters results in a new model that presents a radically different account of the nucleus and its behaviour. Hence in describing nuclear processes, we are left with a number of models that are inconsistent with each other.

There are many ways that models can 'represent' economic or physical systems with different levels of abstraction appropriate in different contexts. In some cases abstract representations simply cannot be improved upon; but this in no way detracts from their value. When we want to understand nuclear fission we use the liquid drop model which gives us an account that is satisfactory for mapping the model's predictions onto a technological/experimental context. Yet we know this model cannot be an accurate representation of nuclear structure. Similarly we often use many different kinds of models to represent a single system. For example, we find a range of models being used for different purposes within the analytical/research departments at central banks. They are all designed to help understand and control the monetary and financial systems, but they range from theoretical small-scale micro-models representing individual behaviour, to empirical models which track financial markets, to large-scale macroeconometric models representing the whole economy. Sometimes they are used in conjunction, other times they are used separately. We do not assess each model based on its ability to accurately mirror the system, rather the legitimacy of each different representation is a function of the model's performance in specific contexts.

2.3.2 Simulation and representation

There is another and increasingly popular sense in which a model can provide representations, that is through the process of simulation. Sometimes simulations are used to investigate systems that are otherwise inaccessible (e.g. astrophysical phenomena) or to explore extensions and limitations to a model itself. A simulation, by definition, involves a similarity relation yet, as in the case of a model's predictions mapping onto the world, we may be able to simulate the behaviour of phenomena without necessarily knowing that the simulated behaviour was produced in the same way as it occurred in nature. Although simulation and modelling are closely associated it is important to isolate what it is about a model that enables it to 'represent' by producing simulations. This function is, at least in the first instance, due to certain structural features of the model, features that explain and constrain behaviour produced in simulations. In the same way that general theoretical principles can constrain the ways in which models are constructed, so too the structure of the model constrains the kinds of behaviour that can be simulated.

R. I. G. Hughes' discussion of the Ising model provides a wealth of information about just how important simulation is, as well as some interesting details about how it works. He deals with both computer simulations of the behaviour of the Ising model and with simulations of another type of theoretical model, the cellular automaton. The Ising model is especially intriguing because despite its very simple structure (an array of points in a geometrical space) it can be used to gain insight into a diverse group of physical systems especially those that exhibit critical point behaviour, as in the case of a transition from a liquid to a vapour. If one can generate pictures from the computer simulation of the model's behaviour (as in the case of the two-dimensional Ising model) it allows many features of critical behaviour to be instantly apprehended. As Hughes notes however, pictorial display is not a prerequisite for simulation but it helps.

His other example of simulation involves cellular automata models. These consist of a regular lattice of spatial cells, each of which is in one of a finite number of states. A specification of the state of each cell at a particular time gives the configuration of the cellular automata (CA) at that time. It is this discreteness that makes them especially suited to computer simulations because they can provide exactly computable models. Because there are structural similarities between the Ising model and the CA it should be possible to use the CA to simulate the behaviour of the Ising model. His discussion of why this strategy fails suggests some interesting points about how the structural constraints on these simple models are intimately connected to the ways in which simulations can provide knowledge of models and physical systems. Hughes' distinction between computer simulation of the model's behaviour and the use of computers for calculational purposes further illustrates the importance of regarding the model as an active agent in the production of scientific knowledge.

The early theoretical business-cycle models of Frisch (discussed by Boumans) were simulated to see to what extent they could replicate generic empirical cycles in the economy (rather than specific historical facts). This was in part taken as a test of the adequacy of the model, but the simulations also threw up other generic cycles which had empirical credibility, and provided a prediction of a new cycle which had not yet been observed in the data. In a different example, the first macroeconometric model, built by Tinbergen to represent the Dutch economy (discussed by van den Bogaard and more fully in Morgan 1990), was first estimated using empirical data, and then simulated to analyse the effects of six different possible interventions in the economy. The aim was to see how best to get the Dutch economy out of the Great Depression and the simulations enabled Tinbergen to compare the concrete effects of the different proposals within the world represented in the model. On this basis, he advocated that the Dutch withdraw from the gold standard system, a policy later adopted by the Dutch government.

Consequently we can say that simulations allow you to map the model predictions onto empirical level facts in a direct way. Not only are the simulations a way to apply models but they function as a kind of bridge principle from an abstract model with stylised facts to a technological context with concrete facts. In that sense we can see how models are capable of representing physical or economic systems at two distinct levels, one that includes the higher level structure that the model itself embodies in an abstract and idealised way and the other, the level of concrete detail through the kinds of simulations that the models enable us to produce. Hence, instead of being at odds with each other, the instrumental and representative functions of models are in fact complementary. The model represents systems via simulations, simulations that are possible because of the model's ability to function as the initial instrument of their production.

Because of the various representative and investigative roles that models play, it is possible to learn a great deal from them, not only about the model itself but about theory, the world and how to connect the two. In what follows we discuss some ways that this learning takes place.

2.4 LEARNING

2.4.1 Learning from construction

Modelling allows for the possibility of learning at two points in the process. The first is in constructing the model. As we have pointed out,

there are no rules for model building and so the very activity of construction creates an opportunity to learn: what will fit together and how?
Perhaps this is why modelling is considered in many circles an art or
craft; it does not necessarily involve the most sophisticated mathematics
or require extensive knowledge of every aspect of the system. It does
seem to require acquired skills in choosing the parts and fitting them
together, but it is wise to acknowledge that some people are good model
builders, just as some people are good experimentalists.

Learning from construction is clearly involved in the hydrodynamics
case described by Margaret Morrison. In this case, there was visual
experimental evidence about the behaviour of fluids. There were also
theoretical elements, particularly a set of intractable equations supposed
to govern the behaviour of fluids, which could neither account for nor
be applied directly to, the observed behaviour. Constructing a mathematical model of the observed behaviour involved a twofold process of
conceptualising both evidence and the available theories into compatible terms. One involved interpreting the evidence into a form that could
be modelled involved the 'conceptualisation' of the fluid into two areas.
The other required developing a different set of simplifications and
approximations to provide an adequate theoretical/mathematical
model. It is this process of interpreting, conceptualising and integrating
that goes on in model development which involves learning about the
problem at hand. This case illustrates just how modelling enables you to
learn things both about the world (the behaviour of fluids) and the
theory (about the way the equations could be brought to apply).

A similar process of learning by construction is evident in the cases
that Marcel Boumans and Stephan Hartmann describe. In Boumans'
example of constructing the first generation of business-cycle models,
economists had to learn by trial and error (and by pinching bits from
other modelling attempts) how the bits of the business-cycle theory and
evidence could be integrated together into a model. These were essentially theoretical models, models designed to construct adequate business-cycle theories. Thereafter, economists no longer had to learn how
to construct such theoretical models. They inherited the basic recipe for
the business-cycle, and could add their own particular variations. At a
certain point, a new recipe was developed, and a new generation of
models resulted. In Hartmann's examples, various alternative models
were constructed to account for a particular phenomenon. But in the
MIT-Bag Model and the NJL model, both of which he discusses in
detail, we see that there is a certain process by which the model is

gradually built up, new pieces added, and the model tweaked in response to perceived problems and omissions.

2.4.2 Models as technologies for investigation

The second stage where learning takes place is in using the model. Models can fulfil many functions as we have seen; but they generally perform these functions not by being built, but by being used. Models are not passive instruments, they must be put to work, used, or manipulated. So, we focus here on a second, more public, aspect of learning from models, and one which might be considered more generic. Because there are many more people who use models than who construct them we need some sense of how 'learning from using' takes place.

Models may be physical objects, mathematical structures, diagrams, computer programmes or whatever, but they all act as a form of instrument for investigating the world, our theories, or even other models. They combine three particular characteristics which enable us to talk of models as a technology, the features of which have been outlined in previous sections of this essay. To briefly recap: first, model construction involves a partial independence from theories and the world but also a partial dependence on them both. Secondly, models can function autonomously in a variety of ways to explore theories and the world. Thirdly, models represent either aspects of our theories, or aspects of our world, or more typically aspects of both at once. When we use or manipulate a model, its power as a technology becomes apparent: we make use of these characteristics of partial independence, functional autonomy and representation to learn something from the manipulation. To see how this works let us again consider again some of the examples we discussed already as well as some new ones.

We showed earlier (in section 2.2) how models function as a technology that allows us to explore, build and apply theories, to structure and make measurements, and to make things work in the world. It is in the process of using these technologies to interrogate the world or our theory that learning takes place. Again, the pendulum case is a classic example. The model represents, in its details, both the theory and a real world pendulum (yet is partially independent of both), and it functions as an autonomous instrument which allows us to make the correct calculations for measurements to find out a particular piece of information about the world.

The general way of characterising and understanding this second way of 'learning from using' a model is that models are manipulated to teach us things about themselves. When we build a model, we create a kind of representative structure. But, when we manipulate the model, or calculate things within the model, we learn, in the first instance, about the model world – the situation depicted by the model.

One well-known case where experimenting with a model enables us to derive or understand certain results is the 'balls in an urn' model in statistics. This provides a model of certain types of situations thought to exist in the world and for which statisticians have well-worked out theories. The model can be used as a sampling device that provides experimental data for calculations, and can be used as a device to conceptualise and demonstrate certain probability set ups. It is so widely used in statistics, that the model mostly exists now only as a written description for thought experiments. (We know so well how to learn from this model that we do not now even need the model itself: we imagine it!) In this case, our manipulations teach us about the world in the model – the behaviour of balls in an urn under certain probability laws.

The Ising model, discussed by Hughes, is another example of the importance of the learning that takes place within the world of the model. If we leave aside simulations and focus only on the information provided by the model itself, we can see that the model had tremendous theoretical significance for understanding critical point phenomena, regardless of whether elements in the model denote elements of any actual physical system. At first this seems an odd situation. But, what Hughes wants to claim is that a model may in fact provide a good explanation of the behaviour of the system without it being able to faithfully represent that system. The model functions as an epistemic resource; we must first understand what we can demonstrate in the model before we can ask questions about real systems. A physical process supplies the dynamic of the model, a dynamic that can be used to generate conclusions about the model's behaviour. The model functions as a 'representative' rather than a 'representation' of a physical system. Consequently, learning about and from the model's own internal structure provides the starting point for understanding actual, possible and physically impossible worlds.

Oftentimes the things we learn from manipulating the world in the model can be transferred to the theory or to the world which the model represents. Perhaps the most common example in economics of learning about theory from manipulation within a model, is the case of Edgeworth-Bowley boxes: simple diagrammatic models of exchange between two people. Generations of economics students have learnt exchange theory by manipulations within the box. This is done by tracing through the points of trade which follow from altering the starting points or the particular shape of the lines drawn according to certain assumptions about individual behaviour. But these models have also been used over the last century as an important technology for deriving new theoretical results not only in their original field of simple exchange, but also in the more complex cases of international economics. The original user, Edgeworth, derived his theoretical results by a series of diagrammatic experiments using the box. Since then, many problems have found solutions from manipulations inside Edgeworth-Bowley box diagrams, and the results learnt from these models are taken without question into the theoretical realm (see Humphrey 1996). The model shares those features of the technology that we have already noted: the box provides a representation of a simple world, the model is neither theory nor the world, but functions autonomously to provide new (and teach old) theoretical results via experiments within the box.

In a similar manner, the models of the equation of exchange described by Mary Morgan were used to demonstrate *formally* the nature of the theoretical relationship implied in the quantity theory of money: namely how the cause–effect relationship between money and prices was embedded in the equation of exchange, and that two other factors needed to remain constant for the quantity theory relation to be observable in the world. This was done by manipulating the models to show the effects of changes in each of the variables involved, constrained as they were by the equation. The manipulation of the alternative mechanical balance version of the model prompted the theoretical developments responsible for integrating the monetary theory of the economic cycle into the same structure as the quantity theory of money. It was because this analogical mechanical balance model represented the equation of exchange, but shared only part of the same structure with the theoretical equation of exchange, that it could function autonomously and be used to explore and build new theory.

The second model built by Fisher, the hydraulic model of the monetary system incorporated both institutional and economic features. It was manipulated to show how a variety of real world results might arise from the interaction of economic laws and government decisions and to 'prove' two contested theoretical results about bimetallism within the world of the model. But, these results remained contested: for although the model provided a qualitative 'explanation' of certain historically observed phenomena, it could not provide the kind of quantitative representation which would enable theoretically-based prediction or (despite Fisher's attempts) active intervention in the monetary system of the time.

These manipulations of the model contrast with those discussed by Adrienne van den Bogaard. She reports on the considerable arguments about the correct use of the models she discusses in her essay. Both the barometer and the econometric model could be manipulated to predict future values of the data, but was it legitimate to do so? Once the model had been built, it became routine to do so. This is part of the econometrics tradition: as noted above, Tinbergen had manipulated the first macroeconometric model ever built to calculate the effects of six different policy options and so see how best to intervene to get the Dutch economy out of the Great Depression of the 1930s. He had also run the model to forecast the future values for the economy assuming no change in policy. These econometric models explicitly (by design) are taken to represent both macroeconomic theory and the world: they are constructed that way (as we saw earlier). But their main purpose is not to explore theory, but to explore past and future conditions of the world and perhaps to change it. This is done by manipulating the model to predict and to simulate the outcomes which would result if the government were to intervene in particular ways, or if particular outside events were to happen. By manipulating the model in such ways, we can learn things about the economy the model represents.

2.5 CONCLUSION

We have argued in this opening essay that scientific models have certain features which enable us to treat them as a technology. They provide us with a tool for investigation, giving the user the potential to learn about the world or about theories or both. Because of their characteristics of autonomy and representational power, and their ability to effect a relation between scientific theories and the world, they can act as a powerful agent in the learning process. That is to say, models are both a means to and a source of knowledge. This accounts both for their broad applicability, and the extensive use of models in modern science.

Our account shows the range of functions and variety of ways in which models can be brought to bear in problem-solving situations. Indeed, our goal is to stress the significance of this point especially in light of the rather limited ways that models have, up to now, been characterised in the philosophical literature. They have been portrayed narrowly as a means for applying theory, and their construction was most often described either in terms of 'theory simplification' or derivation from an existing theoretical structure. These earlier views gave not only a limited, but in many cases an inaccurate, account of the role of models in scientific investigation. Our view of models as mediating instruments, together with the specific cases and detailed analyses given in these essays, go some way toward correcting the problem and filling a lacuna in the existing literature.

A virtue of our account is that it shows how and why models function as a separate tool amongst the arsenal of available scientific methods. The implication of our investigations is that models should no longer be treated as subordinate to theory and data in the production of knowledge. Models join with measuring instruments, experiments, theories and data as one of the essential ingredients in the practice of science. No longer should they be seen just as 'preliminary theories' in physics, nor as a sign of the impossibility of making economics a 'proper' science.

NOTE

Two earlier users of the term 'mediators' in accounts of science should be mentioned. Norton Wise has used the term in various different contexts in the history of science, and with slightly different connotations, the most relevant being his 1993 paper 'Mediations: Enlightenment Balancing Acts, or the Technologies of Rationalism'. His term 'technologies' is a broad notion which might easily include our 'models'; and they mediate by playing a connecting role to join theory/ideology with reality in constructing a rationalist culture in Enlightenment France. Our focus here is on using models as instruments of investigation about the two domains they connect. The second user is Adam Morton (1993) who discusses mediating models. On his account the models are mathematical and mediate between a governing theory and the phenomena produced by the model; that is, the mathematical descriptions generated by the modelling assumptions. Although our account of mediation would typically include such cases it is meant to encompass much more, both in terms of the kinds of models at issue and the ways in which the models themselves function as mediators.

REFERENCES

Cartwright, N. (1983). *How the Laws of Physics Lie.* Oxford: Oxford University Press.

Hesse, M. (1966). *Models and Analogies in Science.* Notre Dame: University of Notre Dame Press.

Hughes, R. I. G. (1997). 'Models and Representation', *Philosophy of Science* 64, S325–336

Humphrey, T. M. (1996). 'The Early History of the Box Diagram', *Federal Reserve Board of Richmond Economic Quarterly*, 82:1, 37–75.

Morgan, M. S. (1990). *The History of Econometric Ideas.* Cambridge: Cambridge University Press.

(1997a). 'The Technology of Analogical Models: Irving Fisher's Monetary Worlds', *Philosophy of Science* 64, S304–314.

(1997b). 'The Character of "Rational Economic Man"', *Dialektik*, 78–94.

Morrison, M. (1998). 'Modelling Nature: Between Physics and the Physical World', *Philosophia Naturalis* 38:1, 65–85.

Morton, Adam (1993). 'Mathematical models: Questions of Trustworthiness', *British Journal for the Philosophy of Science*, vol. 44, 4, 659–74.

Wise, M. N. (1993). 'Mediations: Enlightenment Balancing Acts, or the Technologies of Rationalism' in P. Horwich (ed.), *World Changes: Thomas Kuhn and the Nature of Science*, Cambridge, MA: MIT Press.

CHAPTER 3

Models as autonomous agents

Margaret Morrison

3.1 INTRODUCTION

Perhaps the key philosophical question regarding the nature of models concerns their connection to concrete physical systems and the degree to which they enable us to draw conclusions about these systems. This presupposes, of course, that models can sometimes be understood as representative of objects or systems in the world. But how should we understand this presupposition? It seems not quite correct to say that models accurately describe physical systems since in many cases they not only embody an element of idealisation and abstraction, but frequently represent the world in ways that bear no similarity to physically realisable objects, e.g. the electron as a point particle.[1] Hence, we need a reformulation of the philosophical question; more specifically, since models are sometimes deliberately based on characterisations we know to be false how can they provide us with information about the world.

There are different answers to this latter question, each of which depends first, on how one views the nature and role of models and secondly, how one understands the philosophically problematic issue of what it means to accept a model as providing reliable information about real systems, as opposed to simply successful predictions. I will say something about both of these issues and how they relate to each other in what follows but let me begin by mentioning two different and rather surprising characterisations of models given by two different physicists,

[1] I want to distinguish between idealisation and abstraction in much the same way that Nancy Cartwright (1989) does. An idealisation is a characterisation of a system or entity where its properties are deliberately distorted in a way that makes them incapable of accurately describing the physical world. By contrast, an abstraction is a representation that does not include all of the systems properties, leaving out features that the systems has in its concrete form. An example of the former is the electron as a point particle and the latter is the omission of intermolecular forces from the ideal gas law.

Dirac, the esteemed theoretician, and Heinrich Hertz, the equally esteemed experimentalist. Each view recognises but deals differently with the epistemological issues that surround the use of models in physics. And they do so in a way that sheds some light on why allegiance to fundamental theory as the explanatory vehicle for scientific phenomena distorts our view of how models function in scientific contexts. The core of my argument involves two claims (1) that it is models rather than abstract theory that represent and explain the behaviour of physical systems and (2) that they do so in a way that makes them autonomous agents in the production of scientific knowledge. After drawing the comparison between Hertz and Dirac (which provides some nice stage setting) I go on to discuss some examples of different kinds of models and modelling techniques, as well as the relation between models and theory. The examples draw attention to a secondary point which is nevertheless important for sorting out how models work, they show that the distinction between phenomenological and theoretical models proves to be neither a measure of the model's accuracy to real systems nor its ability to function independently of theory. In other words, it is not the case that phenomenological models provide a more accurate account of physical systems than theoretical models. Hopefully the details of my argument will highlight not only the explanatory power of models but the ways in which models can remain autonomous despite some significant connections with high level theory.

3.2 THE LIMITS OF MODELS

In the *Principles of Mechanics* (*PM*) Hertz claims that 'when from our accumulated previous experience we have once succeeded in deducing images of the desired nature, we can then . . . develop by means of them, as by means of models, the consequences which in the external world only arise in a comparatively long time' (*PM*, p.1). In other words, our ability to draw inferences based on the structure of our models enables us to 'be in advance of the facts' and extend our knowledge in productive ways. In that same work he stated that 'the relation of a dynamical model to the system of which it is regarded as a model, is precisely the same as the relation of the images which our mind forms of things to the things themselves' (p. 428). Hertz has quite a bit to say about images in the introduction to *PM*, much of it centring on the fact that the agreement between the mind and the world ought to be likened to the agreement of two systems that are models of each other. In fact in his discussion of dynamical models later in the book Hertz claims that we can account for the similarity between the mind and nature by assuming that the mind is capable of making actual dynamical models of things, and of working with them. The notion of a model is introduced in the context of determining the natural motion of a material system. However, he makes a rather strong claim about the kind of knowledge we get from models:

> If we admit . . . that hypothetical masses can exist in nature in addition to those which can be directly determined by the balance, then it is impossible to carry our knowledge of the connections of natural systems further than is involved in specifying models of the actual systems. We can then, in fact, have no knowledge as to whether the systems which we consider in mechanics agree in any other respect with the actual systems of nature which we intend to consider, than in this alone, – that the one set of systems are models of the other. (p. 427)

Even though models enable us to expand our knowledge in the form of predictions we seem to be restricted to the domain of the model when it comes to making claims about specific processes. This is rather notable since it implies that the philosophical question I raised at the beginning cannot be answered if we expect the answer to tell us something about correspondences between models and concrete systems. In that sense 'knowing from models' involves a particular and limited kind of knowing.

Dirac, on the other hand, was unsympathetic to the idea that pictorial, or other kinds of models had any place in physics. He wanted relativistic quantum mechanics to be founded on general principles rather than on any particular model of the electron. His contemporaries, Pauli and Schrodinger, assumed that the problem of integrating spin and relativity would require a sophisticated model of the electron; but for Dirac the notion of model building seemed an inappropriate methodology for physics, especially since it left unanswered what for him was a fundamental question, namely, '. . . why nature should have chosen this particular model for the electron instead of being satisfied with a point charge'. His only concession to the traditional models of classical physics was the claim in *Principles of Quantum Mechanics* that one may extend the meaning of the word picture to include any way of looking at the fundamental laws which makes their self-consistency obvious. In other words, a deduction of these laws from symmetry principles, for example, might enable us to acquire a picture (here taken to be an understanding) of atomic phenomena by becoming familiar with the laws of quantum theory.

At first glance this characterisation seems somewhat odd; one would typically expect (especially in this period of the development of physics) the theoretician to be the proponent of model building while assuming that the experimentalist would view models as unfortunate abstractions from concrete systems. But it is exactly because a theoretician like Dirac sees the world as governed by fundamental laws and invariances that the need for models becomes not only otiose but creates the unnecessary philosophical problem of determining whether nature is actually like one's chosen model. Hertz, the experimentalist, perceives nature as filled with enormous complexity and it is the desire to understand how it might possibly be constructed, that motivates his reliance on models.

What is perhaps most interesting in this comparison is the fact that Hertz, in his own way, was as sceptical as Dirac about whether one could acquire certainty about physical systems from models; yet each dealt with the problem in a different way. One emphasised that models were in some sense the limit of our knowledge while the other eschewed their use altogether. Because models can incorporate a significant amount of detail about what real systems might be like, and because there is some-times no way to directly verify the accuracy of this detail, Dirac prefers to rely on broad sweeping theoretical laws and principles in characteris-ing natural phenomena; principles such as the fact that the spacetime properties of the relativistic equation for the electron should transform according to the theory of relativity. Hertz, however, thinks that we need to have a 'conception' of things in order for us to draw inferences about future events. This conception is furnished by a model which may be much simpler than the system it represents but nevertheless provides enough detail that we are able to use its structural and material con-straints for predictive purposes.

But, are these differences simply methodological preferences or is there some systematic disparity between the use of models rather than theoretical principles? And, perhaps more importantly, need the two be mutually exclusive. In some philosophical accounts of models described in the literature, for example, the semantic view as outlined by Giere (1988) and van Fraassen (1980) as well as Redhead's (1980) account of the relationship between models and approximations, the latter question need not arise since models are typically thought to bear a close relation-ship to high level theory. According to the semantic view a theory is simply a family of models; for example, in classical mechanics we have a variety of models that supposedly represent actual physical systems, the pendulum, the damped harmonic oscillator etc. All obey the funda-mental laws of mechanics and in that sense can be seen to be instantia-tions or applications of these laws in conjunction with specific conditions relevant to the system, which together define the model.

Michael Redhead (1980) also stresses the relation to theory but makes an important distinction between models and what he calls approximations. Theories in physics are usually expressed via a system of equations whose particular solutions refer to some physically possible coordination of variables. An approximation to the theory can be considered in two different senses. We can either have approximate solutions to the exact equations of the theory as in the case $dy/dx - \lambda y = 0$ where the solution might be expanded as a perturbation series in λ with the nth order approximation being

$$\Upsilon_n = 1 + \lambda x + \lambda^2 x^2/2! + \ldots + \lambda^{n-1} x^{n-1}/(n-1)! \qquad (3.1)$$

if the boundary condition $y = 1$ at $x = 0$; or, we can look for exact solutions to approximate equations by simplifying our equation before solving it. Here Υ_n is an exact solution for

$$dy/dx - \lambda y + \lambda^n x^{n-1}/(n-1)! = 0 \qquad (3.2)$$

which, for small λ, is approximately the same as (3.1).

The main idea is that approximations are introduced in a mathematical context and the issue of physical interpretation need not arise. However, in the case of models the approximation involves simplifying, in both a mathematical and physical sense, the equations governing a theory before solutions are attempted; that is, one is concerned with solutions to a simplified theory rather than approximate solutions to an exact theory. In that sense every model involves some degree of approximation in virtue of its simplicity but not every approximation functions as a model.

Both of these accounts tend to undermine Dirac's desire for general principles to the *exclusion* of models. Redhead points to cases where models are necessary because we simply cannot work with our theory while the semantic view suggests that we need models where we want to fit the theory to concrete situations. For example, if we want to consider cases of linear restoring forces we can do so by constructing a model of the system that behaves like a spring and mass, a vibrating string, a simple pendulum, etc.; all of which are oscillators involving harmonic motion. Although the model may be a gross simplification of an actual physical system it nevertheless functions by showing us how Newton's laws apply in specific cases. We can see then that models do play a central and powerful role in 'doing physics'.

But, I want to make a stronger and somewhat different claim about models than either the semantic or approximation view. Specifically I want to argue for an account of models that emphasises their role as autonomous agents in scientific activity and inquiry. The autonomy is the result of two components (1) the fact that models *function* in a way that is partially independent of theory and (2) in many cases they are *constructed* with a minimal reliance on high level theory. It is this partial independence in construction that ultimately gives rise to the functional independence. But, as we shall see below, even in cases where there is a close connection with theory in developing the model it is still possible to have the kind of functional independence that renders the model an autonomous agent in knowledge production. One aspect of this functional autonomy is the role models play in both representing and explaining concrete processes and phenomena. In what follows I want to show how different kinds of models fulfil these functions in different sorts of ways. Because I claim that the representative and explanatory capacities of models are interconnected it is important to display how the representative power of models differs not only with respect to the kind of system we wish to model but also with respect to the resources available for the model's construction. In that sense my view differs from Dirac's notion that we gain an understanding from abstract theoretical principles. Rather, I want to claim that the latter furnish constraints on the class of allowable models; and when we want to find out just how a physical system actually functions we need to resort to models to tell us something about how specific mechanisms might be organised to produce certain behaviour.

3.3 MODELS AND MODELLING: KINDS AND TECHNIQUES

In looking at specific examples of the ways models represent physical systems one of the things I want to stress is that the classification of theoretical models suggested by the semantic view does not seem rich enough to capture the many ways in which models are constructed and function. The semantic view characterises theoretical models as 'models of theory' – there is a basic theoretical structure that does more than simply constrain the acceptable models, it provides the fundamental building blocks from which the model is constructed. In that sense even though we may have several different models to represent the linear restoring force there is a basic structural unity provided by the theory that serves to unite the models in a single family or system. However, there is reasonable evidence to suggest that this represents only a very limited picture of model building in physics; many models are constructed in a rather piecemeal way making use of different theoretical concepts in nothing like a systematic process. (A similar claim can be made for models in economics; see for example Marcel Boumans' paper in this volume where he shows how model construction involves a complex activity of integration.)

Some recent work by Cartwright, Suarez and Shomar (1995) calls attention to the problems with the semantic view by emphasising the importance of 'models of phenomena' as structures distinct from 'models of theory'. They discuss the model constructed by the London brothers to take account of a superconductivity phenomenon known as the Meissner effect. What is significant about the Londons' model is that the principle equation had no theoretical justification and was motivated on the basis of purely phenomenological considerations. From this example Cartwright et. al. stress the need to recognise independence from theory in the method and aims of phenomenological model building. Although I agree in spirit with this approach the problem is one of spelling out what constitutes 'independence from theory'. In fact, if we look at the distinction between phenomenological and theoretical models we see that it is sometimes difficult to draw a sharp division between the two, especially in terms of their relation to theory. Not all theoretical models are 'models of theory' in the sense of being derivable from theory nor is every phenomenological model free from theoretical concepts and parameters. A common misconception is that the existence of well-established background theory automatically facilitates the construction of models and provides the backdrop for distinguishing between theoretical and phenomenological models. It is also commonly thought that the theoretical/phenomenological distinction reflects a difference in descriptive accuracy, that theoretical models are more abstract and hence sacrifice descriptive accuracy while phenomenological models are more accurate/realistic accounts of actual physical systems.

But, this is not to suggest that we cannot, or that it is unhelpful in some contexts, to distinguish between theoretical and phenomenological models; rather, my point is simply that the distinction is not one that is especially useful for a philosophical characterisation since many of the significant features of models cannot be accurately reflected on the basis of that contrast. And, as the examples below show, claims to descriptive accuracy cannot be drawn along the axis that supposedly distinguishes theoretical from phenomenological models. As I mentioned at the beginning, these points about classification of models are significant for my main thesis about models as independent sources of knowledge; it is partly because models have a rather hybrid nature (neither theory nor simple descriptions of the world) that they are able to mediate between theory and the world and intervene in both domains. What this intervention consists of will, I hope, become clear in section four where I examine in more detail the construction and function of specific kinds of models.

There are many reasons why models are necessary in scientific practice. In some cases it is impossible to arrive at exact solutions because the equations are nonlinear or involve partial differentials hence the construction of models is crucial. But there are also cases where there is no consistent or unified treatment of a specific phenomenon, as in the case of the atomic nucleus where a variety of different models is required to account for various types of experimental results – one is used to describe scattering and another to describe fission etc. Part of the reason for this diversity of models is that the atomic nucleus exhibits many different types of behaviour, from the classical where it behaves like a liquid drop to the quantum mechanical where it shows a shell structure similar to that of atoms. The models themselves are not strictly 'theoretical' in the sense of being derived from a coherent theory, some make use of a variety of theoretical and empirical assumptions. For example, for the bulk properties of the nuclei, like size and behaviour of binding energy, the liquid drop model is used; this model also explains fission. In the formula for the mass of the nucleus, the part that gives the form of the dependence of the nuclear binding energy on the number of neutrons and protons is taken directly from theory, but the coefficients are adjusted to give the best fit for observed binding energies. The shell model, on the other hand, treats nucleons in nuclei as moving independently in a central potential which has roughly the shape of a square well and incorporates the quantum behaviour that is inexplicable using the liquid drop model.

Although Quantum Chromodynamics (QCD) is a theory about strong interactions, one of which is the nuclear force, because of the theory's structure it is applicable only to high energy domains; hence, models of the nucleus are required when we want to model low energy phenomena. Nor does QCD tell us how to derive or construct models of the nucleus. In fact, nuclear models are usually characterised as phenomenological since they were developed in an attempt to classify and explain different experimental results like scattering, neutron capture, etc. Yet they obviously incorporate a significant amount of theoretical structure, specifically, aspects of classical and quantum mechanics that are crucial features of the model's design.

Phenomenological models then should not be seen as completely

independent of theory, as being purely descriptive rather than explanatory, nor do they necessarily provide more realistic accounts of the phenomena. Although it may provide a 'model of the phenomena' a
phenomenological model can also be reliant on high level theory in
another way – for its *applicability*. An example is the model of the boundary layer describing the motion of a fluid of low viscosity past a solid
surface. The hydrodynamic equations describing the motion of a fluid are
nonlinear so in order to get a solvable set of equations we need to divide
the fluid conceptually into two regions; a thin boundary layer near the
surface where viscosity is dominant and the remainder where viscosity is
negligible and the fluid can be treated as ideal. Here very different approximations are used for the same homogeneous fluid in different parts of the
system and the model itself requires two different theoretical descriptions;
that is, it relies on two different theories for its applicability.

So far we have mentioned briefly two different ways in which phenomenological models relate to theory. Given this relation how should
we understand the model as maintaining the kind of independence I
spoke of earlier? This issue is especially important in cases of theoretical models where there are even closer ties between high level theoretical structure and models. As a way of illustrating exactly how theory
functions in these cases (i.e. how it contributes to the representative and
explanatory role of models) and how models can retain a level of independence I want to look closely at two different kinds of models. The
first is the rather straightforward model of the pendulum. Although this
is a theoretical model in the sense that it can be *derived* from theory, it is
nevertheless capable of providing, with the appropriate corrections, an
extremely realistic account of the instrument itself and its harmonic
motion. In other words, the abstractness of theory is not inherited by
the model, but neither can the theory (Newtonian mechanics) itself
provide the kind of explanatory information about real pendulums that
is useful in say, measuring the force of gravity. Instead, we have a well
established theory, Newtonian mechanics, that still requires models for
particular applications. Although in this case the construction of the
model is closely aligned with theory there is a functional independence
that gives the model a 'life of its own'.[2] The model can function as an
autonomous source of knowledge and an instrument for measuring
gravitational acceleration. Next I want to discuss in more detail the
phenomenological modelling in hydrodynamics that I mentioned

---

[2] The phrase is borrowed from Ian Hacking's (1983) description of experimental entities. I want
to argue for a similar characterisation for models as independent entities.

above. In both of these examples theory plays some role but the model can be seen to function independently. The important difference between the two is that in the hydrodynamic case the model is required because the theory itself does not provide a feasible way to solve problems or explain why the fluid behaves as it does; in the pendulum case the model functions as a kind of measuring instrument thereby acting in a way that is independent of theory. In each context not only is the representative function of the model different, that is to say each model embodies a different notion of what it means to 'represent' a physical system, but the sense in which the models are 'realistic' (an accurate description of the system) has little to do with whether they are phenomenological or theoretical.

3.4 WHAT KINDS OF MODELS REPRESENT REAL SYSTEMS?

In his chapter on models Giere (1988) discusses the pendulum as an illustration of simple harmonic motion. That is, we can represent this kind of motion using an idealised theoretical model of a real pendulum. To construct such a model we assume that the pendulum has a mass m in a uniform gravitational field with no friction. In the case of the real pendulum the downward gravitational force is partially balanced by the tension along the string S which can in turn be resolved into a vertical and horizontal component. The idealised 'model' of the pendulum considers only horizontal motion leaving out the vertical component. However, the equation describing horizontal motion cannot, by itself, describe simple harmonic motion unless we assume a small angle of swing, then we can easily calculate the solutions for the equation of motion in the horizontal direction that do give us harmonic motion. The equation describing the horizontal motion of S also embodies another idealising condition, namely a uniform gravitational force.[3] And, although one might be able to think of ways to improve the approximation one cannot complete the process so that the equation can correctly represent all the existing gravitational forces. In addition to gravitational forces there are also frictional forces such as air resistance on the pendulum bob. The point is that the law of the pendulum does not represent a real pendulum in any 'true' sense; what Giere claims is important for physicists is being close enough for the problem at hand.

---

[3] This is false due to the fact that the earth is not a uniform homogeneous sphere and is influenced by the motion of the sun, moon and planets.

The theory (Newtonian mechanics) provides us with well-established techniques for introducing corrections into the model. From Giere's description we get no real sense of how the model functions except as a device to apply theory; in other words, we need the model as a way of seeing how Newtonian laws govern the motion of the pendulum. Despite these close links with theory there is an important sense in which we can see the pendulum model functioning independently, as an instrument that is more than simply a vehicle for theory. What Giere refers to as 'being close enough for the problem at hand' may require one to have an extremely realistic account of how the 'model' pendulum described by the law differs from the physical object. That is, we might be required to add on a great many correction factors to make the model as close a copy as possible of the real pendulum. In such contexts the accuracy of the representation can have a significant impact on whether the model can function in the desired way. The following example shows how the addition of correction factors contributes to the model's independence by illustrating how a very accurate theoretical model can function in an autonomous way.

3.4.1 Theoretical models meet the world

As we saw above when we want to gain information about the behaviour of a real pendulum we can't simply apply laws of Newtonian mechanics directly; rather we need a way of describing an idealised pendulum – a model – and then applying the laws to the model. Although the pendulum model is a highly abstract version of the real thing it is possible, using the plane pendulum and its accompanying model, to measure local gravitational acceleration to four significant figures of accuracy.[4] A great many corrections are necessary but the important point is that as a model whose structure is derived from theory (a theoretical model) it is capable of absorbing corrections that provide a highly realistic description of the actual apparatus.

If one knows the cord length and period one can solve for the acceleration of gravity where

$$g = 4\pi^2 \, l / T_0^2.$$

A large number of corrections, depending on the desired level of accuracy, must be applied to the model since it will always deviate from the

[4] For a detailed discussion see Nelson and Olsson (1986).

actual pendulum in a variety of ways. For example, in addition to the corrections for finite mass distribution and amplitude there are the effects of air resistance, elastic corrections due to wire stretching and motion of support. If we have a point pendulum supported by a mass-less, inextensible cord of length L the equation of motion for oscillations in a vacuum is

$$\ddot{\theta} + (g/l)\sin\theta = 0$$

where $\theta \equiv d\theta/dt$. If we have small angular displacements the equation above can be solved by a perturbation expansion. The correction to the period is

$$\Delta T/T_0 = \sum_{n=1}^{\infty} \left(\frac{(2n)!}{2^{2n}(n!)^2} \right)^2 \sin^{2n}\left(\frac{\theta_0}{2} \right) = \frac{1}{16}\theta_0^2 + \frac{11}{3072}\theta_0^4 + \ldots$$

where θ_0 is the maximum angular displacement in radians. Besides lengthening the period the finite pendulum displacement introduces an admixture of higher harmonics that can be observed by a Fourier analysis of the time-dependent displacement.

When we consider the mass distribution corrections we need to take into account that the pendulum bob is a finite size, the suspension wire has a mass, there may be additional wire connections to the bob and one may have to take account of the structure that supports the pendulum. We can describe these effects using the physical pendulum equation

$$T = 2\pi \, (I/Mgh)^{1/2}$$

where I is the total moment of inertia about the axis of rotation, M is the total mass and h is the distance between the axis and the centre of mass.

One also needs to take into account the various ways in which air can change the measured period. The first is buoyancy. Archimedes' principle states that the apparent weight of the bob is reduced by the weight of the displaced air. This has the effect of increasing the period since the effective gravity is decreased. The correction is

$$\Delta T/T_0 = \left(\tfrac{1}{2}\right) (m_a/m)$$

where m_a is the mass of the air displaced by the bob. There is also a damping correction; air resistance acts on the pendulum ball and the wire causing the amplitude to decrease with time and increasing the period of oscillation. The force law for any component of the system is determined by the Reynolds number R for that component. R is defined as

$$R = \rho VL / \eta$$

where ρ and η are the fluid density and viscosity, V is a characteristic velocity and L is a characteristic length. The drag force is usually expressed in the form

$$F = \tfrac{1}{2} C_D A \rho v^2$$

where the drag coefficient C_D is a dimensionless number which is a function of the Reynolds number.[5] Because the damping force is a combination of linear and quadratic damping the function is one that contains both effects simultaneously. One can compute the decrease in amplitude using the work-energy theorem. The damping force is defined by

$$F = b|v| + cv^2$$

where c and b are physical damping constants and the work done by F acting on the centre of mass over the first half period is

$$W = -\int_0^{\pi/w_0} Fv\,dt = -\int_0^{\pi/w_0} (-bv + cv^2)v\,dt$$

To determine the correction to the period one must consider the differential equation of motion. Since both damping forces are small we can take them as independent perturbations. For linear damping with small oscillations the equation of motion has an exact solution but not for the case of quadratic damping. In fact, in the latter case the equation of motion is not analytic since the sign of the force must be adjusted each half-period to correspond to a retarding force. The problem is solved by using a perturbation expansion applied to an associated analytic problem where the sign of the force is not changed. In this case the first half-period is positively damped and the second half-period is negatively damped. The resulting motion is periodic. Although only the first half-period corresponds to the damped pendulum problem the solution can be reapplied for subsequent half-periods. The correction to the period is

---

[5] The Reynolds number is defined as UL/V where U is the typical flow speed, L is the characteristic length of the flow and V is viscosity. A Reynolds number is a dimensionless quantity and for a steady flow through a system with a given geometry the flowlines take the same form at a given value of the Reynolds number. Hence, air flow through an orifice will be geometrically similar to that of water through a similar orifice if the dimensions and velocity are chosen to give identical values of the Reynolds number.

obtained from the requirement that the solution must be periodic. Air resistance also has an indirect effect on the period through correction for finite amplitude.

The variation of the bob's motion during the pendulum cycle causes a further air correction for the motion of the air surrounding the bob. The kinetic energy of the system is partly that of air with the effective mass of the system therefore exceeding the bob mass. The kinetic energy of the air can be taken into account by attributing an added mass to the bob's inertia proportional to the mass of the displaced air.

Yet another set of corrections involves elasticity since a real pendulum has neither an inextensible wire nor is it mounted on a perfectly rigid support. The length of the pendulum can be increased by stretching of the wire due to the weight of the bob or there may also be a dynamic stretching from the apparent centrifugal and Coriolis forces acting on the bob during its motion. The period can be either increased or decreased depending on the nature of the support. A very massive support is required if the period is to be independent of the support, so, for example, for four figure accuracy the support should be at least 10^4 times more massive than the bob.

This kind of modelling is interesting because it presents the paradigm case of what we usually think of as the goal of model building. We start with a background theory from which we can derive an idealised model that can then be corrected to provide an increasingly realistic representation (model) of a concrete physical phenomenon or system. The idealised structure/model (this may also occur with certain kinds of laws) can be corrected in a variety of ways depending on the level of accuracy we want. We know the ways in which the model departs from the real pendulum, hence we know the ways in which the model needs to be corrected; but the *ability* to make those corrections results from the richness of the background theoretical structure. Despite the fact that the pendulum model is 'theoretical' and abstract, for the task we want it to fulfil it can be made more or less true of a real phenomenon.

One might think that the reason this case is so successful is our direct access to the system we are modelling. We can observe, independently of the model, behaviour of the real system and then use the appropriate physics to calculate the necessary corrections to the ideal case. Although this might make some aspects of model construction slightly easier it in no way guarantees success or accuracy for even here it is sometimes impossible to get exact solutions. An example is the case of air corrections to the period due to quadratic damping, we need to use a perturbation expansion but this is rather deceptive since only the first few terms converge and give good approximations. In fact the series diverges asymptotically yielding no solution. Moreover, as we shall see in the hydrodynamic case, access to the 'real system' does not necessarily result in the ability to provide more precise correction factors.

Moreover, it is not just that we can add correction factors – it is the *way* in which the factors are added that enhances the model's power. In the pendulum case the sophistication of the apparatus (the real pendulum) determines the precision of the measurement of gravitational acceleration but it is the appropriate analysis of all the correction factors that gives us the *accuracy* of the measurement of G. In that sense the model allows us to represent and calculate to a reasonably accurate extent each of the factors that together yield the value for gravitational acceleration. And, if we were asked why the force is what it is we could answer by providing an explanation in terms of the model and its corrections.

As we saw in the example of nuclear models, however, these kinds of detailed and cumulative corrections may not always be possible. There we have contradictory models governing the same system and any attempt to add corrections to one usually results in it being no longer valid in its intended domain (see Morrison 1998). The obvious question to ask is whether this in any way results from the lack of a comprehensive background theory governing certain kinds of nuclear reactions. The answer, I believe, is no. As we shall see in the hydrodynamics example discussed below the existence of a well established background theory in no way *guarantees* that one can even provide models of particular systems, let alone the kind of accuracy displayed in the pendulum case.

In modelling the pendulum we do not have the kind of independence of construction that we have in the case of nuclear models. In the latter case each model makes use of different theoretical constraints, some classical, some quantum mechanical and some from both domains. The models are constructed piecemeal to account for and classify theoretical results, with no particular theory governing their development. In that sense they exhibit a great deal more independence in construction. Nevertheless, with both the nuclear models and the pendulum model (which is theoretical) we have a functional independence that allows the model(s) a degree of autonomy regardless of its relation to theory. In the hydrodynamic example I now want to consider there are well established theories governing fluids yet they are incapable of yielding models that enable us to calculate particular kinds of flows. The model exhibits both independence of construction and functional independence despite the availability of background theory.

3.4.2 The boundary layer – phenomenological abstraction

Around the end of the nineteenth century the science of fluid mechanics began to develop two divergent ways. One was theoretical hydrodynamics which evolved from Euler's equations for a frictionless, non-viscous fluid and the other was the empirical science of hydraulics, formulated as a way to account for the marked conflict with experiment due to the omission of fluid friction from the theory. Although a complete set of equations for flows with friction, the Navier-Stokes equations, had been known for some time their mathematical intractability prevented any kind of systematic treatment of viscous fluid motion. However, what made the situation even more awkward was the fact that for both water and air, the two most important fluids (from a practical point of view), viscosity is very small and hence the frictional forces are small in comparison to gravity and pressure forces. Hence, it was difficult to understand why the classical theory was in such glaring conflict with experiment in these cases and why hydraulics was unable to yield solutions. The answer was provided by Ludwig Prandtl in 1904 with his model of a fluid with a very thin boundary layer. It allowed for the possibility of treating viscous flows and also gave approximate solutions to the Navier-Stokes equations.

Prandtl's model is interesting for a number of reasons, most importantly it shows us the sequence in developing a successful account of fluid dynamics. From a water tunnel which was itself a kind of physical model that provided a visualisation of fluid flows, Prandtl was able to represent and explain various aspects of the flow. From there he developed the physical/conceptual model of the fluid incorporating a boundary layer. The way the boundary layer was conceived allowed him to formulate a mathematical model that integrated the Navier-Stokes equations and the equations of motion for ideal fluids. Although this is an example of phenomenological modelling we are nevertheless given a physical explanation of the importance of viscous flows and how one could understand this process from the point of view of both the Navier-Stokes equations of motion and those of classical hydrodynamics. The model was explanatory because it supplied a representation of different kinds of flows in different regions of the fluid thereby allowing one to understand the nature of the conflict with experiment. Not only was this a way of conceptualising the fluid so that one could treat the flows mathematically but the small water tunnel provided a visualisation of these different areas in the fluid; hence it served as the source for the phenomenological model. The understanding produced by the tunnel in turn furnished the understanding necessary for developing the model. The model is phenomenological not because there is no theory from which to draw but because it is motivated solely by the phenomenology of the physics; in fact, once the model is developed theory does play an important role in its application.

In order to understand how the model of the boundary layer worked we need to first look at the model of a perfect fluid suggested by the classical theory. Here we have a frictionless and incompressible fluid whose motion involves no tangential forces (shearing stresses) when it comes in contact with another body; that is, it offers no internal resistance to changes in shape. The two contacting layers act on each other with normal forces (pressures) only. In the absence of these tangential forces on the boundary between a perfect fluid and a solid wall there is a difference in relative tangential velocities, i.e. there is a slip. The problem with this account is that there is no way to explain the drag of a body.

The falsity of the picture stems from the fact that the inner layers of a real fluid transmit tangential (frictional) as well as normal stresses associated with their viscosity. In real fluids the existence of intermolecular attractions causes the fluid to adhere to a solid wall and thus gives rise to shearing stresses. In most cases like that of water and air there are very small coefficients of viscosity resulting in small shearing stresses. Hence the theory of perfect fluids neglects viscosity which allows for far reaching simplification of the equations of motion. The problem, however, is that even in these cases of very small viscosity (air and water) the condition of no slip prevails near a solid boundary. This condition is the physical origin of the extremely large discrepancy between the value of drag in a real and perfect fluid.

What Prandtl demonstrated with his water tunnel was the existence of this no slip effect in cases of the flow of a fluid with small viscosity past a solid body; that is, there was no slip on the boundary or wall of the body itself. His machine consisted of a hand operated water tank where the water was set in motion by means of a paddle wheel (figure 3.1). The flow was made visible to the naked eye by using a mineral consisting of microscopically small, reddish and very lustrous scales. When the water undergoes rapid strains the scales are oriented and exhibit a lustre allowing one to recognise the formation of vortices. This formation is caused by the frictional or boundary layers. The apparatus

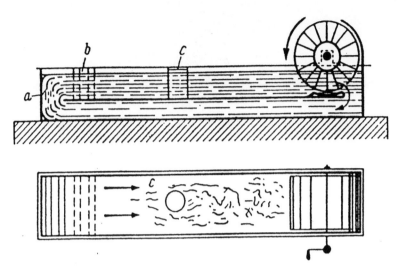

Figure 3.1 Prandtl's water-wheel showing the formation of vortices beyond the circular cylinder (Prandtl 1927, p. 761).

allowed him to examine flow past a wall, flow past a circular arc at zero incidence and flow past a circular cylinder, showing how the vortices formed in each case. Although the existence of frictional forces were known, what the experiments indicated was that the flow about a solid body can be divided into two parts, the thin layer in the neighbourhood of the body where friction plays an essential role (the boundary layer) and the region outside this layer where it can be neglected.

The reason viscous effects become important in the boundary layer is that the velocity gradients there are much larger than they are in the main part of the flow due to the fact that a substantial change in velocity is taking place across a very thin layer. In this way the viscous stress becomes significant in the boundary layer even though the viscosity is small enough for its effects to be negligible elsewhere in the flow. Prandtl's experiments showed that two kinds of flows can exist in the boundary layer; laminar, which involves steady flow with the fluid moving in parallel layers or laminae, and turbulent flow which involves non-regular motion such that the velocity at any point may vary in both direction and magnitude with time. This latter type of motion is accompanied by the formation of eddies and the rapid interchange of momentum in the fluid. Turbulent flow is thought to be caused by instability developed by the laminar boundary layer; the change from

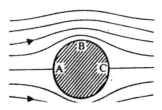

Figure 3.2 Pressure differences at various points around the cylinder.
Depiction of Prandtl's photographs in (1927)

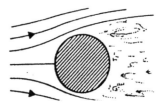

Figure 3.3 Boundary layer separation.

laminar to turbulent flow occurs at a critical value of Reynolds number.

An important consequence of this fact is that in certain circumstances involving flow near blunt bodies, like circular cylinders or spheres, boundary layers can actually separate from the boundary causing the entire flow of a low viscosity fluid to be significantly different from that predicted by the classical theory. The phenomenon of separation is governed by the transition from laminar to turbulent flow. What happens is that flow in the immediate neighbourhood of a solid wall becomes reversed due to external pressure or increase in the velocity of the fluid causing the boundary layer to separate from it. This is accompanied by a pronounced formation of eddies in the wake of the body. Thus the pressure distribution is changed and differs from that in the frictionless stream.

In the case of a circular cylinder, for example, the classical theory predicts a variation of pressure p along the boundary with a local maximum at the forward stagnation point A which falls to a minimum at the midpoint B and increases again to maximum at the end (C) (figure 3.2). This implies that between the mid and end points there is a substantial increase in pressure along the boundary in the direction of flow. It is this severe pressure gradient along the boundary that causes vortices to form and the boundary layer to separate as indicated in figure 3.3. A fluid layer which is set into rotation by friction from the wall pushes itself out into the free fluid causing a complete transformation in the motion.

Although motion in the boundary layers is regulated by the pressure gradient in the mainstream flow, the latter is also markedly influenced by any separation that may occur. The flow is rotational throughout the turbulent region whereas in the absence of separation it would be rotational only in the boundary layer where the viscosity is important; the vorticity would be zero in the mainstream. Hence, we can say that separation causes the vorticity to penetrate from the boundary layer into the fluid.[6] Mathematically what this implies is that what happens in the limit as the viscosity μ goes to 0 can be quite different from what happens when $\mu = 0$. Practically then, boundary layer theory gives an answer to the important question of what shape a body like an aerofoil must have in order to avoid separation.

To see how we get from the conceptual model derived from Prandtl's water tunnel to a mathematical account (for the case of laminar flow) we begin with an incompressible fluid with a constant viscosity coefficient η. The velocity $\mathbf{v}$ of an element of fluid must satisfy the Navier-Stokes equation

$$\partial \mathbf{v}/\partial t + (\mathbf{v} \cdot \nabla)\mathbf{v} = - \nabla p/\rho + (\eta/\rho)\,\nabla^2\,\mathbf{v} \qquad (3.1)$$

where p is the pressure and ρ is the density. The continuity equation is

$$\partial \rho/\partial t \qquad (3.2)$$

and so $\nabla \cdot \mathbf{v} = 0$. A problem is completely defined by this set of equations in addition to the appropriate boundary conditions. These include (1) the fact that at a solid surface the fluid cannot penetrate so that if $\mathbf{n}$ denotes the normal to the surface we can represent this condition as $\mathbf{v} \cdot \mathbf{n} = 0$; and (2) the attractive intermolecular forces between a viscous fluid and a solid surface will cause the tangential component of the velocity to vanish at the surface, i.e. $\mathbf{v} \times \mathbf{n} = 0$.

In some cases we can neglect the viscosity of the fluid in which case it can be considered ideal. This is possible only if the viscosity term $(\eta/\rho)\nabla^2 \mathbf{v}$ in equation 3.1 is much smaller than the inertia term $(\mathbf{v} \cdot \nabla)\mathbf{v}$. So for an ideal incompressible fluid equation 3.1 becomes

$$\partial \mathbf{v}/\partial t + (\mathbf{v} \cdot \nabla)\mathbf{v} = - \nabla p/\rho \qquad (3.3)$$

while equation 3.2 remains the same. For this set of equations only boundary condition (1) applies since the lack of viscous forces prevents

---

[6] Difficulties with theoretical treatment of turbulent flows are discussed in Schlichting (1968). A more general discussion of boundary layer problems can be found in Landau and Lifshitz (1959).

us from making any claims about the tangential velocity of the fluid at the surface. Since 3.3 is a first order partial differential while 3.1 is second order, the solutions for the equations of ideal fluids cannot be required to satisfy condition (2) and in general will not do so. But, as we have seen, the neglect of the viscous forces means that the theory cannot explain the resistive forces acting on a body moving uniformly through a fluid. However, if we consider only equation 3.1 we cannot find an exact solution due to its nonlinearity, hence we need to resort to a model in an attempt to represent the phenomenology of the fluid and find a solvable equation.

As we saw above even a fluid with a very low viscosity cannot be treated as an ideal case since boundary condition (2) is violated. In the model of the fluid containing a boundary layer, if we assume that the layer is very thin, the Navier-Stokes equation can be simplified into a solvable form; condition (2) can then be satisfied and an analytical continuation of the solution in the boundary layer to the region of ideal flow can be given. So, for the steady flow of a fluid over a flat plate we choose the x axis in the direction of flow and the origin so that the plate occupies the plane $y = 0$, $x \geq 0$. (Assume the system is infinite in the z direction so that velocity is independent of z.) The Navier-Stokes and continuity equations are:

$$v_x + \partial v_x/\partial_x + v_y \partial v_y/\partial_y = (1 - \rho)\partial p/\partial_x + (\eta/\rho)(\partial^2 v_x/\partial_x^2 + \partial^2 v_x/\partial_y^2) \tag{3.4}$$

$$v_x + \partial v_y/\partial_x + v_y \partial v_y/\partial_y = (-1/\rho)\partial p/\partial_y + (\eta/\rho)(\partial^2 v_y/\partial_x^2 + \partial^2 v_y)\partial_y^2) \tag{3.5}$$

$$\partial v_x/\partial x + \partial v_y/\partial_y = 0 \tag{3.6}$$

The thinness δ of the boundary layer compared with the linear dimensions of the boundary allows for the possibility of certain approximations in the equations of motion thereby enabling one to determine the flow in the boundary layer.[7] In other words, the boundary layer thickness is supposed to be everywhere small compared with distances parallel to the boundary over which the flow velocity changes appreciably. Across the boundary layer the flow velocity changes from the value 0 at the boundary to some finite value characteristic of an inviscid fluid.

---

[7] If we take l as the length of the plate in the real system, in the boundary layer $\partial/\partial x$ will usually be of the order of $1/l$ and hence much smaller than $\partial/\partial y$ which will be of the order $1/\delta$.

Equation 3.4 can then be replaced by the much simpler boundary layer equation

$$v_x \, \partial v_x / \partial x + v_y \, \partial v_x / \partial y = (\eta/\rho) \, \partial^2 v_x / \partial^2 y \qquad (3.7)$$

which is solved in conjunction with the continuity equation. The boundary conditions for this system are that at the surface of the plate where velocity equals 0

$$v_x = v_y = 0 \text{ for } y = 0 \text{ and } x \geq 0$$

while at large distances from the plate the velocity must be that in the mainstream of the fluid so that

$$v_x \to U \text{ as } y \to \pm\infty$$
$$v_y \to 0 \text{ as } y \to \pm\infty$$

where U is the x-component of velocity just outside the boundary layer. (U is not a rapidly varying function of y so that the impossibility of locating exactly the 'edge' of the boundary layer is of no consequence.) This set of equations can then be solved by defining a stream function to solve the continuity equation. Then using a change to dimensionless variables we can reduce the boundary layer equation to an ordinary nonlinear differential equation (the Blausius equation) which can be solved numerically (see Gitterman and Halpern (1981)).

Once the fluid is divided conceptually into two regions the model allows for a number of approximations to the Navier-Stokes equations that apply in the boundary layer. One then obtains a set of solvable equations for that area while treating the rest of the fluid as ideal, thereby allowing for a different set of solvable equations for the remainder of the fluid. Hence, we have a macroscopic system where the phenomenology of the physics dictates the use of very different approximations for the same homogenous fluid in different parts of the system. Although the solutions for these problems are ultimately dependent on the structural constraints provided by classical hydrodynamics and the Navier-Stokes equations, the important point is that the approximations used in the solutions come not from a direct *simplification* of the mathematics of the theory but from the *phenomenology* of the fluid flow as represented by the model. In other words, neither theory is capable of suggesting how to provide a model that can give solvable equations for viscous fluids, and it is impossible to reconcile the physical differences between real and ideal fluids simply by retaining the no-slip condition in a mathematical

analysis of inviscid fluids. If one drops the viscous force from the equation of motion the order of the differential is reduced by one resulting in
one of the boundary conditions becoming redundant.

What the model of the boundary layer enables us to do is represent
the fluid in a way that facilitates a mathematical solution to the problem
of viscous flow and explains conceptually why neither the theory of ideal
fluids nor the Navier-Stokes equations alone were sufficient for solving
the problem. Of special importance however is the fact that in laminar
flows the behaviour of the fluid outside the boundary layer is governed
by the classical theory so, even though the model is constructed on the
basis of phenomenological considerations, it nevertheless involves a mix
of phenomenological and theoretical components. Both the Navier-
Stokes equations and the classical theory figure in the boundary layer
model, yet the model was constructed and functions independently of
both giving it an autonomous role in the understanding and explanation
of viscous flows.

3.5 REPRESENTATION AND EXPLANATION

Each of the models I have discussed offers a different kind of representational strategy. The pendulum case is an attempt to provide a reasonably
accurate measure of *g*, hence we need a fairly accurate account (model) of
the measuring apparatus. This is possible to the extent that we have a well
established theory that tells us how to calculate for certain kinds of parameters. However, even then there are specific kinds of corrections for which
we cannot get exact solutions, hence we are forced to remain at a level of
abstraction within the model when making the calculations. Nevertheless,
the pendulum model is one of the closest examples of the traditional
notion of representing as mirroring. Although there exist cases like this
they are relatively rare in modern physics which is why we need, ultimately, to revise our ideas about the way that models are constructed and
function. The pendulum example represents an 'ideal case' in the sense
that most modelling in physics does not approximate the picture it presents. That is, we have a theory that allows us to construct a model which
we can then correct to get an increasingly realistic picture of the actual
object or physical system. Sometimes we don't have a well-established
background theory from which to derive our models and even when we
do there is no guarantee it can furnish the models we need. Since many
of the philosophical questions about the nature and function of models
clearly presuppose a close alignment between models and theory it would

seem that a proper understanding of models requires that we begin by changing the kinds of questions we ask. For example, the question of whether a model is realistic or heuristic is structurally too simple to capture the different ways in which models can provide information about the world. The hydrodynamic case illustrates why because it shows not only a rather different model-theory relation but the diversity of ways that models can yield knowledge. Similarly, nuclear models are not realistic in any strong sense, yet they function as an epistemic resource for dealing with specific kinds of nuclear phenomena.

In developing the model of the boundary layer Prandtl began by building a small water tunnel that produced flows past particular kinds of boundaries. With this tank he was able to demonstrate the no-slip condition for viscous fluids and the formation of vortices which led to separation of the boundary layer from particular kinds of bodies. In other words he could physically show that the fluid was in some sense divisible into two separate parts, the boundary layer where frictional forces were significant and the other where the fluid could be treated as non-viscous. Hence the tank itself functioned as a model of fluid flow demonstrating various kinds of phenomena via the set up of a specific kind of idealised structure. The kind of understanding produced by observing the behaviour of fluid in the tank enabled Prandtl to develop a mathematical model that was grounded in the phenomenology of the physics yet made use of theoretical constraints governing fluid mechanics. The conceptual understanding provided by the tank model was transferred to the mathematical model in a way that would have been impossible from either theory alone or other kinds of experimental methods. This was due to the fact that there was no direct route from ordinary experimental findings in hydrodynamics to a way of understanding the conflict with theory.

But, there is another important point about this model – its phenomenological status in no way guaranteed a realistic account of the system. Although the mathematical model described the two areas within the fluid it embodied other abstract assumptions that made it a less than accurate representation of the actual phenomena. As we saw above, in order to give a mathematical treatment of the boundary layer phenomenon it is necessary to resort to the Navier-Stokes equations as well as the equations for an ideal fluid. We know that the ideal fluid ignores viscosity effects but it also takes no account of energy dissipation in a moving fluid due to heat exchange in its various parts. The fluid is regarded as a continuous medium, meaning that any small volume element in the fluid is always supposed so large that it still contains a great number of molecules. In that sense it builds in idealising or abstract assumptions that extend beyond the boundary layer problem. Although the Navier-Stokes equations account for the force of friction they too embody false assumptions about the nature of the liquid. For instance, in the derivation of these equations it is assumed that the fluid is a continuum with density defined at each point and therefore they can be used to represent cases where the density varies continuously from point to point. However in real fluids the density outside an atom or molecule is zero and maximum inside. And, between these values there is a discontinuous transition. Although the Navier-Stokes equations are mathematically intractable, the thinness of the boundary layer allows for their simplification to a good approximation. In other words, the motion of the liquid in Prandtl's flow chamber is represented mathematically by two sets of equations neither of which gives a 'realistic' account of the fluid. However, his model (both the physical tank and mathematical representations) enables one to account for a phenomenon that was untreatable within the theoretical framework provided by the classical and Navier-Stokes equations; but, the idealising assumptions present in the model cannot be concretised beyond their present form – that is just how you represent a fluid in hydrodynamics.

In this case the phenomenological model, rather than producing a descriptively accurate account of fluid flow, yields an abstract representation using different approximations for the same homogenous fluid in different parts of the system. The theoretical model of the pendulum is more straightforward in that many but not all of the abstractions or idealisations can be corrected. Both models are thought to be highly successful from the point of view of their goal and both are thought to provide legitimate representations of physical systems, yet only one (the theoretical model of the pendulum) is realistic in the sense usually defined by philosophical analysis; namely, the product of ever-increasing realistic approximations. Nuclear models 'represent' the nucleus in yet another way. It is not possible to develop a coherent theory of nuclear structure based on knowledge of nuclear forces. Instead we have a number of contradictory models each of which gives a different account of the structure of the nucleus, each of which involves different approximations. In this case what it means to represent the nucleus is context specific and determined by the particular kinds of experimental results that need to be connected with each other. Like the boundary layer case the models are phenomenologically based yet they involve bits of theoretical structure as well.[8]

The important lesson here is that the descriptive accuracy with which models represent physical systems is not simply a function of whether the model is theoretical or phenomenological. Nor does that distinction indicate a reliance on or independence from theory. To that extent the contrast between theoretical and phenomenological models, although important for recognising particular aspects of model construction, ceases to have philosophical importance as a way of isolating realistic models from those that are more abstract.

At the beginning I stressed the relation between the representative capacity of models and their explanatory function. We saw that there are many acceptable ways in which one can represent a system using models, the legitimacy of the representation depends partially on how the model is used. We need a very accurate representation if the pendulum model is going to enable us to measure the force of g but we can settle for a rather abstract characterisation of fluid flow in areas remote from the boundary layer. The reason that models are explanatory is that in representing these systems they exhibit certain kinds of structural dependencies. The model shows us how particular bits of the system are integrated and fit together in such a way that the system's behaviour can be explained.

For example, in hydrodynamics we can model a system that explains a variety of fluid motions provided certain dynamical relations hold, relations that depend on geometrical boundaries and velocity fields. In other words, we can have two flows around geometrically similar bodies, say spheres, that involve different fluids, different velocities, and different linear dimensions yet, if at all geometrically similar points the forces acting on a particle bear a fixed ratio at every instant of time, that is sufficient for the model to explain the streamlines of different flows. The relations expressed in the model are such that they explain 'how possibly' certain kinds of behaviour take place. The reason the model can do this is that it incorporates more detail about structural dependencies than high level theory. For instance, the corrected pendulum model tells us about how the real pendulum behaves, in a way that Newton's laws of motion do not. Similarly, the boundary layer model tells us about

---

[8] One might be tempted to conclude that it is the lack of a unifying theoretical structure that results in the disunity among nuclear models. But this is too simple since the well-established kinetic theory of gases yields several different models of the molecular structure of a gas, one of which contradicts a fundamental theorem of statistical mechanics. See Morrison (1990).

laminar flow in ways that classical fluid dynamics and the Navier-Stokes equations do not. The model explains the behaviour of the system because it contextualises the laws in a concrete way. That is, it shows how they are applied in specific circumstances. The structural dependencies I spoke of above are just those law-like relations that the phenomena bear to one another. The model makes these evident in a way that abstract theory cannot.

This explanatory feature is one of the ways in which we can come to see how models act as autonomous agents. In the examples of both nuclear and hydrodynamic models they provide the only mechanism that allows us to represent and understand, to the extent that we do, experimental phenomena. In the case of boundary layer phenomena the model was able to unify theory and experiment in fluid dynamics in a way that was impossible from a strictly theoretical point of view. In both of these cases the construction of the model involves varying degrees of independence from theory which results in a functional independence for the model itself. But, even in the pendulum case where the model is closely aligned with theory, it is the representation provided by the model in terms of correction factors rather than the sophistication of the apparatus that creates a level of confidence in the accuracy of the measurement. Although the explanatory role is a function of the representational features of the model there is no one way to characterise the nature of that representation. By acknowledging the complexities and diverse natures of models we can begin to appreciate the unique and autonomous position they occupy, one that is separate from both theory and experiment yet able to intervene in both domains.[9]

REFERENCES

Cartwright, Nancy (1989). *Nature's Capacities and their Measurements*. Oxford: Clarendon Press.

Cartwright N., T. Shomar, and M. Suarez (1995). 'The Tool Box of Science', in L. Nowack (ed.), *Poznan Studies in the Philosophy of Science*, Amsterdam: Rodopi.

Dirac, P. A. M. (1958). *Principles of Quantum Mechanics*. 4th edn. Oxford: Oxford University Press.

Fraassen, Bas van (1980). *The Scientific Image*. Oxford: Clarendon Press.

[9] I would like to thank the Wissenschaftskolleg zu Berlin, the Social Sciences and Humanities Research Council of Canada and the Centre for the Philosophy of The Natural and Social Sciences at LSE for support of my research. I am grateful to Mary Morgan for valuable comments on earlier drafts of this chapter and for conversations on the issues I discuss.

Giere, Ronald (1988). *Explaining Science: A Cognitive Approach.* Chicago: University of Chicago Press.

Gitterman, M. and V. Halpern (1981). *Qualitative Analysis of Physical Problems.* New York: Academic Press.

Hertz, Heinrich (1958). *The Principles of Mechanics.* Trans. D. E. Jones and J. T. Walley, New York: Dover Books.

Hacking, Ian (1983). *Representing and Intervening.* Cambridge: Cambridge University Press.

Landau, L. D. and E. M. Lifshitz (1959). *Fluid Mechanics.* Oxford: Pergamon Press.

Morrison, Margaret (1990). 'Unification, Realism and Inference', *British Journal for the Philosophy of Science,* 41, 305–32.

(1998). 'Modelling Nature: Between Physics and The Physical World' *Philosophia Naturalis,* 38, 64–85.

Nelson, Robert and M. G. Olsson (1986). 'The Pendulum – Rich Physics from a Simple System', *American Journal of Physics,* 54, 112–21.

Prandtl, Ludwig (1904). 'Uber Flussigkeitsbewegung bei sehr kleiner Reibung' reprinted in *Gesammelt Abhandlungen.* 3 vols. Berlin: Springer (1961).

(1927). 'The Generation of Vortices in Fluids of Small Viscosity' reprinted in *Gesammelt Abhandlungen* (1961).

Redhead, Michael (1980). 'Models in Physics', *British Journal for the Philosophy of Science,* 31, 145–63.

Schlichting, H. (1968). *Boundary Layer Theory.* 6th edn. New York: McGraw-Hill.

CHAPTER 4

Built-in justification

Marcel Boumans

4.1 INTRODUCTION

In several accounts of what models are and how they function a specific view dominates. This view contains the following characteristics. First, there is a clear-cut distinction between theories, models and data and secondly, empirical assessment takes place after the model is built. In other words, the contexts of discovery and justification are disconnected. An exemplary account can be found in Hausman's *The Separate and Inexact Science of Economics* (1992). In his view, models are definitions of kinds of systems, and they make no empirical claims. Although he pays special attention to the practice of working with a model – i.e. conceptual exploration – he claims that even then no empirical assessment takes place. 'Insofar as one is only working with a model, one's efforts are purely conceptual or mathematical. One is only developing a complicated concept or definition' (Hausman 1992, 79). In Hausman's view, only theories make empirical claims and can be tested. Above that, he does not make clear where models, concepts and definitions come from. Even in Morgan's account 'Finding a Satisfactory Empirical Model' (1988), which comes closest to mine and will be dealt with below, she mentions a 'fund' of empirical models of which the most satisfactory model can be selected. This view in which discovery and justification are disconnected is not in accordance with several practices of mathematical business-cycle model building. What these practices show is that models have to meet implicit criteria of adequacy, such as satisfying theoretical, mathematical and statistical requirements, and be useful for policy. So in order to be adequate, models have to integrate enough items to satisfy such criteria. These items

I gratefully thank Mary Morgan and Margaret Morrison for their encouraging involvement in writing this paper. Thanks also go to the participants of the 'Models as Mediators' workshop at the Tinbergen Institute, 19 March 1996, when this paper was presented and to Hsiang-Ke Chao for his constructive comments.

include besides theoretical notions, policy views, mathematisations of the cycle and metaphors also empirical data and facts. So, the main thesis of this chapter is that the context of discovery is the successful integration of those items that satisfy the criteria of adequacy. Because certain items are empirical data and facts, justification can be built-in.

4.2 THE PROCESS OF MODEL BUILDING

To clarify the integration process, it is very helpful to compare model building with baking a cake without having a recipe. If you want to bake a cake and you do not have a recipe, how do you take the matter up? Of course you do not start blank, you have some knowledge about, for example, preparing pancakes and you know the main ingredients: flour, milk, raising agent and sugar. You also know how a cake should look and how it should taste. You start a trial and error process till the result is what you would like to call a cake: the colour and taste are satisfactory. Characteristic for the result is that you can not distinguish the ingredients in the cake any more.

Model building is like baking a cake without a recipe. The ingredients are theoretical ideas, policy views, mathematisations of the cycle, metaphors and empirical facts. The model-building process will be explored in three case studies. Each case study discusses a classic paper, in the sense that it contains a new recipe that initiated a new direction in business-cycle research. Each recipe is a manual for building a business-cycle model, but in each case the set of ingredients is different. The integration of a new set of ingredients demands a new recipe, otherwise the result will fall apart. However, a recipe is not unique in the sense that it is the one and only way to integrate a certain set of ingredients. Thus a new recipe is a manual for *a* successful integration of a new set of ingredients.

4.3 INTRODUCTION TO THE CASE STUDIES

The first two business-cycle models that are dealt with, Frisch's and Kalecki's, have some features in common. Both were so-called Aftalion models. That is, the cyclic behaviour of the models is mainly based on the characteristic that a period is needed for the production of capital goods (see Aftalion 1927). Secondly, the cycle itself was mathematically represented by the harmonic oscillation. Thirdly, the models were very simple, containing just a few variables and equations, at the most four. Fourthly, Frisch's and Kalecki's models were presented at the Leiden meeting of the Econometric Society in 1933 (Marschak 1934). Both papers had an important impact in the field of the mathematical study of business cycles. They probably presented the first two rigorous business-cycle models ever published (Andvig 1985, p. 85). The main difference between both models is that Kalecki was looking for a fully endogenous explanation of the undamped cycle, because he considered the economic system to be unstable. According to Frisch the economic system was stable and the endogenous cycle was a damped one, so he was looking for an exogenous completion of the explanation of the observed cycle which was of rather stable amplitude.

The third and last case study, Lucas' business cycle model, is a more recent example of a successful integration. Lucas is the central figure of 'new classical economics'. His papers, some of them jointly written with others, published at the beginning of the 1970s were the take-off of this new approach in macroeconomics. His model provided a general-equilibrium account of the Phillips curve, to provide a real challenge to the then dominant neo-Keynesian economics.

Every case study is summarised in a figure, where the relevant model is put in the centre. Around that model the ingredients are placed that are transformed into the model. These transformations are indicated by arrows. Although the list of ingredients should cover the reference list and all the graphs and figures that can be found in the original paper in question, only the most dominant ingredients in the model-building process are discussed here.

4.4 MICHAL KALECKI[1]

The first mathematical business-cycle model in the history of economics was Kalecki's 'Proba teorii koniunktury' (1933).[2] This Polish essay was read in French as 'Essai d'une théorie des mouvements cycliques construite à l'aide de la mathématique supérieure' at the Econometric Society in Leiden in 1933. Its essential part was translated into English

---

[1] Kalecki was born in 1899 in Poland. From 1917 to 1918 he studied at the Polytechnic of Warsaw and from 1921 to 1923 at the Polytechnic of Gdansk. His studies were interrupted by military service and were brought to an end by his father's unemployment, when Kalecki was forced to find work in order to support his family. Kalecki's studies were in engineering. He became a member of the Research Staff of the Institute for the Study of Business Cycles and Prices (ISBCP) in Warsaw in late 1929. In his political views, Kalecki always considered himself a socialist (see Feiwel 1975).

[2] The original Polish text of 1933 is translated and published as 'Essay on the Business Cycle Theory' (Kalecki 1990, pp. 65–108) with only minor editorial corrections.

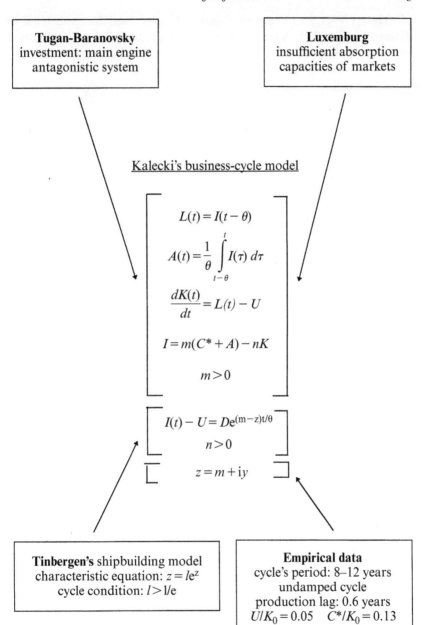

Figure 4.1 Kalecki's business-cycle model.

as 'Macrodynamic Theory of Business Cycles', and was published in
Econometrica in 1935. A less mathematical French version was published
in *Revue d'Economique Politique,* also in 1935. The English paper was discussed in Tinbergen's survey (1935), appearing in the same issue of
Econometrica. The model ingredients are summarised in figure 4.1.

4.4.1 The model

Kalecki's model contained four equations describing 'the functioning of
the economic system as a whole', using Frisch's term 'macrodynamic' to
denote this ambition. Three relations described capitalistic production:

$$L(t) = I(t - \theta) \tag{4.1}$$

$$A(t) = \frac{1}{\theta} \int_{t-\theta}^{t} I(\tau) \, d\tau \tag{4.2}$$

$$\frac{dK(t)}{dt} = L(t) - U \tag{4.3}$$

where I denotes total of investment orders, A, total production of capital
goods, L, volume of deliveries of industrial equipment, and, K, volume
of existing industrial equipment. U denotes demand for restoration of
equipment used up and was assumed to be constant. θ denotes the
average production lag. The fourth relation resulted from the interdepen-
dence between investment and yield of existing enterprises (see below):

$$I = m(C^* + A) - nK \tag{4.4}$$

where C^* denotes autonomous consumption, m and n are positive.
  Kalecki arrived at this equation in the following way. B, total real
income of capitalists is equal to the sum of their consumption, C, and
production of capital goods, A:

$$B = C + A \tag{4.5}$$

Consumption, C, is composed of an autonomous part, C^*, and a part
proportional to income, B:

$$C = C^* + \lambda B \tag{4.6}$$

From this we get:

$$B = \frac{1}{1 - \lambda}(C^* + A) \tag{4.7}$$

The ratio of volume of investment orders, I, to the volume of existing industrial equipment, K, is an increasing function of the gross yield, B/K:

$$\frac{I}{K} = f\left(\frac{B}{K}\right) \tag{4.8}$$

We already saw that B is proportionate to $C^* + A$, we thus obtain:

$$\frac{I}{K} = \phi\left(\frac{C^* + A}{K}\right) \tag{4.9}$$

Kalecki assumed ϕ to be linear, i.e.

$$\frac{I}{K} = m\left(\frac{C^* + A}{K}\right) - n \tag{4.10}$$

where the constant m is positive. Normally, n is not restricted to any range of values. But it will be shown below that to get the main cycle Kalecki had to assume n to be positive too.

4.4.2 The ingredients

4.4.2.1 Theoretical ingredients: Marxist economics

Kalecki's essay was not inspired by the contemporary mainstream of orthodox economics but first and foremost by Marxist economics, whose theorists of crises were N. Tugan-Baranovsky and R. Luxemburg. Tugan-Baranovsky's ideas were developed at the beginning of the twentieth century by, among others, A. Aftalion. Kalecki's model was one of the two first attempts at a mathematisation of verbal theories of the business cycle.

Tugan-Baranovsky. The role of investment as the main factor of reproduction in capitalism was an element of Kalecki's theory which he owed to Tugan-Baranovsky. Many years later, in 1967, Kalecki wrote that he regarded Tugan-Baranovsky's argument on problems of realisation in capitalism as his lasting contribution to the analysis of how capitalism functions in its various phases (Osiatynski 1990, 439). Tugan-Baranovsky was possibly the first interpreter of Marx's schemes of reproduction to stress investments as the main engine of capitalistic economic development in Marx's theory (see Reuten's chapter in this volume). Tugan-Baranovsky believed that capitalism was not a 'harmonious' but an 'antagonistic' system. In an antagonistic system consumption is neither the ultimate goal nor the criterion of economic activity. Production which only serves further production is entirely justified, provided that it is profitable. Hence he did not regard as absurd the assumption that capitalism is based on investments that serve only further investment and so, with the appropriate inter-industry proportions, the development of capitalism did not depend on sales outlets.

   Luxemburg. The fact that Kalecki did not aim at an equilibrium model was inspired by the contemporary debates about the theory of capital accumulation put forward by Luxemburg. She emphasised the difficulties of realising production because of the insufficient absorption capacity of markets, which she believed was a barrier to expanded reproduction under capitalism. Kalecki himself several times pointed to his ties with Luxemburg's theory and, through it, with the Marxist school of thought (Osiatynski 1990, 439).

4.4.2.2 Mathematical moulding

To mould the above Marxist views into a mathematical model of a cycle, Kalecki used Tinbergen's ship-building paper (1931). Tinbergen's ship-building model, in itself an example of a very important investment cycle, provided the idea of how to construct a mathematical model of the cycle and also the mathematical techniques used for that purpose.

   Tinbergen's ship-building model. In an empirical study, Tinbergen found that the increase in tonnage, $f'(t)$, is an inverse function of total tonnage two years earlier, $f(t - \theta)$, i.e.

$$f'(t) = -af(t - \theta) \tag{4.11}$$

where $a > 0$, $\theta \approx 2$ years. This equation was analysed by Tinbergen by solving the characteristic equation

$$z = a\theta e^z \tag{4.12}$$

where z is a complex number, $z = x + iy$.

   As a result, the general solution is a sum of trigonometric functions:

$$f(t) = \sum_{k=1}^{\infty} D_k e^{-x_k t} \sin(y_k t + \omega_k) \tag{4.13}$$

where the amplitudes D_k and the phases ω_k are determined by the shape of the movement in an initial period. It followed from Tinbergen's analysis that only one sine function had a period longer than the delay, θ, and that this cycle only exists when $a\theta > e^{-1}$. According to Tinbergen this cycle was the only cycle with an economic meaning, because the other sine functions had a period shorter than the delay θ. The parameter a had a value somewhere between $\frac{1}{2}$ and 1, so that $a\theta > e^{-1} (= 0.37)$

and thus the main cycle in Tinbergen's model existed and had a period of about 8 years.

4.4.2.3 Empirical data

The observed business cycle was a rather stable cycle: 'In reality we do not observe a clear *regular* progression or digression in the amplitude of fluctuations' (Kalecki 1990, 87; see also Kalecki 1935, 336). By 'statistical evidence' the cycle's period was given to be between 8 to 12 years. The average production lag was determined on the basis of data of the German *Institut für Konjunkturforschung*. The lag between the beginning and termination of building schemes was 8 months, the lag between orders and deliveries in the machinery-making industry was 6 months. So, Kalecki assumed the average duration of θ to be 0.6 years.

Two other important empirical values (see below) were U/K_0 and C^*/K_0, where K_0 is the average value of K. The 'rate of amortisation' U/K_0, determined on the basis of combined German and American data, was about 0.05. For the evaluation of C^*/K_0, he used only American data, and fixed on 0.13.

Clearly Kalecki did not use Polish data, although he worked at the *Polish Institute for Economic Research*, which affiliation was printed at the bottom of his *Econometrica* paper. The most probable reason for using German and American data, and not Polish data, was that one of his model assumptions was that total volume of stocks remains constant all through the cycle. This assumption was justified by existing 'totally or approximately isolated' economic systems like that of the United States.

4.4.3 The integration

The integration of the ingredients (discussed above) had to be done in such a way that mathematisation of the Marxist views resulted in a reduced form equation which resembled Tinbergen's cycle equation and fulfilled its cycle criterion. Beyond that, the cycle which resulted from that equation should meet the generic empirical facts. How far one is able to reconstruct the integration process that actually took place is, of course, very difficult but with the aid of Kalecki's published works one can lift a corner of the veil. Namely, one part of mathematical moulding is calibration: the values of certain parameters have to be chosen in such a way to make the integration successful.

From Kalecki's four-equation model (equations (4.1) to (4.4)) one can derive an equation in one variable, the so called reduced form equation.

The reduced form equation of this four-equation model was a mixed differential-difference equation of both differential order and difference order one:

$$(m + \theta n)\mathcal{J}(t - \theta) = m\mathcal{J}(t) - \theta\mathcal{J}'(t) \tag{4.14}$$

where $\mathcal{J}(t)$ is the deviation of $I(t)$ from U, $\mathcal{J}(t) = I(t) - U$. To use Tinbergen's ship-building results to discuss his own macro-model, Kalecki transformed this equation into the equation Tinbergen analysed in his ship-building paper (cf. equation 4.12), by assuming that $\mathcal{J}(t) = De^{(m-z)t/\theta}$:

$$z = le^z \tag{4.15}$$

where $l = e^{-m}(m + \theta n)$. One result was that the main cycle only exists when the following inequality is satisfied:

$$l > e^{-1} \tag{4.16}$$

which is equivalent to:

$$m + \theta n > e^{m-1} \tag{4.17}$$

The parameter m is already assumed to be positive (see equation 4.10). It can be shown that the above inequality is satisfied only when n is positive too. In other words the main cycle exists only when n is positive.

In the original Polish essay of 1933 Kalecki tried to prove that n must be positive. This proof was questioned by Rajchman (Kalecki 1990, 471), who in conclusion, rightly, accepted the condition $n > 0$ as an additional assumption by Kalecki. In his 'Macro-dynamic Theory' Kalecki (1935) asserts only that the constant m is positive, but adds that a necessary condition for a cyclical solution is the positive value also of the coefficient n.

As in Tinbergen's case, z is a complex number: $x + iy$. The solution of the reduced form equation 4.14 was:

$$\mathcal{J}(t) = ae^{(m-x)t/\theta}\sin yt/\theta \tag{4.18}$$

where a is a constant. Kalecki also chose x to be equal to m, so that the main cyclical solution became constant, which was in accordance with reality: 'This case is especially important because it corresponds roughly to the real course of the business cycle' (Kalecki 1990, 87; see also Kalecki 1935, 336):

$$\mathcal{J}(t) = a \sin yt/\theta \tag{4.19}$$

By taking into consideration this 'condition of a constant amplitude' ($x = m$), Kalecki derived from equation 4.15 the following equations:

$$\cos y = m/(m + \theta n) \qquad (4.20)$$

and

$$y/tgy = m \qquad (4.21)$$

Between m and n there was another dependency for they are both coefficients in the equation 4.4. That equation must also hold true for the one-cycle-averages of I and A equal to U, and for the average value of K equal to K_0:

$$U = m(C^* + U) - nK_0 \qquad (4.22)$$

Hence:

$$n = (m-1)U/K_0 + mC^*/K_0 \qquad (4.23)$$

Using his empirical values for θ, U/K_0 and C^*/K_0, the result was that the model generated a cycle with a period of 10 years, which was in accordance with the empirical business cycle period, ranging from 8 to 12 years, so 'the conclusions from our theory do not differ very much from reality' (Kalecki 1990, 91; see also Kalecki 1935, 340).

4.4.4 Comments

By building his business-cycle model, Kalecki was able to integrate a list of ingredients: Marxist's theoretical ideas on the role of investment and reproduction in capitalistic economies, Tinbergen's mathematical model of an investment cycle and generic data of the business cycle. To make the integration of these ingredients satisfactory, two parameters, n and m, had played a crucial but controversial role. The choice of n to be positive was not suggested by economic nor by empirical considerations but was only justified by the motive of integration: it made the model fulfil Tinbergen's cycle condition. The choice of the real part x of the complex number z to be equal to m, enabled Kalecki to integrate into the model the fact that the cycle is rather stable. This choice was not suggested by economic theory nor by Tinbergen's cycle model. Thanks to the integration of the cycle condition and the characteristic of a stable cycle the ingredients could be combined together to make a model with a resulting cycle period of 10 years. This was seen by Kalecki as an empirical justification of the model and thus as a justification for choice of both *n* and *x*. Kalecki's case shows that integration and justification are both sides of the same coin.

4.5 RAGNAR FRISCH[3]

Frisch's model presented at the Leiden meeting was published in the Cassel volume (Frisch 1933), and was also discussed by Tinbergen in his 1935 survey. The model, known as the Rocking Horse Model consisted of two elements each solving a 'fundamental' problem in business-cycle theory: the 'propagation' problem and the 'impulse' problem.[4] The 'propagation' model, taking the largest part of the paper, showed how an initial shock is propagated through the system, by explaining the character of the cycle by the structural properties of the system. The 'propagation' model did not show how the cycles are maintained: 'when an economic system gives rise to oscillations, these will most frequently be damped. But in reality the cycles we have occasion to observe are generally not damped' (Frisch 1933, 197). The 'impulse model' solved that problem. In his discussion of Kalecki's paper, Frisch was very critical about Kalecki's 'condition of a constant amplitude'. Frisch's point was that this condition is very sensitive to variations in the data. In order to have a stable cycle, the parameter concerned had to have exactly that particular value and 'since the Greeks it has been accepted that one can never say an empirical quantity is exactly equal to a precise number' (Frisch quoted in Goodwin 1989). Thus, Frisch sought a new recipe, not so dependent on exact values of parameters, and one which was more in accordance with the kind of statistical data actually observed: the business cycle is not a smooth harmonic oscillator but more like a changing harmonic, a cycle of which the period lengths and amplitudes are varied. To start again with the complete picture, the model ingredients are summarised in figure 4.2.

---

[3] Frisch was born in Norway in 1895. He studied economics from 1916 to 1919. In 1921 he received a scholarship to go abroad to study economics and mathematics. He received his PhD in 1926 on a mathematical statistical subject. Frisch was appointed to 'privat docent' at the University of Oslo in 1923. In the following year he was advanced to assistant professor. In 1928 he became a lecturer ('docent') and was appointed full professor in economics at the University of Oslo in 1931. Frisch became more and more left wing with the years. During the interwar period he was not active in any political party, but his views changed roughly from supporting the liberal party to supporting the right wing of the Labor Party.

[4] Frisch's well-known 'rocking horse model' is treated at length in Andvig (1985), and fully discussed from the econometric point of view in Morgan (1990). In this chapter I only deal with those aspects of interest in relation to the other models.

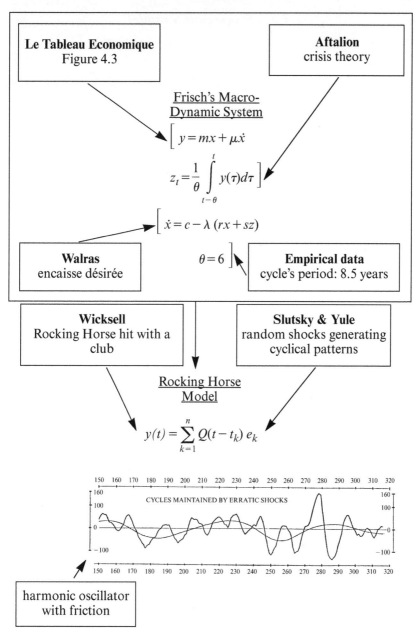

Figure 4.2 Frisch's Rocking Horse Model (Frisch 1933, 203).

4.5.1 Propagation model

Frisch's propagation model is like a nutcake, after baking the cake some of the ingredients are still recognisable (see figure 4.2). Therefore, in the discussion of this model the ingredients and their transformations into the model are jointly dealt with.

Frisch's propagation model consisted of three equations. The first equation, called the 'production policy equation', was Frisch's version of the accelerator principle: the need for an increase in total capital stock was caused by the fact that annual consumption was increasing. The accelerator principle was derived from his version of the *Tableau Economique* giving the relations between 'the human machine' (i.e. consumption and labour), land, stock of capital goods, stock of consumers' goods, production of capital goods and production of consumer goods.

$$y = mx + \mu \dot{x} \tag{4.24}$$

where x is consumption and y is production started.

In fact the equation was derived from a simplified version (figure 4.3) of his *Tableau Economique* in which there is no land and no stock of consumers' goods.

The stock of capital goods, Z, was taken as an essential element of the analysis. When the yearly depreciation on the capital stock is $hx + ky$, the rate of increase of this stock will be:

$$\dot{Z} = y - (hx + ky) \tag{4.25}$$

The capital stock needed to produce the yearly consumption, x, is equal to:

$$Z = \eta x \tag{4.26}$$

By defining $m = h/(1 - k)$ and $\mu = \eta/(1 - k)$ we obtain the above production equation (4.24).

The second equation of the model was Frisch's mathematical interpretation of Aftalion's insight into the production delay, namely that the discrepancy in time between the moment of starting the production of a capital good and the moment of the completion of that good may provoke cyclical oscillations. This equation was mathematically equivalent to Kalecki's second production equation (4.2):

$$z_t = \frac{1}{\theta} \int_{t-\theta}^{t} y(\tau) d\tau \tag{4.27}$$

where z is carry-on activity.

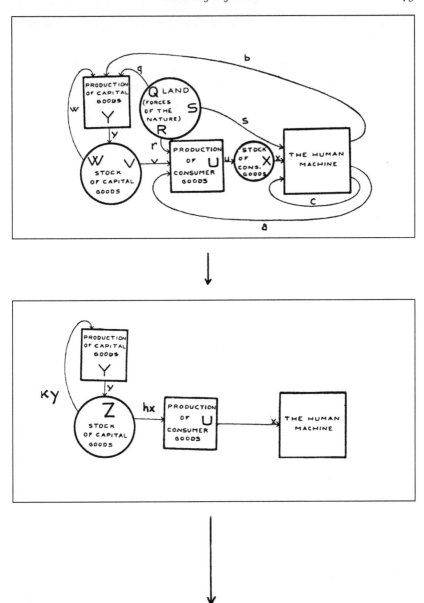

Figure 4.3 Frisch's *Tableau Economique* (Frisch 1933, 174).

The third equation, called the 'consumption equation', contained a 'monetary brake mechanism'.

$$\dot{x} = c - \lambda \, (rx + sz) \qquad (4.28)$$

This mechanism was based on the notion of Walras' *'encaisse désirée'*, depending on consumption and production, and diminishing the rate of increase of consumption in boom periods, because of 'limitations of gold supply, the artificial rigidity of the monetary systems, psychological factors, and so on' (Frisch 1933, 179).

This last equation was considered very controversial. Tinbergen in his survey stated that 'it is desirable that these relations be analysed economically and investigated statistically before they are made a basis for a theory, especially when the number of competing possibilities is as large as here. It might be that the "brake" included in the mechanism of equation . . . works in a different way' (Tinbergen 1935, 271).

4.5.2 The integration of the propagation model

For the same reasons as in Kalecki's case, the discussion will concentrate on the calibration part.

The reduced form equation of Frisch's three-equation model is a mixed differential-difference equation:

$$\theta \ddot{x}_t + (\theta \lambda r + \lambda s \mu)\dot{x}_t + \lambda s m x_t - \lambda s m x_{t-\theta} - \lambda s \mu \dot{x}_{t-\theta} = 0 \qquad (4.29)$$

which is much more complicated than Tinbergen's ship-building equation, because of the higher differential order. Frisch didn't solve the equation analytically but used the method of numerical approximation. As a consequence, the values of the model parameters had to be chosen in advance. The result was a set of three damped cycles: 8.57 years, 3.5 years and 2.2 years, of which the first two had a period in accordance with the observed cycle periods. Both the damping character and the cycle periods found convinced Frisch that 'the results here obtained, in particular those regarding the length of the primary cycle of $8\frac{1}{2}$ years and the secondary cycle of $3\frac{1}{2}$ years, are not entirely due to coincidence but have a real significance' (Frisch 1933, 190).

The parameter values were not chosen at random. Frisch admitted that they were the outcome of very rough guessing. Some values 'may in a rough way express the magnitudes which we would expect to find in actual economic life' (Frisch 1933, 185), other values were even 'rougher' estimates (Frisch 1933, 186). But, better estimates would leave the damping characteristic and the period lengths unaltered, provided that these values fulfil certain weak conditions. However, this did not apply for the length of the construction period of capital goods, which Frisch assumed to be 6 years. According to Tinbergen the chosen value was too high and at the same time the lengths of the solution periods were highly dependent on it (Tinbergen 1935, 271–2). Thus, although the damping character was not dependent on a specific value of the parameters, the period lengths as solutions of the reduced form equation (4.29) were almost fully determined by the construction period, θ. The high value for θ was probably chosen to obey the empirical criteria of the cycle lengths.

4.5.3 The propagation and impulse model

Frisch explained the maintenance of the dampening cycles and the irregularities of the observed cycle by the idea of erratic shocks regularly disturbing the economic system. He brought the propagation and impulse elements together in the following way:

$$y(t) = \sum_{k=1}^{n} Q(t - t_k) e_k \qquad (4.30)$$

where $Q(t)$ is a damped oscillation determined by the propagation model and e_k's are the random shocks.

4.5.4 The ingredients of the propagation and impulse model and their integration

4.5.4.1 The integrating metaphor
The propagation and impulse model integrated 'the stochastical point of view and the point of view of rigidly determined dynamical laws' (Frisch 1933, 198). The idea to solve the integration problem was delivered by Knut Wicksell by means of a 'perfectly simple and yet profound illustration' (Frisch 1933, 198): a rocking horse hit by a club.

4.5.4.2 Stochastical moulding
Eugen Slutsky and George Udny Yule had shown that cycles could be caused by a cumulation of random shocks. Slutsky's (1937, originally 1927 in Russian), 'Summation of Random Causes as the Source of Cyclic Processes' dealt with the problem of whether it is possible 'that a definite structure of a connection between random fluctuations could form them into a system of more or less regular waves' (Slutsky 1937,

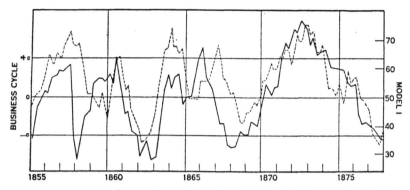

-------- An index of English business cycles from 1855 to 1877; scale
on the left side. ------- Terms 20 to 145 of Model I; scale on the right side.

Figure 4.4 Slutsky's random series juxtaposed with business cycles (Slutsky 1937, 110).

106). Economic waves, like waves on the sea, never repeat earlier ones
exactly, either in duration or in amplitude. So he considered harmonic
analysis an unsatisfactory way to analyse them. He showed for a cyclic
process that it is not necessary that the system itself generates a cycle,
any weight system to be used in the cumulation of erratic shocks could
generate cyclic behavior:

$$y(t) = \Sigma_k Q_k e_{tk} \qquad (4.31)$$

where Q_k's are constants.

Slutsky studied experimentally the series obtained by performing iter-
ated differences and summations on random drawings (lottery drawings,
etc.). For example he plotted the following series, called Model I:

$$y_t = x_t + x_{t-1} + \ldots + x_{t-9} + 5 \qquad (4.32)$$

where the x_t's are obtained by taking the last digits of the drawings of a
government lottery loan. Without further comment he plotted in the
same figure an index of the English business cycles for 1855–1877 in jux-
taposition, implying that one could describe business cycle data with
such an equation.

4.5.4.3 Mathematical moulding
The ingredient that was used for the mathematical moulding was the
equation which describes a pendulum hampered by friction:

$$\ddot{y} + 2\beta\dot{y} + (\alpha^2 + \beta^2)y = 0 \qquad (4.33)$$

The solution of this equation can be written in the form

$$y(t) = P(t - t_0)y_0 + Q(t - t_0)\dot{y}_0 \qquad (4.34)$$

where $P(t) = pe^{-\beta t}\sin(v + \alpha t)$ and $Q(t) = qe^{-\beta t}\sin\alpha t$ and where p, q and v are functions of α and β.

When the pendulum is hit at the points of time t_k $(k = 1 \ldots n)$ by shocks of strength e_k the velocity is suddenly changed from $\dot{y}_k$ to $\dot{y}_k + e_k$. So that after n shocks $y(t)$ will be

$$y(t) = P(t - t_0)y_0 + Q(t - t_0)\dot{y}_0 + \sum_{k=1}^{n} Q(t - t_k)e_k \qquad (4.35)$$

After a while the influence of the first two terms will be negligible because they will be almost damped out, so that the result will be the model equation 4.30.

4.5.4.4 Empirical facts
Like Slutsky, Frisch produced experimentally a graph as a cumulation of random shocks, the weight function being now a damped sine function. The result was 'the kind of curves we know from actual statistical observation' (Frisch 1933, 202).

4.5.5 Comments

It is already obvious from figure 4.2 that Frisch's case is a much more complex and sophisticated example of the integration process. Because the model had to explain a more complicated phenomenon: a regular irregular cycle – a changing harmonic – the model not only contains more ingredients, but also has more layers. Each layer solved a specific problem. Slutsky's and Yule's model of the business cycle (see figure 4.4) was explicitly non-theoretical. There was no economic explanation of the choice of the weights in the summation of the random shocks. Frisch's propagation model provided such an economic theoretical explanation of the weights in the cumulation of the random shocks that generated these nice realistic cyclical patterns. Thus, empirical evidence was explicitly built-in: it only needed theoretical foundation.

The lengths of the cycle's periods generated by the propagation model contributed to its empirical adequacy but were clearly built in by the choice of the production delay. The other ingredients provided in their turn the theoretical foundation of the propagation model. Frisch's case shows how, in each layer, theoretical foundation and empirical justification were integrated, but only the combination provided a model with could be considered adequate by all the criteria.

4.6 ROBERT LUCAS[5]

Lucas' 1972 paper discussed here, 'Expectations and the Neutrality of Money', established a general-equilibrium account of the Phillips curve and showed that this 'non-neutrality of money' is caused by misperception. The belief in the neutrality of money is the idea that changes in the money supply affect the price level only and are neutral with respect to output and employment. The Phillips curve is a phenomenological relation between inflation and output (or employment), named after its discoverer. It was an empirical relation and an important Keynesian policy instrument, but it could not be backed up by Keynesian theories. Another problem was its instability and the theoretical explanation of this instability. Its instability became dramatically apparent at the beginning of the 1970s. Lucas' account of the Phillips curve and its instability – in 1972 – was therefore a real challenge to the then still dominant Keynesian standards.

After the Second World War the business cycles became less strong in amplitude and the cyclic character became less apparent. Gradually the conception of the business cycles shifted from component cycles to stochastic fluctuations in output. A model that integrated a harmonic oscillation as the basic concept of the business cycle did not satisfy anymore. As was mentioned in section 4.2, new ingredients ask for new recipes.

4.6.1 The ingredients

In Lucas' case we have the advantage that we have an explicit account of the ingredients that Lucas wanted to integrate and how they were chosen. These ingredients are summarised in figure 4.5.

In the introduction of his *Studies in Business-Cycle Theory* (1981) Lucas recollects how Edmund Phelps hosted a conference imposing as a ground rule that no author could discuss his own paper, but instead certain basic questions presented by Phelps in the form of an agenda.

---

[5] Lucas was born in 1937 in the United States. In 1964, Lucas received his PhD from the University of Chicago. From 1963 to 1974, he stayed at Carnegie-Mellon University. He is now professor in economics at the University of Chicago. In his political views he considered himself on the right.

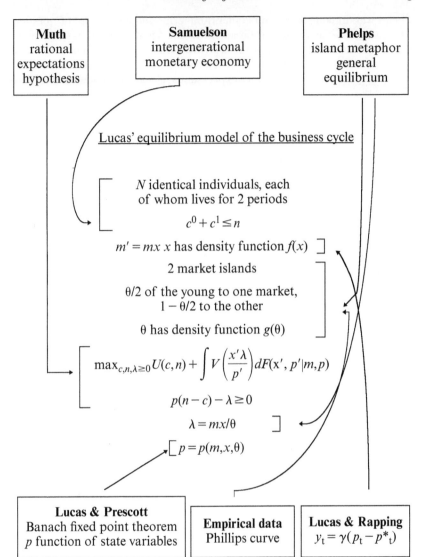

Figure 4.5 Lucas' equilibrium model of the business cycle.

Phelps was thinking in general-equilibrium terms and used his list of questions to focus our discussion in this direction. If one agent is fooled into imagining relative prices are moving in his favor, is his action not offset by another with the opposite misapprehension? . . . Why formulate price expectations as adaptive in the *levels*. . .? How can the price expectations relevant for labour-market decisions differ from those relevant for bond-market decisions? Is it possible, in short, to describe an entire economy operating in a mutually consistent way that is led into large-scale employment fluctuations via informational imperfections alone? (Lucas 1981, 6–7)

Lucas' model was an attempt to answer these questions in a mathematical form.

The initial idea simply was to situate some Lucas-Rapping households in an elementary economy, subject the system to stochastic shocks to the money supply, and see what would happen. Samuelson's intergenerational monetary economy offered a convenient analytical setting because it both fit the two-period consumer maximization problem Rapping and I had studied and the function of 'money' in it was so clearly defined. The idea of defining an equilibrium as a point in a space of functions of a few 'state variables' was one that Prescott and I had utilized in 'Investment under Uncertainty'. This analytical device had forced Prescott and me to be precise as to the meaning of terms like *information and expectations* and led us to formulate and utilize Muth's rational-expectations hypothesis in exactly the way I then used it in 'Expectations and the Neutrality of Money'. In short, the needed ingredients for a general-equilibrium formulation seemed to be readily at hand. (Lucas 1981, 7)

Lucas and Rapping (1969) had conducted an econometric study of the labour market, which showed that expectational errors on the part of workers are mainly responsible for fluctuations in real output (presented in the most simple way):

$$y_t = \gamma(p_t - p_t^*) \tag{4.36}$$

where p_t^* denotes price expectations. In that paper they had used the adaptive expectations hypothesis (again in the most simple way presented):

$$p_t^* = p_{t-1} + \delta(p_{t-1} - p_{t-1}^*) \tag{4.37}$$

In the 'Neutrality' paper, Lucas used Muth's rational expectations hypothesis:

$$p_t^* = E(p|I_{t-1}) \tag{4.38}$$

where I_{t-1} denotes all available information at $t-1$. But the first time Lucas used the rational expectations hypothesis was in a joint paper with Edward Prescott (1971) on investment behavior. To determine the characteristics of the equilibrium path they had defined equilibrium as a function of state variables and used the Banach fixed point theorem. It is thus from this paper Lucas transported the main methods used in his 'Neutrality' model.

But these ingredients were not enough to obtain a Phillips curve. In a one-sector system with monetary shocks as the only source of uncertainty, the monetary shocks acted like neutral monetary movements. 'At this point, it became clear to me why Phelps had imagined an island economy, with traders scattered and short on useful, system-wide information' (Lucas 1981, 7). This 'island' feature as we see below, became a key element in getting nonneutrality.

4.6.2 The integration in the model

The model is built up as follows. Each period, N identical individuals are born, each of whom lives for two periods. The production-consumption possibilities for any period are described (in per capita terms) by:

$$c^0 + c^1 \leq n \tag{4.39}$$

where n is the output produced by a member of the younger generation, c^0 output consumed by its producer, and c^1 output consumed by a member of the older generation.

Exchange occurs at two physically separated markets, between which there is no communication within a trading period. The older generation is allocated across these two markets so as to equate total monetary demand between them. The young are allocated stochastically across both markets, fraction $\theta/2$ going to one and $1-\theta/2$ to the other, introducing fluctuations in relative prices between the two markets. θ has density function g. A second source of disturbance arises from stochastic changes in the quantity of money, which in itself introduces fluctuations in the nominal price level:

$$m' = mx \tag{4.40}$$

where m is the pre-transfer money supply, per member, of the older generation; m' is the post-transfer balance; and x is a random variable with density function f. Thus the state of the economy is entirely described by three variables m, x, and θ.

All individuals evaluate current consumption c, current labour supplied, n, and future consumption, c', according to the common utility function

$$U(c,n) + \mathrm{E}\{V(c')\} \tag{4.41}$$

where the functions U and V fulfil certain conditions.

Because future consumption cannot be purchased directly by an age-0 individual, a known quantity of nominal balances λ is acquired in exchange for goods. If next period's price level is p' and if next period's transfer is x', these balances will then purchase $x'\lambda/p'$ units of future consumption. Then the decision problem facing an age-0 person is:

$$\max_{c,n,\lambda \geq 0} U(c,n) + \int V\left(\frac{x'\lambda}{p'}\right) dF(x', p'|m,p) \qquad (4.42)$$

subject to:

$$p(n-c) - \lambda \geq 0$$

The Kuhn-Tucker conditions for this maximisation problem are:

$$U_c(c,n) - p\mu \leq 0, \qquad \text{with equality if } c>0$$
$$U_n(c,n) + p\mu \leq 0, \qquad \text{with equality if } n>0$$
$$p(n-c) - \lambda \geq 0, \qquad \text{with equality if } \mu>0$$

$$\int V'\left(\frac{x'\lambda}{p'}\right)\frac{x'}{p'} dF(x',p'|m,p) - \mu \leq 0 \qquad \text{with equality if } \lambda>0$$

where μ is a non-negative multiplier. When the three first equations are solved for c, n, and $p\mu$ as functions of λ/p, it appears that the solution value for $p\mu$ is a positive, increasing function, $h(\lambda/p)$.

Money supply, after transfer, in each market is $Nmx/2$. Because the market receives a fraction $\theta/2$ of the young, the quantity supplied per demander is $(Nmx/2)/(\theta N/2) = mx/\theta$. Equilibrium requires that

$$\lambda = mx/\theta \qquad (4.43)$$

Since $mx/\theta > 0$, substitution into the last Kuhn-Tucker condition gives the equilibrium condition:

$$h\left(\frac{mx}{\theta p}\right)\frac{1}{p} = \int V'\left(\frac{mxx'}{\theta p'}\right)\frac{x'}{p'} dF(x',p'|m,p) \qquad (4.44)$$

It was remarked earlier that the state of the economy is fully described by the three variables m, x and θ. So, one can express the equilibrium price as a function $p(m,x,\theta)$ on the space of possible states. The probability distribution of next period's price, $p' = p(m',x',\theta') = p(mx,x',\theta')$ is known, conditional on m, from the known distributions of x, x', and θ'. Thus the equilibrium price is defined as a function $p(m,x,\theta)$, which satisfies the above equilibrium equation.

Using the Banach fixed point theorem, Lucas showed that the equilibrium equation has an unique solution. Further, he showed that this unique equilibrium price function has the form $m\varphi(x/\theta)$. In other words, the current price informs agents only of the *ratio* x/θ. 'Agents cannot discriminate with certainty between real and monetary changes in demand for the good they offer, but must instead make inferences on the basis of the known distribution $f(x)$ and $g(\theta)$ and the value of x/θ revealed by the current price level' (Lucas 1972, 114).

A consequence of the fact that agents cannot discriminate between real and monetary demand shifts is that monetary changes have real consequences. In other words, the model suggests that there exists 'a nonneutrality of money, or broadly speaking a Phillips curve, similar in nature to that which we observe in reality' (Lucas 1972, 103): He considered the following variant of an econometric Phillips curve

$$y_t = \beta_0 + \beta_1(p_t - p_{t-1}) + \epsilon_t \qquad (4.45)$$

where y_t is the log of real GNP (or employment) in period t, p_t the log of the implicit GNP deflator for t, and $\epsilon_1, \epsilon_2, \ldots$ a sequence of independent, identically distributed random variables with 0 mean.

From the model one can derive that the approximate probability limit of the estimated coefficient $\hat{\beta}_1$ is positive:

$$\hat{\beta}_1 = \frac{\eta_n \eta_\varphi}{1 - 2\eta_\varphi + 2\eta_\varphi^2} > 0 \qquad (4.46)$$

where η_n and η_φ are the elasticities of the functions n and φ, respectively, evaluated at $E[\ln x]$. This positive estimate for β_1 was interpreted as 'evidence for the existence of a "trade-off" between inflation and real output' (Lucas 1972, 118), in other words, the Phillips curve.

4.6.3 Comments

Lucas' case nicely supports the integration thesis, because in his own recollection of the model building, Lucas used the same terminology to describe this process. Lucas had already some experience, shared with Prescott and Rapping, as a model-builder before he made his famous recipe. That earlier experience was ingeniously used to arrive at his 'Neutrality' model. The successful integration of the ingredients (see figure 4.5) convinced Lucas that most ingredients to build a business-cycle model were elementary and they remain in his later models.

Exceptions are: Samuelson's intergenerational monetary economy and the Banach fixed point theorem. The first one because there are other ways to introduce the uncertain future, and the last one because, once he had proved the existence of an equilibrium price as a function of state variables with the aid of this theorem, existence could then be assumed in his later work.

4.7 MATHEMATICAL MOULDING

An important element in the modelling process is mathematical moulding. Mathematical moulding is shaping the ingredients in such a mathematical form that integration is possible, and contains two dominant elements. The first element is moulding the ingredient of mathematical formalism in such a way that it allows the other elements to be integrated. The second element is calibration, the choice of the parameter values, again for the purpose of integrating all the ingredients.

As a result, the choice of the mathematical formalism ingredient is important. It determines the possibilities of the mathematical moulding. In the 1930s, to people such as Tinbergen, Frisch and Kalecki, mixed difference-differential equations seemed to be the most suitable formalism for a business-cycle model. With hindsight, one can doubt the validity of the choice of that ingredient.

In general, it is difficult to solve mixed differential-difference equations. Moreover, at the time Tinbergen, Kalecki and Frisch were studying them, there were hardly any systematic accounts available. Systematic overviews of the subject have appeared only since the beginning of the 1950s.[6] As a consequence, they were studied by Tinbergen, Kalecki and Frisch as if they were the same as the more familiar differential equations. Differential equations are solved by assuming the solution to be a complex exponential function. From this assumption the characteristic equation can be derived. The roots of this characteristic

---

[6] E.g., R. Belmann, and K. Cooke, *Differential-Difference Equations*, New York: Academic Press, 1963. However, the various mathematical aspects of Kalecki's analysis already attracted attention in the 1930s. In the first place, there is Frisch and Holme's (1935) treatment of Kalecki's equation, 'The Characteristic Solutions of a Mixed Difference and Differential Equation Occurring in Economic Dynamics', *Econometrica*, 3, 225–239, but also three papers by R. W. James and M. H. Beltz: 'On a Mixed Difference and Differential Equation', *Econometrica*, 4, 1936, 157–60; 'The Influence of Distributed Lags on Kalecki's Theory of the Trade Cycle', *Econometrica*, 6, 1938, 159–62; 'The Significance of the Characteristic Solutions of Mixed Difference and Differential Equations, *Econometrica*, 6, 1938, 326–43.

equation can be real or complex. The order of the equation determines the number of roots. Complex roots lead to periodic solutions (trigono-metric functions) and real roots to exponential solutions. The general solution is a finite weighted sum of these solutions. The weights are determined by the initial conditions.

Tinbergen, Kalecki and Frisch used the same method to solve mixed difference-differential equations. As a result the general solution is again a weighted sum of harmonic functions. But now the sum consists of an infinite number of terms (see, for example, equation 4.13). The total sum of harmonic functions determine the ultimate behaviour of the variables, and from Fourier analysis we know that this can be any arbitrary movement when the weights are not further spec-ified. Tinbergen and Kalecki considered only the first cycle and Frisch only the first three cycles in the harmonic series to be important, and thus disregarded the rest. The question is whether this is justified. In Zambelli's (1992) 'Wooden Horse That Wouldn't Rock' Frisch's model is analysed and worked out with computer simulations which show that when Frisch's model is subjected to an external shock, it evolves back to the equilibrium position in a non-cyclical manner.

4.8 TWO RELATED ACCOUNTS

In the paper I have argued that the model-building process is the integration of several ingredients in such a way that the result – the model – meets certain *a priori* criteria of quality. And because some of the ingredients are the generic data the model is supposed to explain, justification is built-in. Both theses differ from two related accounts, namely Morgan's (1988) account of what satisfactory models are and Cartwright's (1983) simulacrum account of how models are built. Morgan's view is that econometricians of the 1930s 'have been primar-ily concerned with finding satisfactory empirical models, not with trying to prove fundamental theories true or untrue' (Morgan 1988, 199). The ideas about assessing whether the models were 'satisfactory' were rarely expressed clearly and depended on the purpose of the models (Morgan 1988, 204). Morgan gives five ideas that cover the aims and criteria of the early econometricians:

1. To measure theoretical laws: Models must satisfy certain theoreti-cal requirements (economic criteria).
2. To explain (or describe) the observed data: Models must fit observed data (statistical or historical criteria).

3. To be useful for policy: Models must allow the exploration of policy options or make predictions about future values.
4. To explore or develop theory: Models must expose unsuspected relations or develop the detail of relationships.
5. To verify or reject theory: Models must be satisfactory or not over a range of economic, statistical, and other criteria. (Morgan 1988, 205)

Morgan (see also Kim, De Marchi and Morgan 1995) interprets these early econometricians' idea of testing as something like quality control testing. Criteria were applied to empirical models: 'Do they satisfy the theoretical criteria? Do they satisfy standard statistical criteria? Can they be used to explore policy options? Do they bring to light unexpected relationships, or help us refine relationships? A model found to exhibit desired economic-theoretical and statistical qualities might be deemed satisfactory' (Kim, De Marchi and Morgan, 83). The empirical models were matched both with theory and with data, to bridge the gap between both.

Another view that is related to mine is Nancy Cartwright's simulacrum account of models in her *How the Laws of Physics Lie* (1983). Her account deals with the problem of bridging the gap between theory and phenomena in physics. Her aim is to argue against the facticity of fundamental laws, they do not picture the phenomena in an accurate way. For this we need models: 'To explain a phenomenon is to find a model that fits it into the basic framework of the theory and thus allows us to derive analogues for the messy and complicated phenomenological laws which are true of it' (Cartwright 1983, 152). The striving for too much realism in the models may be an obstacle to explain the relevant phenomenon. For that reason she introduces an 'anti-realistic' account of models: models are simulacra: 'the success of the model depends on how much and how precisely it can replicate what goes on' (Cartwright 1983, 153).

To fulfill this bridge function, Cartwright argues that models consist partly of genuine properties of the objects modelled, but others will be merely properties of convenience or fiction. The properties of convenience are introduced into the model to bring the objects modelled into the range of the theory. These latter properties play an important role in her argument that fundamental explanatory laws cannot be interpreted realistically. To bridge the gaps on the route from phenomena to models to theory, properties of convenience or fiction have to be introduced.

The main difference between the view of this chapter and Morgan's and Cartwright's accounts is that they conceive models as instruments to bridge the gap between theory and data.

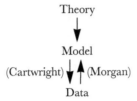

This view is too one-dimensional; this chapter maintains the view that models integrate a broader range of ingredients than only theory and data.

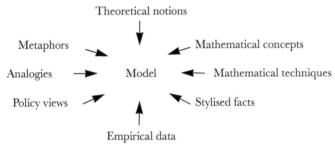

Cartwright's account is on how models are built to fit theory to data. Her conception of models is a subcase of the view developed in this paper. In the first place because it is one-dimensional (see above). In her view, theory is true of the objects of the model and the model is true of the objects in reality (Cartwright 1983, 4). In my account a broader range of ingredients are integrated and the truth relation has a different direction (see the arrows in both figures above): the model is true for all of these ingredients. Secondly, when she talks about theories, she assumes these already provide a mathematical framework, in contrast to the verbal theories used in the business-cycle cases, where the mathematical formalism is one of the ingredients that has to be integrated. But more importantly, the introduction of properties of convenience in the model is a special case of mathematical moulding. Of course, Cartwright's account was designed to clarify her position in a realism debate, so she does not go further into the meaning of the concept of convenience.

The emphasis in Morgan's account of satisfactory empirical models is on what good models are: models are matched both with theory and data to satisfy certain criteria. She compares this way of satisfying with

quality control testing. Her story is about econometric models of the
1930s, but her conception of quality control testing can be extrapolated
to the classical mathematical business-cycle models. Morgan's account
of satisfactoriness is for that matter more clarifying and broader. Models
not only have to meet theoretical requirements but also other 'qualities'.
The case studies of the business-cycle models show that the success of a
model-building process depends on the fulfilment of a broader range of
requirements than only those from theories.

4.9 CONCLUSIONS

The main thesis of this chapter is that the model-building process of
first-generation mathematical models is the integration of several ingredients in such a way that the result – the model – meets certain *a priori*
criteria of quality. The ingredients are elements of the set containing
theoretical notions, metaphors, analogies, mathematical concepts and
techniques, policy views, stylised facts and empirical data. The integration is satisfactory when the model meets a number of *a priori* criteria.
These criteria are a selection of the following not necessarily complete
list of requirements. The model could be

1. a solution to theoretical problems: e.g. Frisch's model provided a
    theoretical account of the weights in the cumulation of the random
    shocks;
2. an explanation of empirical phenomena: e.g. Lucas' model estab-
    lished a theoretical account of the Phillips curve and its instability;
3. an indication of the possibilities of economic policy: e.g. Kalecki's
    model provided an instrument for economic policy, whereas Lucas'
    model showed the impossibilities of Keynesian policy;
4. a provider of a mathematical conception of the relevant phenom-
    ena: e.g. Kalecki's model conceptualised a mathematical frame-
    work which generated a constant cycle with a realistic period
    length.

   The above thesis contains two subtheses: One is about 'discovery', the
other about 'justification', but both are closely connected.

THE CONTEXT OF DISCOVERY   In the model-building process mathematical moulding plays a key role. The integration takes place by transforming the ingredients into a mathematical form to merge them into
one framework. One aspect of this moulding is the choice of the most
adequate mathematical formalism. Another aspect is calibration: the

choice of the parameters in such a way that the model not only fits the generic data, but also integrates the other ingredients.

THE CONTEXT OF JUSTIFICATION   When the set of ingredients contains the generic facts the model is supposed to explain, then justification is built-in.

The three classic models discussed in this chapter, ranging from fully endogenous to fully exogenous explanations of the business cycle, were satisfactory to the model builders because they covered the theoretical, mathematical and statistical requirements of the builders, fitted the data they saw as most characteristic of the business cycle and mirrored their theoretical views on the economic system.

The characteristics of the cycle covered by each model were in accordance with the observed business cycle and made the three men believe that their model had real significance. In each case, these characteristics were presented at the end of their paper, giving the strong suggestion that the justification of their model was disconnected from the building process. This is how it was presented in each publication. Such presentation hides the actual process of model building, which is more like a trial and error process till all the ingredients, including the empirical facts, are integrated. In other words, justification came not afterwards but was built-in.

REFERENCES

Aftalion, A. (1927). 'The Theory of Economic Cycles Based on the Capitalistic Technique of Production', *Review of Economic Statistics*, 9, 165–70.

Andvig, J. C. (1985). *Ragnar Frisch and the Great Depression*. Oslo: Norwegian Institute of International Affairs.

Cartwright, N. (1983). *How the Laws of Physics Lie*. Oxford: Clarendon Press.

Feiwel, G. R. (1975). *The Intellectual Capital of Michal Kalecki*. Knoxville: University of Tennessee Press.

Frisch, R. (1933). 'Propagation Problems and Impulse Problems in Dynamic Economics', in *Economic Essays in Honour of Gustav Cassel*. London: Allen & Unwin.

Goodwin, R. M. (1989). 'Kalecki's Economic Dynamics: A Personal View', in *Kalecki's Relevance Today*. London: Macmillan.

Hausman, Daniel M. (1992). *The Inexact and Separate Science of Economics*. Cambridge: Cambridge University Press.

Kalecki, M. (1935). 'A Macrodynamic Theory of Business Cycles', *Econometrica*, 3, 327–44.

(1990). 'Essay on the Business Cycle Theory', in Osiatynski (1990).

Kim, J., N. De Marchi and M. S. Morgan (1995). 'Empirical Model Particularities and Belief in the Natural Rate Hypothesis', *Journal of Econometrics*, 67, 81–102.

Lucas, R. (1972). 'Expectations and the Neutrality of Money', *Journal of Economic Theory*, 4, 103–24.

    (1981). *Studies in Business-Cycle Theory*. Cambridge MA: MIT Press.

Lucas, R. and E. Prescott (1971). 'Investment under Uncertainty', *Econometrica*, 39, 659–81.

Lucas, R. and L. Rapping (1969). 'Real Wages, Employment, and Inflation', *Journal of Political Economy*, 77, 721–54.

Marschak, J. (1934). 'The Meeting of the Econometric Society in Leyden, September-October, 1933', *Econometrica*, 2, 187–203.

Morgan, M. S. (1988). 'Finding a Satisfactory Empirical Model', in Neil De Marchi (ed.), 199–211, *The Popperian Legacy in Economics*. Cambridge: Cambridge University Press.

    (1990). *The History of Econometric Ideas*. Cambridge: Cambridge University Press.

Osiatynski, J. (ed.) (1990). *Collected Works of Michal Kalecki, Volume I, Capitalism, Business Cycles and Full Employment*. Oxford: Clarendon Press.

Slutsky, E. (1937). 'Summation of Random Causes as the Source of Cyclic Processes', *Econometrica*, 5, 105–46.

Tinbergen, J. (1931). 'Ein Schiffbauzyklus?', *Weltwirtschaftliches Archiv*, 34, 152–64.

    (1935). 'Annual Survey: Suggestions on Quantitative Business Cycle Theory', *Econometrica*, 3, 241–308.

Zambelli, S. (1992). 'The Wooden Horse That Wouldn't Rock: Reconsidering Frisch', 27–54, in K. Velupillai (ed.), *Nonlinearities, Disequilibria and Simulation*. Basingstoke: Macmillan.

CHAPTER 5

The Ising model, computer simulation, and universal physics

R. I. G. Hughes

So the crucial change of emphasis of the last twenty or thirty years that distinguishes the new era from the old one is that when we look at the theory of condensed matter nowadays we inevitably talk about a 'model.'

Michael Fisher (1983, 47)

It is a curious fact that the index of *The New Physics* (Davies 1989), an anthology of eighteen substantial essays on recent developments in physics, contains only one entry on the topics of computers and computer simulation. Curious, because the computer is an indispensable tool in contemporary research. To different degrees, its advent has changed not just the way individual problems are addressed but also the sort of enterprise in which theorists engage, and hence the kind of theory that they propose. Consider, for example, chaos theory. Although the ideas underlying the theory were first explored by Poincaré at the turn of the century,[1] their development had to await the arrival of the computer. In *The New Physics* the beautiful pictures of fractal structures that illustrate the essay on chaos theory (Ford 1989) are, of course, computer generated. Yet, perhaps because it runs counter to the mythology of theoretical practice, that fact is mentioned neither in the captions that accompany the pictures nor elsewhere in the text. The indispensable has become invisible.

The solitary entry on computers in the index takes us to the essay, 'Critical Point Phenomena: Universal Physics at Large Length Scales', by Alastair Bruce and David Wallace, and, within that essay, to a description of the so-called *Ising model* and the computer simulation of its behaviour. The model is at the same time very simple and very remarkable. It is used to gain insight into phenomena associated with a diverse group

---

[1] See Poincaré (1908). A summary is given in Ruelle (1991, 48–50); note Ruelle's references in the last sentence of this chapter to 'the essential role [of computer simulation] in the modern theory of chaos'.

of physical systems – so diverse, in fact, that the branch of physics that explores what is common to them all is called 'universal physics'.

This paper has two sections. In section 5.1 I set out the relations between the phenomena, the Ising model, and the general theory of critical point behaviour; and then outline the role played by computer simulations of the Ising model's behaviour. In section 5.2 I show how the Ising model in particular, and computer simulations in general, can be accommodated within a philosophical account of theoretical representation.

5.1 CRITICAL POINT PHENOMENA AND THE ISING MODEL

5.1.1 The phenomena

Various apparently dissimilar physical systems – magnets, liquids, binary alloys – exhibit radical changes in their properties at some critical temperature.[2]

(a) Above the Curie temperature, T_c(770°C), a specimen of iron will exhibit paramagnetic rather than ferromagnetic behaviour; that is to say, above T_c it can be only feebly magnetised, below T_c its magnetic susceptibility is very high.[3]

(b) At the boiling point of H_2O two phases, liquid and vapour can co-exist. The boiling point increases smoothly with pressure until the critical point is reached ($p_c = 218$ atmospheres, $T_c = 374$°C). At this point the two phases cannot be distinguished; to quote Thomas Andrews, lecturing on similar behaviour in CO_2, at this point, 'if anyone should ask whether it is now in the gaseous or liquid state, the question does not, I believe, admit of a positive reply.'[4]

(c) Within a narrow range of temperatures around its critical temperature a colourless fluid may exhibit critical opalescence, 'a peculiar appearance of moving or flickering striae throughout its entire extent'.[5] By 1900 this effect had been observed in a large number of fluids; subsequently Smulakowsky and Einstein attributed it to fluctuations in the refractive index of the fluid caused by rapid local changes in its density.

---

[2] The four I list are discussed in detail in Domb (1996), a book which is simultaneously a text-book of critical point physics and an internal history of its development. It is very comprehensive, if somewhat indigestible. As an introduction to critical point physics I recommend Fisher (1983).

[3] The elementary treatment of ferromagnetism given in chapter 7 of Lee (1963) is still useful.

[4] Andrews (1869), quoted by Domb (1996, 10).

[5] Andrews (1869), quoted by Domb (1996, 10).

(d) In the 1920s and 1930s x-ray diffraction experiments on various
binary alloys (e.g. copper-gold, copper-zinc) indicated that a transition from order to disorder within an alloy's crystal lattice could
occur at a critical temperature which was well below the alloy's
melting point.

The list is far from exhaustive – transitions to and from a superconducting or a superfluid phase are obvious other examples[6] – but the
four above indicate the diversity of critical point phenomena.

5.1.2 The Ising model

An Ising model is an abstract model with a very simple structure. It
consists of a regular array of points, or *sites*, in geometrical space.
Bruce and Wallace consider a square lattice, a two-dimensional array
like the set of points where the lines on a sheet of ordinary graph
paper intersect each other; the dimensionality of the lattice, however,
is not stipulated in the specification. The number N of lattice sites is
very large. With each site I is associated a variable s_i, whose values are
$+1$ and -1. An assignment of values of s_i to the sites of the lattice
is called an *arrangement, a*, and we write '$s_i(a)$' for the number assigned
to the site I under that arrangement. The sum $\Sigma_i s_i(a)$ then gives us the
difference between the number of sites assigned $+1$ and -1 by a. We
may express this as a fraction of the total number N of sites by
writing,

$$M_a = (1/N)\Sigma_i s_i(a) \tag{5.1}$$

If we think of an arrangement in which each site in the lattice is assigned
the same value of s_i as *maximally ordered*, then it is clear that M_a gives us
the degree of order of an arrangement and the sign of the predominant
value of s_i. For the two maximally ordered arrangements, $M_a = \pm 1$; for
any arrangement which assigns $+1$ to exactly half the lattice sites,
$M_a = 0$.

So far the specification of the model has no specific physical content.
Now, however, two expressly physical concepts are introduced. First,
with each adjacent pair of sites, $<j,k>$, is associated a number $-Js_j s_k$,
where J is some positive constant. This quantity is thought of as an interaction energy associated with that pair of sites. It is a function of s_j and
s_k, negative if s_j and s_k have the same sign, positive if they have opposite

---

[6] See, e.g., Pfeuty and Toulouse (1977, 151–2 and 166–7, respectively).

signs. The total interaction energy E_a of the arrangement a is the sum of these pairwise interaction energies:

$$E_a = \Sigma_{<j,k>} - \mathcal{J}s_j(a)s_k(a) \qquad (5.2)$$

where $<j,k>$ ranges over all pairs of adjacent sites. Clearly, E_a is at a minimum in a maximally ordered arrangement, where $s_j = s_k$ for all adjacent pairs $<j,k>$.

Secondly, the lattice is taken to be at a particular temperature, which is independent of the interaction energy. Between them, the interaction energy E_a and the absolute temperature T determine the probability of p_a of a given arrangement; it is given by Boltzmann's law, the fundamental postulate of statistical thermodynamics:

$$p_a = \mathcal{Z}^{-1}\exp(-E_a/kT) \qquad (5.3)$$

(k is Boltzmann's constant; $\mathcal{Z}$ is called the *partition function* for the model; here its reciprocal $\mathcal{Z}^{-1}$ acts as a normalising constant, to make sure that the probabilities sum to one. To use the terminology of Willard Gibbs, equation 5.3 expresses the fact that the set of arrangements forms a *canonical ensemble*.) The probability of a given arrangement decreases with E_a, and hence with the number of adjacent pairs of opposite sign. As T approaches absolute zero, disordered arrangements have a vanishing probability, and it becomes virtually certain that the system will be in one or other of the fully ordered minimal energy arrangements (which one depends on the system's history). On the other hand, an increase in temperature will flatten out this dependence of p_a on E_a. As a result, since there are many more strongly disordered arrangements than there are strongly ordered arrangements, at high temperatures there will be a very high probability of nearly maximal disorder, in which roughly equal numbers of sites will be assigned positive and negative values of s_i.

The probable degree of order, so to say, is measured by the *order parameter*, M. This parameter is obtained by weighting the values of M_a for each arrangement a by the probability p_a of that arrangement:

$$M = \Sigma_a p_a M_a \qquad (5.4)$$

Since we are keeping $\mathcal{J}$ (and hence E_a) constant, and summing over all possible arrangements, M is a function of the temperature, and is at a maximum at absolute zero. Between the extremes of low and high temperature, the Ising model exhibits the analogy of critical point behaviour. There is a critical region, a range of values of T in which the value of M drops dramatically, and also a critical temperature, T_c, above which M is effectively zero.

Another significant temperature-dependent parameter is the *correlation length*. It is defined indirectly. Although the couplings between adjacent sites act to correlate the values of s_i, at sites several spacings apart, thermal agitations tend to randomise them. The result is that the correlation coefficient $\Gamma(r)$ for the lattice falls off exponentially with distance r.[7] We write, $\Gamma(r) = \exp(-r/\xi)$. Since ξ in this equation has the dimensions of length, we may regard the equation as an implicit definition of the *correlation length*.[8] Effectively this parameter provides a measure of the maximum size of locally ordered regions – islands, as it were, within the lattice as a whole.[9] At low temperatures, where the lattice is highly ordered, these islands are locally ordered regions within which the value of s_i is opposite to the predominant one; thus, if M_a for a typical arrangement is close to -1, they are regions where s_i is uniformly positive. At high temperatures the relevant islands exist within a sea of general disorder, and can be of either sign. At both high and low temperatures ξ is small. Near the critical temperature, however, these islands can be very large; indeed, in the infinite Ising model, as T approaches T_c from either direction, ξ tends to infinity. It is worth emphasising that ξ gives a measure of the *maximum* size of locally ordered regions. As will appear in section 5.1.7, the emergence of critical point phenomena depends crucially on the fact that, near the critical temperature, islands of all sizes up to the correlation length co-exist, and participate in the behaviour of the model.

5.1.3 Interpretations of the Ising model

Although Bruce and Wallace focus on the two-dimensional Ising model for expository reasons, notably the ease with which computer simulations of its behaviour can be displayed, there are, in fact, various kinds

[7] A correlation function for a pair of sites at separation r is defined as follows. For a given arrangement a we denote the values of s_i at the two sites by $s_0(a)$, $s_r(a)$. Mean values of s_0 and the product $s_0 s_r$ are defined by: $<s_0> =_{\mathrm{df}} Z^{-1}\Sigma_a \exp(-H/kT)s_0$; $<s_0 s_r> =_{\mathrm{df}} Z^{-1}\Sigma_a \exp(-H/kT)s_0 s_r$. ($H$ is the total energy of the lattice.) Since, if s_0 and s_r are uncorrelated, we would expect $<s_0 s_r> = <s_0><s_r>$, we define the correlation function $\Gamma(r)$ by $\Gamma(r) =_{\mathrm{df}} <s_0 s_r> - <s_0><s_r>$. When $H = E_a$, $<s_0> = 0 = <s_r>$, and so $\Gamma(r) = <s_0 s_r>$. The more general recipe given here allows for added energy terms, like those associated with an external magnetic field.

[8] Formally, $\xi = -r/\ln\Gamma(r)$. Since $\Gamma(r)$ is never greater than one, its logarithm is never positive.

[9] Strictly, fluctuations allow locally ordered regions to appear and disappear. The correlation length ξ may more properly be thought of as a measure of the size of those locally ordered regions that persist for a certain length of time. We may also define a correlation time that measures the mean lifetime of such regions – see Pfeuty and Toulouse (1977, 5).

of physical systems for which it provides a model. Pfeuty and Toulouse (1977, p. 4) show properly Gallic tastes in listing 'films, adsorbed phases, solids made up of weakly-coupled planes (like *mille-feuille* pastry)'.[10] But, as I mentioned earlier, nothing in the specification of the model is peculiar to the two-dimensional case. In this section I will go back to the phenomena I listed earlier, and see how the abstract three-dimensional Ising model can be interpreted in terms of them.

(a)   Ising himself thought of the site variable in his model as the direction of the magnetic moment of an 'elementary magnet'. As he wrote later, 'At the time [the early 1920s] . . . Stern and Gerlach were working in the same institute [The Institute for Physical Chemistry, in Hamburg] on their famous experiment on space quantization. The ideas we had at that time were that atoms or molecules of magnets had magnetic dipoles and that these dipoles had a limited number of orientations.'[11] The two values of s_i in the abstract Ising model are interpreted as the two possible orientations, 'up' or 'down', of these dipoles, and the coupling between neighbouring dipoles is such that less energy is stored when their moments are parallel than when they are anti-parallel. The order parameter, M, for the lattice is interpreted as the magnetisation of the ferromagnetic specimen, and the correlation length as a measure of the size of magnetic domains, regions of uniform dipole orientation.

(b and c)   A model of liquid-vapour mixture near the critical point that is isomorphic to the Ising model was introduced by Cernuschi and Eyring in 1939, and the term 'lattice gas' first appeared in papers by Yang and Lee in 1952.[12] In the lattice gas, the Ising model is adjusted so that the two values of s_i are 1 and 0. The sites themselves are thought of as three-dimensional cells, and the value, 1 or 0, assigned to a site is taken to denote the presence or absence of a molecule in the corresponding cell. For an adjacent pair of cells, $<j,k>$, $-Js_js_k$ takes the value $-J$ when $s_j = s_k = 1$, and zero otherwise. In other words, there is no interaction between an empty cell and any of its neighbours. The order parameter now depends on the fraction of cells occupied, i.e. the mean density of the liquid-vapour mixture, and the correlation length on the size of droplets within the vapour. Local fluctuations in the density give rise to local variations in refractive index, and hence to critical opalescence.

(d)   A binary alloy is a mixture of atoms of two metals, A and B. More realistic here than a simple cubic Ising model, in which all lattice sites

---

[10] See also Thouless (1989) on condensed matter physics in fewer than three dimensions.

[11] Ernst Ising, letter to Stephen Brush (undated), quoted in Brush (1967, 885–6). On Stern and Gerlach, see Jammer (1966, 133–4).   [12] See Domb (1996, 199–202), Brush (1967, 890).

are equivalent, would be a lattice that reflected the crystal structure of a particular alloy (e.g., a specific brass alloy). Very often such lattices can be decomposed into two equivalent sublattices. For example, if we take a cubic lattice, whose sites can be labelled $a_1, a_2, \ldots a_i, \ldots$, and introduce into the centre of each cube another site b_j, then the set $\{b_j\}$ of sites introduced will constitute another lattice congruent with the first. Which is to say that a body-centred cubic lattice can be decomposed into two equivalent cubic lattices interlocked with each other. The value of s_i that an arrangement assigns to each site I of this composite lattice represents the species of the atom at that site. An arrangement is perfectly ordered when each a-site is occupied by an atom of species A, and each b-site by an atom of species B, and maximally disordered when the atoms are randomly assigned to sites. We may define a parameter S_α analogous to the function M_a defined for the Ising model, and hence derive for the composite lattice an order parameter that depends, not on pair-wise couplings between neighbouring atoms, but on the long-range regularity present in the system as a whole.[13] As in the standard Ising model, the greater the degree of order in the lattice, the less the amount of energy stored in it; it was by analogy with the idea of long-range order in alloys that Landau (1937) introduced the term 'order parameter' into the theory of critical phenomena in general.[14]

Notice that if we take a simple cubic Ising model and label alternate sites along each axis as a-sites and b-sites, then it becomes a composite lattice of the general type I have just described. The set of a-sites and the set of b-sites both form cubic lattices whose site spacing is twice that of the original lattice. As before, we let the value of s_i represent the species of atom at site I. Using an obvious notation, we allow the interaction energies between neighbouring sites to take the values

---

[13] I have assumed that there are as many A-atoms as a-sites, and B-atoms as b-sites. For this special case, let p_α denote the proportion of a-sites occupied by A-atoms under the arrangement α. Then we may write:

$$S_\alpha = {}_{df} [p_\alpha - p_{random}] / [p_{perfect} - p_{random}]$$
$$= [p_\alpha - 1/2] / [1 - 1/2]$$
$$= 2p_\alpha - 1.$$

To confirm that the behaviour of S_α mimics that of M_a in the Ising model described earlier, observe what occurs (i) when all A-atoms are on a-sites, (ii) when A-atoms are randomly distributed over the lattice, and (iii) when all A-atoms are on b-sites. Bragg and Williams (1934, 702–3) generalise this definition to accommodate the general (and more realistic) case, when there are fewer a-atoms than a-sites.

[14] See Domb (1996, 18). Note, however, that many authors (e.g. Amit 1984 and Domb himself) still use 'magnetisation' rather than 'order parameter' as a generic term; see 5.1.6, below.

J_{AA}, J_{AB}, J_{BB}. If $J_{AB} < J_{AA}$ and $J_{AB} < J_{BB}$, then arrangements of maximum order (with A-atoms on *a*-sites and B-atoms on *b*-sites, or conversely) become arrangements of minimum energy. The ordered regions whose size is given by the correlation length are now regions within which the value of s_i alternates from one site to the next.

Simple cubic crystal lattices are rare, and it might appear that, to be useful in any individual case, the Ising model would have to be modified to match the crystal structure of the alloy in question. As we shall see, however, there is an important sense in which the simple cubic Ising model adequately represents them all.

5.1.4 A historical note

The success of the Ising model could scarcely have been foreseen when it was first proposed. The digest of Ising's 1925 paper that appeared in that year's volume of *Science Abstracts* ran as follows:

A Contribution to the Theory of Ferromagnetism. E. Ising. (*Zeits. F. Physik*, 31. 1–4, 253–58, 1925) – An attempt to modify Weiss' theory of ferromagnetism by consideration of the thermal behaviour of a linear distribution of elementary magnets which (in opposition to Weiss) have no molecular field but only a non-magnetic action between neighbouring elements. It is shown that such a model possesses no ferromagnetic properties, a conclusion extending to a three-dimensional field. (W. V. M.)

In other words, the model fails to leave the ground. Proposed as a model of ferromagnetism, it 'possesses no ferromagnetic properties'. Small wonder, we may think, that Ising's paper was cited only twice in the next ten years, on both occasions as a negative result. It marked one possible avenue of research as a blind alley; furthermore, it did so at a time when clear progress was being made in other directions.

In the first place, in 1928 Heisenberg proposed an explanation of ferromagnetism in terms of the newly triumphant quantum theory. His paper was, incidentally, one of the places in which Ising's work was cited. Heisenberg wrote: 'Ising succeeded in showing that also the assumption of directed sufficiently great forces between two neighbouring atoms of a chain is not sufficient to explain ferromagnetism.'[15] In the second, the *mean-field* approach to critical point phenomena was proving successful. At the turn of the century Pierre Curie and Pierre Weiss had explored

---

[15] Heisenberg (1928). The translation is by Stephen Brush (1967, 288). On the Heisenberg model, see Lee (1963, 114–15).

an analogy between the behaviour of fluids and of magnets. Weiss had taken up van der Waals' earlier suggestion that the intermolecular attractions within a gas could be thought of as producing a negative 'internal pressure',[16] and had proposed that the molecular interactions within a magnet might, in similar fashion, be thought of as producing an *added* magnetic field. He wrote, 'We may give $[H_{int}]$ the name *internal field* to mark the analogy with the internal pressure of van der Waals' (Domb 1996, 13). The van der Waals theory of liquid-vapour transitions, and the Weiss theory of ferromagnetism (the very theory that Ising tells us he tried to modify) were examples of *mean-field theories*, so called because in each case an internal field was invoked to approximate the effect of whatever microprocesses gave rise to the phenomena. Other examples were the Curie–Weiss theory of antiferromagnetism, and the Bragg–Williams theory of order-disorder transitions in alloys.[17] The latter was put forward in 1934, the authors noting that 'the general conclusion that the order sets in abruptly below a critical temperature T_c has a close analogy with ferromagnetism', and going on to list 'the many points of similarity between the present treatment and the classical equation of Langevin and Weiss' (Bragg and Williams 1934, 707–8). All these mean-field theories found a place in the unified account of transition phenomena given by Landau in 1937.[18] It was referred to as the 'classical account', not to distinguish it from a quantum theoretic account, but to indicate its status.

Given these developments, how did the Ising model survive? Imre Lakatos, writing on the methodology of scientific research programmes, observes: 'One must treat budding programmes leniently' ([1970] 1978, 92). But in 1935 the Ising approach to ferromagnetism was hardly a budding programme; it had withered on the vine. Any attempt to revive it, one might think, would have been a lurch from leniency into lunacy.

Yet in 1967 Stephen Brush would write a 'History of the Lenz–Ising model', in 1981 a review article entitled 'Simple Ising models still thrive'

---

[16] Van der Waals took the ideal gas law, $PV = RT$, and modified the volume term to allow for the non-negligible volume of the molecules, and the pressure term to allow for their mutual attraction. The resulting equation, $(P - a/V^2)(V-b) = RT$, more nearly approximates the behaviour of real gases.

[17] On mean-field theories in general, see Amit (1984, 6–8) and Domb (1996, 84–6). On antiferromagnetism, see Lee (1963, 202–04).

[18] Landau (1937) himself did not start from the analogy between pressure and magnetic field that I have described. Rather, part of his achievement was to show how that analogy emerged naturally from his analysis. See Domb (1996, 18), Pfeuty and Toulouse (1977, 25–8).

would appear, and its author, Michael Fisher, would announce two years later, '[O]ne model which historically has been of particular importance . . . deserves special mention: this is the Ising model. Even today its study continues to provide us with new insights' (Fisher (1983, 47). Nor was this view idiosyncratic. In 1992, a volume setting out recent results in the area of critical phenomena echoed Fisher's opinion: 'It is of considerable physical interest to study . . . the nearest neighbor ferromagnetic Ising model in detail and with mathematical precision' (Fernandez *et al.* 1992, 6).

Ironically enough, a necessary first step in the resuscitation of the model was the recognition that Ising had made a mistake. While his proof that the linear, one-dimensional model does not exhibit spontaneous magnetisation is perfectly sound, his conjecture that the result could be extended to models of higher dimensionality is not. In 1935 Rudolf Peierls argued that the two-dimensional Ising model exhibited spontaneous magnetisation,[19] and during the late 1930s the behaviour of the model, now regarded as essentially a mathematical object, began to be studied seriously (see Brush 1967, 287). From its subsequent history I will pick out just three episodes: first, the 'Onsager revolution' of the 1940s (the phrase is Domb's); secondly, the theory of critical exponents and universality that emerged in the 1960s, and thirdly, the advent of renormalisation theory in the 1970s.

5.1.5 The Onsager revolution

Lars Onsager's achievement was to produce a rigorous mathematical account of the behaviour at all temperatures – including those in the immediate vicinity of T_c – of the two-dimensional Ising model in the absence of a magnetic field. His was a remarkable feat of discrete mathematics, an exact solution of a non-trivial many-body

---

[19] Peierls (1936). The circumstances that occasioned Peierls' work on this topic were somewhat fortuitous. Peierls tells us (1985, 116) that in 1935 he attended a lecture by a mathematician who derived Ising's result:

But he then claimed that the same result held in two or three dimensions and gave arguments which, he claimed, proved that this was so.

I felt he was wrong. However, rather than look for errors in his arguments, I decided to prove that, at a sufficiently low temperature, a two-dimensional Ising model will have a nonzero average magnetization.

There were, apparently, flaws in Peierls' original proof (see Brush 1967, 887), but it was made rigorous by Griffiths (1964). Peierls' approach is summarised by Thouless (1989, 212), and discussed in more detail by Fernandez *et al.* (1992, 38–44).

problem.[20] He showed, amongst other things, (1) that the two-dimensional model exhibits spontaneous magnetisation, but (2) that it does not occur above a critical temperature, T_c; further, (3) that for the square lattice, $T_c = 2.269 J/k$; and (4) that near T_c physical properties like the specific heat of the lattice show a striking variation with temperature. In his own words, 'The outcome [of this analysis] demonstrated once and for all that the results of several more primitive theories had features which could not possibly be preserved in an exact computation.'[21] I will come back to these results in the next section. Observe, however, that by 1970, when Onsager made these remarks, he could regard the mean-field theories of the classical approach as 'more primitive' than the mathematically secure Ising model. The whirligig of time had brought in his revenges. Concerning the immediate impact of Onsager's work, on the other hand, differing estimates are provided. Thus, whereas Domb declares that, '[The] result was a shattering blow to the classical theory', Fisher talks of 'Onsager's results which, in the later 1950s, could still be discarded as interesting peculiarities restricted to two-dimensional systems', and cite Landau and Lifshitz (1958, p. 438) in corroboration.[22] Landau, of course, was hardly a disinterested observer.

A complete and exact solution for the two-dimensional Ising model in an external field has not yet been achieved,[23] and an analytic solution of the three-dimensional Ising model is not possible. In both cases, however, very precise approximations can be made. With regard to the three-dimensional case we may note that the approach pioneered by Domb, using the technique known as 'exact series expansion', involved comparatively early applications of computer methods to problems in physics.[24]

[20] The essential problem was to calculate the partition function Z for the model. Recall that N is the number of sites in the lattice, assumed large. In any arrangement a there will be a certain number N_+ of sites I such that $s_I(a) = +1$, and a certain number N_{+-} of nearest-neighbour pairs $<j,k>$ for which $s_j(a) = +1$, $s_k(a) = -1$. The energy associated with this arrangement of the model can be expressed as a function $g(N, N_+, N_{+-})$. (See Domb 1996, 112–15.) In order to calculate Z we need to calculate how many arrangements there are for each (N_+, N_{+-}). The solution of this combinatorial problem was Onsager's major achievement. A detailed account of Onsager's work is given in Domb (1996, ch. 5).

[21] Lars Onsager, in autobiographical remarks made in 1970, quoted by Domb (1996, 130).

[22] Both quotations are from Domb (1996), the first from p. 19 of the text, and the second from Fisher's *Foreword*, p. xv.

[23] But see Fernandez *et al.* (1992, 7–11 and ch. 14) for partial results.

[24] See Domb (1996, 22). For an outline of exact series expansion, see Fisher (1983, 54–8); see also Domb (1960), Domb and Sykes (1961).

5.1.6 Critical exponents and universality

I have already mentioned some of the remarkable changes that occur in the physical properties of a system near its critical temperature. Particularly significant are the variations with temperature of four quantities: the specific heat C, the order parameter M, the susceptibility χ, and the correlation length ξ. For expository purposes it is often easier to adopt the idiom of ferromagnetism, to refer, for instance, to the order parameter M as the *spontaneous magnetisation* (the magnetisation in the absence of an external magnetic field). In this idiom χ is the *magnetic susceptibility*, the ratio M/h of the magnetisation to the applied magnetic field h. Of course, one could equally well interpret these quantities in terms of liquid-vapour critical phenomena, where the order parameter is the difference, $\rho-\rho_c$, between the actual density of the liquid-vapour fluid and its density at the critical point, and the susceptibility is the change of this difference with pressure, i.e. the fluid's compressibility. In the liquid-vapour case the relevant specific heat is the specific heat at constant volume (equivalently, at constant density), C_v.

For generality, we describe the variations of these quantities in terms of the *reduced temperature t*, defined as follows:

$$t = (T-T_c)/T_c \tag{5.5}$$

On this scale, $t = -1$ at absolute zero, and $t = 0$ at the critical temperature. Clearly, the number of degrees Kelvin that corresponds to a one degree difference in the reduced temperature will depend on the critical temperature for the phenomenon under study.

Of the four quantities, the spontaneous magnetisation decreases to zero as T_c is approached from below, and remains zero thereafter, but the other three all diverge (go to infinity) as T_c is approached from either direction (see figure 5.1). There is both experimental and theoretical support for the thesis that, close to T_c, the variation with t of each of these four quantities is governed by a power law; that is to say, for each quantity Q we have

$$Q \sim |t|^\lambda \tag{5.6}$$

where, by definition, this expression means that

$$\lim_{t\to 0}[\ln(Q)/\ln(t)] = \lambda \tag{5.7}$$

Given this definition, we see that, if $Q = Q_a|t|^\lambda$, where Q_a is some constant, then $Q \sim |t|^\lambda$. The converse, however, does not hold, a point whose importance will appear shortly.

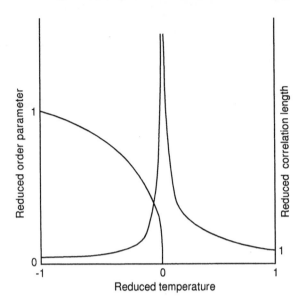

Figure 5.1 The variation with temperature of the order parameter and correlation length of the two-dimensional Ising model. Reproduced from Paul Davies (ed.), *The New Physics*. Cambridge University Press, 1989, 241.

For the magnetisation, M, we have, for $t < 0$,

$$M \sim |t|^{\beta} \tag{5.8}$$

and, for the three diverging quantities,

$$C \sim |t|^{-\alpha} \quad \chi \sim |t|^{-\gamma} \quad \xi \sim |t|^{-\nu}$$

The exponents β, α, γ, and ν are known as *critical exponents*.

The behaviour of the last three quantities is not, as these relations might suggest, symmetrical about T_c where $t = 0$ (see figure 5.1). In fact, at one time it was thought that different exponents were needed on either side of T_c, so that we should write, e.g.

$$\xi \sim |t|^{-\nu} \text{ for } t > 0 \quad \xi \sim |t|^{-\nu'} \text{ for } t < 0 \tag{5.9}$$

There are now, however, strong reasons to think that $\alpha = \alpha'$, $\gamma = \gamma'$, $\nu = \nu'$, and to look elsewhere for the source of asymmetries. In a simple case we might have, for instance,

$$Q \sim Q_+ |t|^{\lambda} \quad Q \sim Q_- |t|^{\lambda} \tag{5.10}$$

where $Q_+ \neq Q_-$, but the exponents are the same.

Table 5.1

Exponent	Ising model ($d=2$)	Ising model ($d=3$)	Mean-field theory
α	0(log)	0.12	0
β	1/8	0.31	1/2
γ	7/4	1.25	1
ν	1	0.64	1/2
δ	15	5.0	3
η	1/4	0.04	0

Source: data from Amit (1984, p. 7).[25]

Two other critical exponents, δ and η, are defined in terms of the variation *at the critical temperature* of M with h, and of the correlation function, $\Gamma(R)$ with R. We have,

$$M \sim h^{1/\delta} \quad \Gamma(R) \sim R^{-(d-2+\eta)} \tag{5.11}$$

In the latter, d denotes the dimension of the physical system.

The exponents have been calculated exactly for the two-dimensional Ising model.[26] In table 5.1 these values are compared with the approximate values obtainable for the three-dimensional version, and the values according to mean-field (Landau) theory.

The fact that $\alpha = 0$ for both the two-dimensional Ising model and the mean-field theory masks a radical difference between the two. In 1944 Onsager proved that in the two-dimensional Ising model C would diverge logarithmically as the temperature approached T_c from either direction; from above ($t>0$) we have:

$$C = A \ln(t) + c \tag{5.12}$$

where c is a finite 'background' term. When the formal definition (5.7) of the $\sim$-relation is applied, this yields:

$$\alpha = \lim_{t \to 0} \ln[\ln(C)]/\ln(t) = 0 \tag{5.13}$$

Hence the use of '0(log)' in the table. In contrast, on the mean-field theory the specific heat remains finite. It rises almost linearly whether T_c

---

[25] Domb (1996, 235) gives the current estimates for β and δ in the three-dimensional Ising model as 0.325 and 4.82 respectively. Unfortunately he does not supply a comprehensive list.

[26] In saying this I am assuming the scaling laws (see below), which were used to establish the exact value for δ; however, precise estimates of the approximate value for δ were available consistent with the exact value shown. See Domb (1996, 175).

is approached from above or below. The zero exponent appears because there is a discontinuity at T_c. Denoting by C^+ (C^-) the limiting value of C approached from above (from below), we have $C^- > C^+$.

Note also that there is no other exponent for which the mean-field value coincides with the value for either Ising model.

These differences are much more threatening to the mean-field theory than might be supposed. One might think that the situation was symmetrical: given two competing models, experiment would decide between them. But in this case the two theories are different in kind. Mean-field theory, as envisioned by Landau, was a universal theory of critical phenomena. He used symmetry considerations and well established thermodynamic principles to generate results which he hoped would be applicable to any system exhibiting critical point behaviour. Now the two-dimensional Ising model may not be an entirely faithful representation of any physical system with which we are familiar. Nonetheless, any system that it *did* represent faithfully would exhibit critical phenomena, and so come under the umbrella of mean-field theory. Thanks to Onsager we now know precisely what the behaviour of such a system would be, and that it would not conform to the mean-field theory's predictions.[27]

The 1960s saw the emergence of a new general approach to critical phenomena, with the postulation of the so-called *scaling laws*, algebraic relationships holding between the critical exponents for a given system. Five of them are given by equations (a)–(e) below. Listed alongside them are the more fundamental inequalities of which the equations are special cases.

(a) $\alpha + 2\beta + \gamma = 2$	(a*) $\alpha + 2\beta + \gamma \geq 2$	(Rushbrooke)	(5.14)
(b) $\alpha + \beta(1 + \delta) = 2$	(b*) $\alpha + \beta(1 + \delta) \geq 2$	(Griffiths)	(5.15)
(c) $v(2 - \eta) = \gamma$	(c*) $v(2 - \eta) \geq \gamma$	(Fisher)	(5.16)
(d) $dv = 2 - \alpha$	(d*) $dv \geq 2-\alpha$	(Josephson)	(5.17)
(e) $d[(\delta - 1)/(\delta + 1)] = 2 - \eta$	(e*) $d[(\delta - 1)/(\delta + 1)] \geq 2 - \eta$	(Buckingham and Gunton)	(5.18)

In equations (d) and (e), and in the corresponding inequalities, d is the dimensionality of the system. The equations (a)–(e) are not independent;

---

[27] The problem with the Landau theory can, in retrospect, be described in two ways. To put it formally, Landau expressed the free energy of a system in terms of a power series of M^2 (the square is used for reasons of symmetry); he then ignored terms above the M^2 and M^4 terms. However, at T_c the series does not converge. To put this physically, what Landau ignored were the fluctuations – regions of varying degrees of order – that occur at T_c. See Amit (1984, 7).

given (a)–(c), (d) and (e) are equivalent. The inequalities, however, are algebraically independent, and indeed were independently proposed by the individuals cited next to them. These inequalities do not all have the same warrant. The first two listed are derivable from the laws of thermodynamics, specifically from the general thermodynamic properties of the free energy of a system, while the others depend on additional assumptions about correlation functions $\Gamma(r)$, which hold for a very broad class of systems.[28] The additional postulates needed to yield the individual equations are known as the *thermodynamic scaling hypothesis* (for (a) and (b)), the *correlation scaling hypothesis* (for (c)), and the *hyperscaling hypothesis* (for (d) and (e)).[29]

All four scaling laws hold exactly for the two-dimensional Ising model, provided that the exact value 15 is assigned to δ (see n. 25). They also hold approximately for the three-dimensional Ising model (the largest error, given the Amit values, is 3%), but that is not surprising, since conformity with these relations was one of the criteria by which the approximations used in calculating the exponents were assessed (see Domb 1996, 175). For the mean-field theory, equations (a), (b), and (c) all hold, but neither the equations (d) and (e) nor the inequalities (d*) and (e*) hold, unless $d = 4$.

Thus in 1970 the situation was this. The differences between the values of the critical exponents calculated for the two-dimensional Ising model, the three-dimensional Ising model, and the mean-field model had suggested, not only that the mean-field theory was not a universal theory of critical phenomena, but also that the search for one was perhaps mistaken. But then, with the postulation of the scaling laws, a new twist had been given to the issue of universality. Unlike the mean-field theory, the laws did not suggest that the critical exponents for different systems would all take the same values; still less did they prescribe what those values might be. What they asserted was that the same functional relations among the critical exponents would obtain, no matter what system was investigated, provided that it exhibited critical point behaviour. More precisely, given a system of a certain dimensionality, if we knew two of these exponents, β and ν, then from the laws we could deduce the other four.

Yet, intriguing though these laws were, in 1970 the status of the various scaling hypotheses, and hence of the laws themselves, was moot.

---

[28] For an extended discussion of the inequalities, see Stanley (1971, ch. 4).

[29] A useful taxonomy of the various inequalities and equations is given by Fernandez *et al.* (1992, 51–2). For a full discussion of the scaling hypotheses, see Fisher (1983, 21–46).

The thermodynamic scaling hypothesis, for instance, could be shown to follow from the requirement that the Gibbs potential for a system exhibiting critical point behaviour be a generalised homogeneous function. (For details, see Stanley 1971, 176–85.) This requirement, however, is entirely formal; it has no obvious physical motivation. As Stanley commented (1971, 18), '[T]he scaling hypothesis is at best unproved, and indeed to some workers represents an *ad hoc* assumption, entirely devoid of physical content.' Nor did physically based justifications fare better; introducing a rationale for adopting the hyperscaling hypothesis, Fisher cheerfully remarks (1983, 41–2), 'The argument may, perhaps, be regarded as not very plausible, but it does lead to the correct result, and other arguments are not much more convincing!'

Experimental verification of the laws was difficult, since the task of establishing precise values for critical exponents was beset by problems. Two examples: first, in no finite system can the specific heat or the correlation length increase without bound; hence their variation with the reduced temperature very close to T_c will not obey a power law, and the relevant critical exponent will not be defined (see Fisher 1983, 14). Secondly, the reduced temperature t that appears in relation 5.8: $M \sim |t|^\beta$ is given by $t = (T - T_c)/T_c$. But T_c may not be known in advance. Hence, if we use a log-log plot of the data to obtain β from $\beta = \lim_{t\to0}(\ln M)/(\ln - t) = \mathrm{d}(\ln M)/\mathrm{d}(\ln - t)$, a series of trial values of T_c may have to be assumed until a straight line is produced. That is to say, prior to obtaining the exponent β, we will need to *assume* that a power law governs the relation between M and t. (See Stanley 1971, 11.)

Again, if the scaling laws are to be justified by appeal to known critical exponents, then those exponents had better not be established via the scaling laws. Yet Domb notes (1996, 174), concerning the calculations performed on various three-dimensional lattices in the 1960s, 'It was usual to make use of the [Rushbrooke] scaling relation . . . and the well-determined exponents γ and β to establish the value of α', and Fisher writes (1983, 41), perhaps with tongue in cheek, 'Experimentally also [the Fisher equality] checks very well. If it is accepted it actually provides the best method of measuring the elusive exponent η!'[30]

Throughout the 1970s and 1980s, however, theoretical and experimental evidence mounted in support of the scaling laws. Furthermore, this evidence pointed to an unanticipated conclusion, that systems

---

[30] If I have lingered unduly on the topic of the scaling laws and their justification, it is because they would provide wonderful material for a detailed case study of justification and confirmation in twentieth-century physics.

exhibiting critical behaviour all fall into distinct classes, within which the values of the critical exponents of every system are the same.[31] These 'universality classes' are distinguished one from the other by two properties: the dimension of the systems involved and the symmetry of its site variable. In the case of the Ising model, for example, where $s_i = \pm 1$, the symmetry is that of the line, which is invariant under reflection; in the case of the Heisenberg model of isotropic ferromagnetism, on the other hand, the site variable is a three-component spin, and thus has spherical symmetry.[32] The study of critical point phenomena had revealed a remarkable state of affairs. While critical temperatures are specific to individual phenomena, as are the amplitudes of quantities like the order parameter and the correlation length, the power laws that govern the variation of these quantities with temperature are determined by dimension and symmetry, properties that radically different systems can have in common.

The study of universality within critical point phenomena gave the Ising model a role different from any it had previously played. In 1967, when Brush wrote his 'History of the Lenz-Ising Model', the model's chief virtues were seen as its simplicity and its versatility. As Brush notes, and we have seen in section 5.1.3, its versatility was evident from the direct, though crude, representations it offered of the atomic arrangements within systems as diverse as ferromagnets, liquid-vapour mixtures, and binary alloys.

Another virtue was its mathematical tractability. At least for the two-dimensional version, exact values for critical exponents were calculable; thus the model could fulfil the negative function of providing a counter-example to the mean-field theory's predictions, and the positive one of confirming the scaling laws. And, *a propos* of results obtained for the three-dimensional Ising model, Domb also suggests, perhaps with hindsight, that 'The idea that *critical exponents for a given model depend on dimension and not on lattice structure* was a first step towards the *universality hypothesis*' (1996, 171; emphasis in the original).

Be that as it may, with the enunciation and acceptance of that hypothesis came a new emphasis on the use of models in condensed matter physics. By definition, universal behaviour supervenes on many different kinds of processes at the atomic or molecular level. Hence its study does

---

[31] For theoretical evidence, see e.g. Griffiths (1970), and Domb (1996, ch. 6). For experimental evidence, see e.g. the work of Balzarini and Ohra (1972) on fluids, discussed by Fisher (1983, 8), and, especially, the review article by Ahlers (1980).
[32] For other examples, see Amit (1984, 8–9).

not demand a model that faithfully represents one of those processes. To quote Bruce and Wallace,

The phenomenon of universality makes it plain that such details are largely *irrelevant* to critical point behaviour. Thus we may set the tasks in hand in the context of simple model systems, with the confident expectation that the answers which emerge will have not merely qualitative but also quantitative relevance to nature's own systems. (1989, p. 242; emphasis in the original)

In other words, in this field a good model acts as an exemplar of a universality class, rather than as a faithful representation of any one of its members.

The virtues we have already remarked enable the Ising model to fit the role perfectly. Its mathematical tractability and – even more – its amenability to computer simulation allow it to be used to explore critical point behaviour with great precision and in great detail. In addition, however crude its representation of ferromagnetism may be at the atomic level, the fact that it was devised with a particular phenomenon in mind provides an added benefit. It enables physicists to use the physical vocabulary of ferromagnetism, to talk of spins being aligned, for example, rather than of site variables having the same value, and it encourages the play of physical intuition, and in this way it facilitates understanding of critical point phenomena.[33] It is not surprising that Kenneth Wilson's 1973 lectures on the renormalisation group include a lecture entirely devoted to the Ising model (Wilson 1975, 797–805).

My discussion of the renormalisation group in the next section is nearly all in terms of the Ising model. It thus offers, *inter alia*, an illustration of how heuristically useful the model can be.

5.1.7 Universality and the renormalisation group

To understand universality, one must remember two things: first, the fact that the term is applied to the behaviour of systems at or near the critical temperature; secondly, that the correlation length ξ has a particular significance at that temperature.

Of the six critical exponents discussed in the previous section, α, β, γ, and ν appear in power laws that describe the behaviour of physical quantities at or near the critical temperature, while δ and η appear in laws describing behaviour at the critical temperature itself. Amongst the

---

[33] On models and understanding, see Fisher (1983, 47); for an account of explanation in terms of the models that theory provides, see Hughes (1993).

former is the relation $\xi \sim |t|^{-\nu}$. It tells us that the correlation length ξ tends to infinity as the system approaches T_c. As I noted in section 5.1.5, in the Ising model the correlation length can be thought of as a measure of the maximum size of a totally ordered island within the lattice as a whole. Not all totally ordered regions, however, are of this size; in fact, at the critical temperature, islands of all sizes will be present, ranging from those whose characteristic length is equal to ξ all the way down to those containing a single site, i.e., those whose characteristic length is given by the lattice spacing.

This prompts the hypothesis that universality results from the fact that islands of all these different sizes are involved in the physics of critical point behaviour. Put in terms of the Ising model, it states that between the lattice spacing and the correlation length there is no privileged length scale. It thus gives rise to two lines of investigation. The first is straightforward: the investigation of what resources the Ising model possesses for providing a description of the lattice in terms of regions rather than individual sites. The second is conceptually subtler: an enquiry into how state variables associated with these larger regions could reproduce the effects of the site variables s_i. At the site level, the tendency for the site variables of adjacent sites to become equal (or, in magnetic terms, for their spins to become aligned) is described in terms of interaction energies between nearest neighbours. By analogy, we may enquire whether there is a description available at the regional level that describes the coupling strengths between one region and another. Whereas we normally think of the nearest-neighbour interactions as ultimately responsible for correlations in the lattice at all ranges from the lattice spacing up to the correlation length, this approach suggests that the coupling strengths between regions of any characteristic length L may be regarded as fundamental, in the sense that they give rise to correlations at all ranges between L and ξ.[34]

Both approaches shed light on the problem of universality. To illustrate the first, consider the example of the Ising model defined in section 5.1.2, which uses a two-dimensional square lattice of sites. This lattice can equivalently be regarded as an array of square cells, or *blocks*, with sides equal in length to the lattice spacing. Now suppose that we obtain a coarse-grained picture by doubling the length of the sides of the blocks. Each block will then contain four lattice sites. Clearly the new

---

[34] The two approaches correspond to Bruce and Wallace's *configurational view* and *effective coupling view* of the renormalisation group, and I have abridged their treatment in what follows.

array of blocks itself constitutes a lattice, and we may define block variables on it by adding the site variables from the sites it contains. These block variables will take one of five values: $-4, -2, 0, +2, +4$. The coarse-graining procedure can be repeated, each step doubling the lengths of the blocks' sides. After n such dilations each block will contain a $2^n \times 2^n$ array of sites. (We can regard the formation of the blocks that contain a single site as step zero.) After n steps, the block variable will take one of $(2^n)^2 + 1$ values.[35]

We may scale these block variables to run from -1 to $+1$; as n increases they tend towards a continuum of values between these limits. If the values of the site variables were independent of one another, as happens at high temperatures, where they are randomised by thermal noise, we would expect the probabilities of these variables over the lattice to form a Gaussian distribution centred on zero. At the critical temperature, however, the effect of correlations is that, as n increases, the probability distribution of the block variables, or *configuration spectrum*, tends towards a bi-modal form with peaks at roughly ± 0.85, as shown in figure 5.2.

The value of n for which the probabilities for a discrete set of values fit this curve is surprisingly small. Bruce and Wallace show (1989, p. 247) how good the fit is for $n = 3$, when there are sixty-five possible values of the block variable, and thereby show that for all lengths greater than eight lattice spacings the probability distribution of (scaled) block variables is a *scale-invariant property*.

They also use the same plot to show an even more striking fact, that this configuration spectrum is independent of the number of values of the site variables in the original model. They do so by taking an Ising model whose site variable can take one of three different values ($-1, 0, +1$) and plotting the probabilities of the 128 different block values obtained when $n = 3$. Whereas the local configuration spectra of the 2-value and the 3-value models are quite different, these differences get washed out, as it were, under coarse-graining.

Taken together, the two results corroborate the link between universality and scale invariance.

Wallace and Bruce extend this analysis by an ingenious argument to show that the way in which the configuration spectrum evolves under coarse-graining is characterised by the critical exponents β and ν. This

---

[35] It is not necessary that we define block variables in this way. Niemeijer and van Leeuwen (1974) consider a triangular lattice, in which the blocks consist of triangles whose sides are multiples of the lattice spacing, and the block variables, like the sites, have two values, to be decided by majority voting.

R. I. G. Hughes

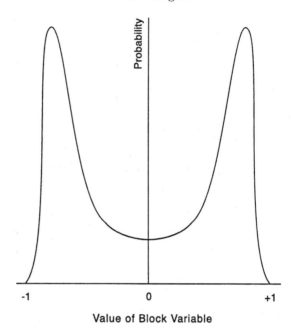

Figure 5.2 Configuration spectrum for two-dimensional Ising model.

result is entirely consistent with the conclusions quoted at the end of the last section, (1) that the values of β and ν suffice to determine the values of all critical exponents for a given universality class, and (2) that the physical properties that define a particular class are the dimensionality of the system and the symmetry of its state variable. Both of the models in question are two-dimensional, and the change from a two-valued to a three-valued state variable is not a change of symmetry, since in each case the symmetry is linear.

The results about configuration spectra that I have quoted were taken from computer simulations of the behaviour of the Ising model. Given the experimentally confirmed fact of universality, we may expect them to hold for any two-dimensional system whose state variable has linear symmetry. But, although these simulations show that the Ising model does indeed display scale-invariant properties at its critical temperature, the question remains how this scale-invariance comes about. To answer this question the second of the two approaches I suggested earlier is called for.

On this approach, any level of description for which the characteristic length of the resolution lies between the lattice spacing and the correlation length could be regarded as fundamental. That is to say, the macroscopic behaviour of the model could be thought of either as produced by nearest-neighbour interactions between sites in the standard way, or as generated by interactions between blocks, provided that the length of the blocks' edges was small compared with macroscopic lengths. Two questions arise. First, what kinds of interaction between blocks would produce the same results as the nearest-neighbour interactions between sites? And secondly, what sort of relation between one set of interactions and the other could be generalised to blocks of different edge-lengths?

Before addressing these questions, let me note that, in the Ising model, the effects of the interactions between sites are attenuated by temperature. In the key equation (3) that governs the statistical behaviour of the lattice, probabilities depend, not just on the interaction energy E_a of an arrangement, but on E_a/kT. This attenuation can be absorbed into the description of the interactions between sites by specifying them in terms of a coupling strength K, where

$$K_1^0 = J/kT \tag{5.19}$$

Here the superscript 0 indicates that we are regarding a site as a block produced at the zero step in the coarse-graining procedure, and the subscript 1 registers the fact that the interactions are between nearest neighbours. The point of the second notational device is this. When we ask what effective interactions exist between blocks produced after n steps of coarse-graining, there is no guarantee that they will be confined to nearest-neighbour interactions. In fact they may have to be represented by a set of effective coupling strengths, K_1^n, K_2^n, K_3^n, etc., where K_1^n represents a coupling strength for adjacent blocks, K_2^n a coupling strength between blocks two block lengths apart, and so on. The superscript n indicates that the blocks have been produced after n steps of the coarse-graining procedure, so that the blocks' sides are of length 2^n lattice spacings. A more economical representation is obtained if we think of a coupling strength as a vector $\mathbf{K}^n$ whose components are $K_1^n, K_2^n, \ldots$. The vector $\mathbf{K}^0$ has only one component, K_1^0.

From the coupling strength $\mathbf{K}^0$, n steps of coarse-graining yields the effective coupling strength $\mathbf{K}^n$. We may represent the effect of this coarse-graining mathematically by an operator T^n on the space of coupling strengths,[36] such that

$$\mathbf{K}^n = T^n(\mathbf{K}^0) \tag{5.20}$$

[36] I distinguish the operator T^n from the temperature T by italicising the latter.

Various physical assumptions have already been built into this mathematical representation. It is assumed that the model is describable in terms of the states of blocks of arbitrary size; furthermore, that whatever level of coarse-graining we choose, the internal energy of the model is expressible in terms of a set of effective coupling strengths associated with interactions between blocks at different separations, these separations always being specified in terms of the block spacing. In brief, built into the representation is the hypothesis we encountered earlier, that between the lattice spacing and the correlation length there is no privileged length scale.

The further requirement, that the interactions between regions of any characteristic length L may be regarded as fundamental, i.e. as generating the correlations at all lengths between L and ξ, constrains the family $\{T^n\}$ of transformations. We require, for all m, n,

$$T^{n+m} = T^m . T^n \qquad (5.21)$$

This equation is satisfied if, for all n,

$$\mathbf{K^{n+1}} = T^1(\mathbf{K^n}) \qquad (5.22)$$

Equation 5.21 is the rule of group multiplication for an additive group. Hence the name 'Renormalisation Group' for $\{T^n\}$.[37] It is a sequence of transformations induced by a sequence of dilations. Like many groups of transformations in physics, the transformations induced by coarse-graining can leave the system looking different, particularly when n is small, while its fundamental physical properties remain undisturbed.[38]

The possibility of universal behaviour appears if the sequence $\mathbf{K^0}$, $\mathbf{K^1}$, $\mathbf{K^2}$, $\mathbf{K^3}$, ... tends to a limit $\mathbf{K^*}$. In this case the transformation T^1 has a *fixed point* $\mathbf{K^*}$ such that

$$T^1(\mathbf{K^*}) = \mathbf{K^*} \qquad (5.23)$$

The fixed point is not determined by a unique sequence $\mathbf{K^0}$, $\mathbf{K^1}$, $\mathbf{K^2}$, ..., but by the transformation T^1. In other words, there are many sequences that tend to the same fixed point; in physical terms, there may be many systems which differ at the atomic level, but which nevertheless manifest the same behaviour at large length scales. Furthermore, when n is large the points $\mathbf{K^n}$, $\mathbf{K^{n+1}}$, $\mathbf{K^{n+2}}$ lie very close together, and are also very close

[37] Strictly, $\{T^n\}$ forms a semi-group; no inverse operation is employed.
[38] Naturally, what counts as fundamental depends on one's preferred group of transformations.

to **K***; in physical terms, systems display scale-invariant behaviour at large length scales.

This skeletal account of renormalisation theory has done no more than show how universality might come about. The theory itself changed the way physicists approach critical point phenomena (see Domb 1996, ch. 7). Its articulation by Kenneth Wilson in the early 1970s earned him a Nobel prize a decade later. It is an unusual theory. It is not mathematically rigorous, and the approximative methods used are chosen on a case by case basis. That is part of the reason why simple mathematical models are so valuable in foundational work. In Wilson's own words, for physicists seeking to apply the theory, 'There is no renormalization cook book.'[39]

5.1.8 Computer simulation of the Ising model

The use of computers to perform calculations has been commonplace in critical point physics for fifty years. I have already mentioned (section 5.1.5) their use by Domb and his co-workers in establishing the behaviour of the three-dimensional Ising model. Some fifteen years later, in applying the renormalisation approach to the two-dimensional Ising model, Wilson also used the computational capacity of the computer. Indeed, in describing what he achieved, he tells us about the computer he used (a CDC 7600) and the details of the calculations involved (Wilson 1975, 802–5).

Yet, although Domb and Wilson both used computers, and although they both worked on the Ising model, to my mind neither of them performed a computer simulation of the model's behaviour.[40] A typical computer simulation of the two-dimensional Ising model involves the representation of an array of sites within the computer in such a way that values of the site variable can be attached to each site. A stochastic algorithm takes one arrangement a of the lattice into another, a', with a specified probability. Repeated applications of the algorithm generate a sequence a, a', a'', . . . of arrangements. The algorithm is so designed that, during a suitably long sequence, arrangements will occur with a relative frequency equal to their probability according to the Boltzmann Law:

$$p_a = \mathcal{Z}^{-1}\exp(-E_a/kT) \tag{5.24}$$

---

[39] Wilson, quoted by Domb (1996, 261).
[40] I treat this distinction at greater length in section 5.2.2.

In brief, under the algorithm the set of arrangements becomes a canonical ensemble. Properties of the model are measured by taking mean values (as determined by the computer) over a suitably large number of arrangements.

This type of simulation is called a *Monte Carlo* simulation because, like a roulette wheel, it chooses outcomes from a set of alternatives with specified probabilities. The simulation of the three-dimensional Ising model devised by Pearson, Richardson and Toussaint (1984) is a nice example.[41] In the basic step of their algorithm, a particular site i is considered, along with the values $s_n(a)$ of the site variables of six nearest neighbours. The value of the variable s_i, is then set to $+1$ with probability

$$p = \exp(-E_n/kT)/[\exp(-E_n/kT) + \exp(E_n/kT)] \qquad (5.25)$$

and to -1 with probability $1 - p$.[42] ($E_n = -J\Sigma_n s_n(a)$, where the summation is over the six neighbouring sites s_n of i.[43]) This step is then reiterated for each site in the lattice in turn, and the sweep over all sites is repeated many times. It can be shown that, whatever the initial arrangement, the simulation will approach an equilibrium situation in which the Boltzmann probabilities are realized.

Pearson and his collaborators modelled a cubic lattice with $64 \times 64 \times 64$ sites. With $1/kT$ set at 0.2212 ($1/kT_c = 0.2217$ for this lattice), the equilibrium situation is reached after about 1700 sweeps through the lattice. As they point out (1983, 242),

This means that the great bulk of the computational effort consists of repeating the simple updating algorithm for all the spins in the lattice. Therefore it is attractive to construct a special purpose device to perform the updatings and make some of the simplest measurements.

The device they constructed performed the Monte Carlo updating on 25 million sites per second; thus each second roughly 100 sweeps of the lattice were performed. 'Despite its modest cost', they claim (1983, 241),

---

[41] For an introduction to Monte Carlo methods, and a discussion of their application to critical point phenomena, see Blinder and Stauffer (1984), and Blinder (1984a), respectively. Both contain sections on the simulation of the Ising model; neither paper, however, does more than mention single-purpose processors of the kind that Pearson *et al.* constructed.

[42] The Monte Carlo process uses a random number generator to produce a number x in the interval [0,1]. This number is then compared with p. If $x < p$, the value $+1$ is selected; otherwise the value -1 is selected (see Blinder and Stauffer, 1984, 6–7).

[43] For generality, Pearson *et al.* include in this energy term an additional h to allow for an external field.

'this machine is faster than the fastest supercomputer on the one particular problem for which it was designed'.[44]

It was possible to adjust the machine so that the coupling strength in
the x-direction was set to zero. It then behaved like an array of two-
dimensional lattices, all of which were updated in one sweep of the
model. In this way the verisimilitude of the simulation could be checked
by comparing the performance of the machine against the exactly known
behaviour of the Ising model. When their respective values for the internal energy of the lattice at various temperatures were compared, they
agreed to within the estimated errors for the values given by the simulation; that is, to within 1 part in 10,000. (See Pearson *et al.*, 1983, 247–8.)

The main function of the processor, however, is to simulate the behaviour of the three-dimensional Ising model, for which exact results are
not available. For this reason, Pearson *et al.* display plots of the variation
with temperature of the magnetisation, susceptibility, and specific heat
of the model as examples of what the simulation can provide.

Computer simulations, in the strict sense, of the behaviour of the Ising
model started to be performed in the 1960s. In his well-known text, Stanley
(1971, facing 6) illustrates his discussion of the lattice-gas model with some
pictures generated by a simulation performed by Ogita and others. Six pictures are displayed, each showing a 64×64 array of square cells, some
black, some white. Stanley's remarks in the caption are a trifle odd, given
that the Ising model can be used to model the lattice gas:

Fig. 1.5. Schematic indication of the lattice-gas model of a fluid system. [There
now follow notes about the individual pictures.] This illustration and the associated temperatures are to be regarded as purely *schematic*. In fact the figure was constructed from a computer simulation of the time-dependent aspects of the
two-dimensional Ising model and actually represents rather different phenomena.

The distinction drawn here, between a *schematic indication* and a *representation* is a nice one. But it may be that Stanley's chief concern is the
difference of dimensionality between the two models.

Obviously, for expository purposes simulations of the two-dimensional
version of the Ising model are ideal, since they can be presented on a
two-dimensional grid or, in other words, as pictures. Yet, as Bruce and
Wallace point out (1989, 239n), pictures like these are in one way
misleading. A picture will show just one arrangement, but the properties

<hr>

[44] In assessing this claim, one should bear in mind when it was made. Note also that while this
machine was being built in San Diego, California, another single-purpose machine for simulating the Ising model was being built in Delft by Hoogland and others; see Hoogland *et al.* (1983).

that concern us are averaged over a sequence of arrangements. It follows that a 'typical' picture will have to be selected for the viewer.

Lending itself to pictorial display is not a prerequisite for a simulation. None the less, it helps. The pictures that illustrate Bruce and Wallace's presentation of renormalisation theory allow many features of critical behaviour to be instantly apprehended, like the almost total disorder above the critical temperature, the high degree of order below it, and, crucially, the presence of ordered islands of all sizes at the critical temperature itself. It also allows the authors to display the results of coarse-graining at these different temperatures, to show that coarse-graining appears to take a model that is not at the critical temperature further away from criticality, so that a model above the critical temperature appears more disordered, and one below it more strongly ordered. At the critical temperature, on the other hand, provided the coarse-grained pictures are suitably scaled, the model looks no more and no less ordered after coarse-graining than before.

Rumination on the different ways in which changes in the pictures occur brings out a perplexing aspect of the renormalisation process. We may distinguish between changes in the pictures that result from altering one of the parameters that govern the algorithm (in particular, the temperature), and changes that result from taking an existing picture and transforming it (in particular, by coarse-graining). When we move from study of the Ising model to the study of physical systems, this raises a puzzling question for the realist: how is she or he to regard the effective coupling strengths that are invoked when the length scale at which we describe the system is increased? On the one hand, they must surely be regarded as real; large regions of the system do have an influence one on the other. On the other hand, they seem like artefacts of the mode of description we choose to employ. A third alternative, which would have delighted Niels Bohr, is that, while there can be no single complete description of a system near to its critical point, renormalisation theory shows how a set of complementary descriptions can be obtained. I will not pursue this issue here.

5.2 THEORETICAL REPRESENTATION: THE ISING MODEL AND COMPUTER SIMULATION

5.2.1 The DDI account of modelling

The Ising model is employed in a variety of ways in the study of critical point phenomena. To recapitulate, Ising proposed it (and immediately

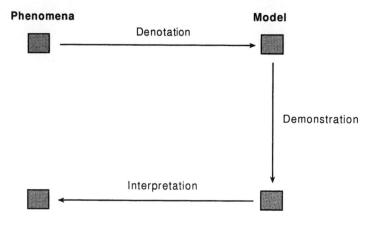

Figure 5.3 A schematic diagram of the DDI account of theoretical representation.

rejected it) as a model of ferromagnetism; subsequently it has been used to model, for example, liquid-vapour transitions and the behaviour of binary alloys. Each of these interpretations of the model is in terms of a specific example of critical point behaviour. But, as we saw in the discussion of critical exponents, the model also casts light on critical point behaviour in general. Likewise, the pictures generated by computer simulation of the model's behaviour illustrate, not only ferromagnetic behaviour close to the Curie temperature, but also the whole field of scale-invariant properties. In this section of the paper and the next I will see how the multiple roles played by the model and its simulation can be accommodated within a very general account of theoretical representation, which I call the *DDI account*.[45]

The account rests on the premise that the function of a model is to provide a representation of a phenomenon, or a cluster of phenomena. The question is how best to characterise the kind of representation that a theoretical model provides. The DDI account invites us to think of a theoretical representation in terms of three components: denotation, demonstration, and interpretation (see figure 5.3). Elements of the subject of the model (a physical system evincing a particular kind of behaviour, like ferromagnetism) are *denoted* by elements of the model; the internal dynamic of the model then allows conclusions (answers to specific questions) to be *demonstrated* within the model; these conclusions can then be *interpreted* in terms of the subject of the model.

---

[45] I discuss this account at greater length in Hughes (1997).

The DDI account thus shares with Goodman's account of pictorial representation the negative thesis that (*pace* Giere) representation does not involve a similarity or resemblance between the representation and its subject, and the positive thesis that elements of the subject are denoted by elements of the representation (see Goodman 1968, ch. 1). On the DDI account, however, theoretical representation is distinguished from pictorial and cartographic representation by the second component in the process, the fact that models, at least as they are used in physics, have an internal dynamic which allows us to draw theoretical conclusions from them. We talk about the behaviour of a model, but rarely about the behaviour of a picture or a map. The conclusions we draw are, strictly speaking, about the behaviour of the model; they need interpretation if they are to be applied to the phenomena under study.

The two-dimensional Ising model, for example, is an abstract mathematical entity. Hence, to paraphrase Goodman (1968, 5), it resembles other mathematical entities much more than it resembles, e.g., a ferromagnetic layer on the surface of a physical object. When the model is used as a model for that kind of physical system, the representation it provides is generated by specific denotations. The values of s_i, in the model denote the orientations of the elementary magnets that are assumed to make up the layer. Likewise, the parameter T in the model denotes the temperature, not of the model, but of the layer being modelled. As T increases, the model does not become hot to the touch.

Notice that, in talking, as I did, of 'the elementary magnets that are assumed to make up the layer', I am already using what Nancy Cartwright has called a 'prepared description' of the model's subject (1983, 133–4), and the use of a prepared description is even more obvious when we employ the (three dimensional) Ising model as a model of a fluid (see (b) in section 5.1.1 above). These prepared descriptions are themselves representations, and so in these cases the Ising model is a representation of a representation. In fact, representational hierarchies of this kind are the norm within physics,[46] and, as we shall see later, they can easily be accommodated within the DDI account.

More problematic for the DDI account than these individual cases is the use of the Ising model in discussions of critical point phenomena in general. The universality thesis suggests that Onsager's analysis of the two-dimensional Ising model has considerable theoretical significance,

---

[46] I discuss these hierarchies in some detail in Hughes (1997).

whether or not the model is a faithful representation of any specific example of critical point behaviour, i.e., whether or not the elements of the model denote elements of any actual physical system. In some discussions (e.g. Fisher 1983) the Ising model comes to represent a large, and very disparate class of physical systems almost in the manner that a congressman represents a district, or member of parliament a riding. It represents a class of systems by being a representative of them. (It is also representative of them, which is not quite the same thing.) In such cases of representation, the concept of denotation seems out of place.[47]

Three possible responses to this problem suggest themselves. The first is to dismiss it, to say that the Ising model occupies a unique position in physics, and that since no other model acts as a representative of such a heterogeneous class of systems, no general account of modelling can be expected to accommodate it. The second response is less cavalier. It follows the line I took earlier in assessing the importance for mean-field theory of Onsager's results. In that case, I argued, since the mean-field theory's claims were purportedly universal, it did not matter whether the two-dimensional Ising model faithfully represented any actual physical system. It was enough that we could describe a physically possible fictional system that the model *could* represent, a fictional system, in other words, whose elements *could* be denoted by elements of the model.

The third response is, I think, more interesting. It requires us to isolate those features or characteristics that physical systems in a given universality class have in common, and which give rise to the scale-invariant behaviour of the members of that class. While these characteristics are not confined to the dimensionality of the systems involved and the symmetry of the physical quantities that govern their internal energy, neither do they include specific details of microprocesses, since these are not shared by all members of the class. To each of them, however, there must correspond a characteristic of the Ising model, if it is to act as a representative of the class. Because of the generality of the predicates involved (e.g., 'forms a canonical ensemble') we may be reluctant to use the term 'denotation' for the relation of correspondence between the characteristics of the model and of the individual physical systems. None the less, we may say that the existence of these correspondences allows us to regard the model as a whole as denoting each system as a whole, without thereby stretching the notion of denotation beyond recognition.

---

[47] Goodman (1968, 4) explicitly excludes these cases from his discussion.

In such a case, the term *interpretation* needs to be similarly generalised. Not every detail of the Ising model can be directly interpreted as denoting a detail of a system in the class. In other words, the Ising model is not a *faithful model* of any such system. Yet it may well be that the behaviour of the model offers the best available explanation of the behaviour of one of the systems. As Fisher writes (1983, 47), immediately before he introduces the Ising model,

> [T]he task of the theorist is to *understand* what is going on and to elucidate which are the crucial features of the problem. So . . . when we look at the theory of condensed matter physics nowadays we inevitably talk about a 'model.' . . . [And] a good model is like a good caricature: it should emphasize those features which are most important and should downplay the inessential details.

We are thus led to a conclusion parallel to Bas van Fraassen's rejection of the principle of inference to the best explanation, and Nancy Cartwright's observation that 'The truth doesn't explain much': from the fact that a model M provides the best explanation of the behaviour of a system S it does not follow that M is a faithful representation of S.

Like many other forms of representation, a model is best understood as an epistemic resource. We tap this resource by seeing what we can demonstrate with the model. This may involve demonstration in the seventeenth-century sense of the word, i.e. mathematical proof. Lars Onsager, for example, used strict methods of mathematical proof to show, for the two-dimensional Ising model in the absence of an external field, that the quantity C (denoting specific heat) diverges logarithmically as $|T - T_c|$ approaches zero; in contrast, Domb and others had perforce to use approximate methods, often supplemented by computer techniques, to show the corresponding result (that $\alpha = 0.12$) for the three-dimensional version. Or it may involve demonstration as the term is currently used. Bruce and Wallace used computer simulations to demonstrate, through a series of pictures, the effects of coarse-graining. In both cases the internal dynamic of the model serves an epistemic function: it enables us to know more, and understand more, about the physical systems we are studying.

5.2.2 *Computer simulation*

In section 5.1.8 I drew a distinction between the use of computer techniques to perform calculations, on the one hand, and computer simulation, on the other. The distinction may not always be clear-cut, but paradigm cases will help to confirm it.

A paradigm case of the former is the use of step-by-step methods to chart the evolution of a system whose dynamics are governed by non-integrable differential equations.[48] Notoriously, in Newtonian gravitational theory, for instance, there is no analytic solution to the three-body problem, even though the differential equations that govern the bodies' motions are precisely defined for all configurations of the system. But from a given configuration of, say, the Sun, Jupiter, and Saturn at time t, we can use these equations to approximate their configuration at time $t + \Delta t$, and then at time $t + 2\Delta t$, and so on. Reiteration of this process yields their configuration at any subsequent time $t + n\Delta t$. (The specification of a configuration here includes not only the positions of the bodies, but also their momenta.) These are just the kinds of calculations that computers can perform more speedily than human beings. Interpolation techniques can then be used to ensure that these approximate solutions fall within acceptable limits. The simplest requires the computer to check that, if it performs two calculations for the successive time intervals from t to $t + \Delta t/2$, and from $t + \Delta t/2$ to $t + \Delta t$, the result is 'sufficiently close' to that obtained for the time interval from t to $t + \Delta t$, and, if it is not, to decrease the value of Δt until an adequate approximation is reached.

A paradigm case of computer simulation is cited by Fritz Rohrlich. It is designed to investigate what takes place at the atomic level when a gold plate is touched by the tip of a nickel pin.[49] To quote Rohrlich,

[T]he simulation assumes 5000 atoms in dynamic interaction, 8 layers of 450 atoms constituting the Au metal, and 1400 atoms the Ni tip. The latter are arranged in 6 layers of 200 atoms followed by one layer of 128 and one of 72 atoms forming the tip. Additional layers of static atoms are assumed both below the Au and above the Ni; their separation provides a measure of the distance during lowering and raising of the Ni tip. The interatomic interactions are modeled quantitatively by means of previously established techniques and are appropriate to a temperature of 300K. To this end the *differential equations of motion* were integrated by numerical integration in time steps of 3.05×10^{-15} sec.
   The tip is lowered at a rate of 1/4Å per 500 time steps. (1991, 511–12; emphasis in the original)

Notice the thoroughly realist mode of description that Rohrlich employs. Aside from his references to the ways interatomic interactions

---

[48] For discussions of differential equations in physics that do not have analytic solutions see Humphreys (1991, 498–500) and Rohrlich (1991, 509–10).
[49] The simulation was originally described in Landman *et al.* (1990).

are modelled and numerical methods are used to integrate the equations of motion, the passage reads like a description of an actual physical process – albeit one which our technology would find it hard to replicate precisely. The impression of a realistic description is heightened by Rohrlich's inclusion of several computer-generated illustrations of the process. As he comments (1991, 511), 'There is clearly a tendency to forget that these figures are the results of a computer simulation, of a calculation; they are not photographs of a material physical model.'

The mimetic function of a simulation is emphasised in the characterisation of a computer simulation given by Stephan Hartmann: '*A simulation imitates one process by another process*. In this definition the term 'process' refers solely to some object or system whose state changes in time. If the simulation is run on a computer, it is called a *computer simulation*' (1996, 83; emphasis in the original). Hartmann contrasts this definition with the 'working definition' proposed by Paul Humphreys (1991, 501): '*Working definition:* A computer simulation is any computer-implemented method for exploring the properties of mathematical models where analytic methods are unavailable.'

Of these two proposals, Hartmann's can clearly be accommodated within the DDI account of representation. Recall, for example, that the account describes a theoretical model as possessing an internal dynamic. The dynamic has an epistemic function: it enables us to draw conclusions about the behaviour of the model, and hence about the behaviour of its subject. But, though the verification of these conclusions may lie in the future, these conclusions are not exclusively about future events, or even events in the future relative to the specification of the model we start from. They can also concern properties of the model coexisting with its specified properties, or general patterns of its behaviour. That is to say, the epistemic dynamic of the model may, but need not, coincide with a temporal dynamic.

As befits a general account of modelling, the DDI account includes as a special case the kind of 'dynamic model' which, according to Hartmann (1996, 83), is involved in simulation, and which 'is designed to imitate the time-evolution of a system'. Whether Hartmann is right to restrict himself to dynamic models of this kind is another question. After all, the prime purpose of the Isling model simulations performed by Pearson *et al.* (see section 5.1.8) was to investigate the variation of the model's properties with temperature, rather than with time. However, I will not press the issue here.

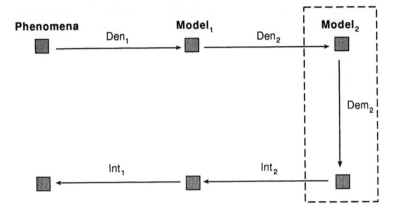

Figure 5.4 Hierarchies of representation on the DDI account of modelling.

I have a more serious reservation about Humphreys' 'working definition', to wit, that it blurs the distinction I have just made between computer
simulation and the use of the computer to solve intractable equations. This
distinction appears very clearly within the DDI account of modelling and
representation. As I emphasised, the dynamic of a model enables us to
demonstrate conclusions. Very often this involves solving equations, and
one way to do so is by the use of computer techniques, as in the case of the
three-body problem. (Notice that it is not the only way. Newton was the
first of a distinguished line of astronomers to apply perturbation theory to
such problems.) Alternatively, however, we may choose to remodel the original representation using a computer, in other words, to nest one representation within another in the manner shown in figure 5.4. The result is
a hierarchy of models of the sort mentioned in section 5.2.1, save that in
this case the whole of the second stage of the modelling process (enclosed
within the dotted line in the diagram, and labelled 'Model 2') is performed
on a computer. In this case, if the process involved in Demonstration 2 is
a temporal process, then it conforms precisely to Hartmann's characterisation of a computer simulation; it imitates the behaviour of the model.

All computer simulations of the Ising model are of this general
kind. They also support one of Hartmann's criticisms (1996, 84) of
Humphreys' 'working definition'. In denying Humphreys' suggestion
that computer simulations are used only when analytic solutions are
unavailable, Hartmann could well have pointed to the two-dimensional
Ising model; computer simulations of its behaviour are used despite the
fact that an exact solution already exists.

Two further points need to be made about the way computer simulation fits within the DDI account of representation. The first is that, while the distinction between computer simulation and the use of computer-driven calculational techniques is conceptually clear-cut, individual cases may be hard to classify, as will appear in section 5.2.4.

The second is that the two alternatives are not exhaustive. Common to my description of both of them was the assumption that the phenomena under study were represented by an abstract mathematical model (Model 1 in figure 5.4). What was at issue was simply the way the behaviour of the model was to be investigated: in one case computer-aided computation was to be used, in the other computer simulation. The assumption may not always hold. A third possibility is that the computer simulation can be a direct representation, so to speak, of a physical phenomenon. At a first level of analysis, for instance, the computer simulation discussed by Rohrlich, of a pin touching a plate, is a direct representation of just this kind. I use the phrase, 'at a first level', because a more fine-grained analysis is possible. In the first place, a 'prepared description' of the pin-plate system is used. The tip of the pin is described as a sequence of layers of Ni atoms, each containing a certain number of atoms. For this reason we may deny that the simulation directly represents the phenomenon. In the second place, even within the simulation itself the interatomic forces are modelled in a particular way, and a specific technique of integrating the relevant differential equations is used. Whereas the latter involves the use of a computer to perform a calculation, the former involves the nesting of one computer simulation within another.

A virtue of the DDI account of representation is that it allows us to articulate the complexity of computer simulation, a complexity over and above those encountered in the writing of the computer programme itself. Of course, whether we need an analysis as fine-grained (not to say picky) as the one I have just offered will depend on the kinds of question we wish to answer.

5.2.3 Cellular automata and the Ising model

Since the seventeenth century developments in physics have gone hand in hand with advances in mathematics. A theoretical model would be of no use to us if we lacked the requisite mathematical or computational techniques for investigating its behaviour. By the same token, it is hardly surprising that the arrival of the digital computer, a mathematical tool of enormous power and versatility, but at the same time a tool of a rather special kind, has stimulated the use of theoretical models that are particularly suited to computer simulation. One such model is the cellular automaton.[50]

A cellular automaton (CA) consist of a regular lattice of spatial cells, each of which is in one of a finite number of states. A specification of the state of each cell at time t gives the *configuration* of the CA at that time. The states of the cells evolve in discrete time steps according to uniform deterministic rules; the state of a given cell I at time $t+1$ is a function of the states of the neighbouring cells at time t (and sometimes of the state of cell I at time t as well). In summary, space, time, and all dynamical variables are discrete, and the evolution of these variables is governed by local and deterministic rules. (As we shall see, the last stipulation is sometimes relaxed to allow probabilistic rules.) We may visualise a CA as an extended checkerboard, whose squares are of various different colours, and change colour together – some, perhaps, remaining unchanged – at regular intervals.

The discreteness that characterises cellular automata makes them ideal for computer simulation. They provide exactly computable models. Indeed, a realisation of a CA is itself a computing device, albeit one with a rather limited repertoire (see Toffoli 1984, 120). The dynamic rules apply to all cells simultaneously, and so the CA enjoys the advantages of parallel processing. To put this another way, if we set out to design a computer that minimises the energy expended in the wiring and the time that signals spend there, rather than in the active processing elements of the computer, we will be led towards the architecture of a cellular automaton.[51]

For the present, however, I will retain the distinction between a cellular automaton, regarded as a model of a physical system, and the computer simulation of its behaviour.[52] As models, cellular automata are

---

[50] The study of cellular automata was initiated by John von Neumann in the late 1940s, with input from Stan Ulam (see Burke 1970). Von Neumann's original project was to model biological processes; indeed the best-known cellular automaton, devised by John Conway in 1970 and described by Martin Gardner in a series of *Scientific American* articles in 1970 and 1971 (see Gardner 1983), was known as 'Life'. In 1984 an issue of *Physica* (10D), edited by Stephen Wolfram, brought together 'results, methods, and applications of cellular automata from mathematics, physics, chemistry, biology, and computer science' (viii). I draw on these essays in what follows.

[51] See Vichniac (1984, 96) and Toffoli (1984, 120), but bear in mind when these essays were written.

[52] A nice example of a cellular automaton as a model, cited by Rohrlich (1991, 513–14), is a model of the evolution of a spiral galaxy, in which the galaxy is represented by a disc formed by concentric rings of cells. The hypothesis modelled is that, if a star is born in a particular cell, there is a probability p that another will be born in an adjacent cell within a time interval $\Delta(=0.01$ billion years!).

radically different from those traditionally used in physics. They are exactly computable; no equations need to be solved, and no numerical calculations need to be rounded off (Vichniac 1984, 118). Demonstration in these models does not proceed in the traditional way, by the use of derivation rules appropriate to the formulae of symbolic representations. Instead, only simple algebraic operations are required (Toffoli 1984, 118). Rohrlich (1991, 515) speaks of the 'new syntax' for physical theory that cellular automata provide, in place of the differential equations characteristic, not only of classical physics, but of quantum theory as well. He suggests that CAs will increasingly come to be used as phenomenological (as opposed to foundational) models of complex systems.

There are obvious structural similarities between cellular automata and the Ising model. And, given the theoretical significance of that model, the question naturally arises: can the behaviour of the Ising model be simulated by a CA? As we saw in section 5.1.8, the task facing the automaton would be that of simulating the wandering of the Ising configuration through the canonical ensemble, so that the frequencies of occurrence of individual arrangements would accord with the probabilities given by Boltzmann's law (equation 5.3). A simple argument, due to Vichniac (1984, 105–106), shows that this cannot be done by a CA that uses fully parallel processing, i.e. that applies its dynamic rules to all cells simultaneously.

The argument runs as follows. Consider a one-dimensional Ising model in a fully ordered spin-up arrangement, where, for every site I, $s_i(a) = +1$. Diagrammatically, we have:

$$\ldots + + + + + + + + + + + + + \ldots$$

Assume that the Ising model acts as a CA, i.e. that each site I can be regarded as a cell, and that the value of $s_i(a)$ denotes its state. Let Ri denote the right-hand neighbour of site I, and Li its left-hand neighbour. In this arrangement, for each site I, $s_{Li}(a) = s_{Ri}(a) = +1$. It follows that, if the CA is fully deterministic, then either all sites will flip at every time step, or none will. This behaviour obtains in the Ising model only at absolute zero, when no transitions occur. Hence, at any other temperature T the CA must behave probabilistically. Since, however, we are still considering the wholly ordered arrangement, the value of the probability function involved will be the same at each site, i.e., each site will be ascribed the same probability $p(T)$ of remaining $+1$, and $1 - p(T)$ of flipping to -1.

Assume, for the sake of argument, that the temperature of the lattice is such that $p(T) = 2/3$. Since N is very large, there is a non-negligible probability that a single transition will leave the lattice in an arrangement whereby, for two thirds of the sites, $s_i(a) = +1$, and for the remaining third, $s_i(a) = -1$. Among the possible arrangements of this kind are these (in which the periodicity shown is assumed to be repeated throughout the lattice):

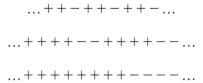

Furthermore, each of these arrangements will be equiprobable.

They will not, however, all have the same total interaction energy, E_a since each has twice as many domain walls (where neighbouring sites are of opposite sign) as the one below. But it is the value of E_a, rather than the proportion of sites where $s_i(a) = +1$, which, by Boltzmann's law, gives the probability of the arrangement a at a given temperature.

The argument, which is not restricted to the one-dimensional case, suggests that the reason why this simple CA simulation of the Ising model gives 'terrible results'[53] is that the rule governing the changes of state of cells of a CA are purely local; each cell's behaviour is governed by the states of its immediate neighbours. While rules could be formulated which reflected the nearest-neighbour interaction energies of the Ising model, no set of CA rules can be sensitive to the *total* interaction energy of the model of the system, as must be the case if the arrangements are to form a canonical ensemble.

We are led here to a deep question in thermodynamics: if interaction energies are due to local interactions, as in the Ising model, whence comes the dependence of the probability of a particular arrangement on a global property, the total interaction energy? I will not try to answer it. Note, however, that an adequate simulation of the Ising model's behaviour can be achieved by abandoning full parallelism, and using either (1) the site-by-site Monte Carlo method described in section 5.1.8, or (2) a CA modified so that its rules are applied alternately to every second cell (Vichniac 1984, 104). This suggests that the non-divisibility

---

[53] Vichniac (1984, 105) reports that 'a spurious checkerboard pattern starts to grow, leading eventually to a state of *maximum* energy' (emphasis in the original).

of time in the simple CA simulation may bear part of the blame for its failure to imitate the Ising model.[54]

This thought leads in turn to a general comment about the use of cellular automata as models of physical systems. As we have seen, they have been hailed as providing 'an alternative to differential equations' (Toffoli 1984), and 'a new syntax' for physical theory (Rohrlich, 1991). Vichniac declares (1984, 97):

> They provide a third alternative to the classical dichotomy between models that are solvable exactly (by analytic means) but are very stylized, and models that are more realistic but can be solved approximately (by numerical means) only. In fact cellular automata have enough expressive power to represent phenomena of arbitrary complexity, and at the same time they can be simulated exactly by concrete computational means . . . In other words, we have here a third class of *exactly computable* models.

Amidst all this enthusiasm, far be it from me to play the role of the designated mourner. But, as the example of the Ising model shows, a word of caution is called for.[55] The reason is straightforward. The constraints that prepared descriptions of physical systems must conform to before they become eligible for modelling by cellular automata are severe. That is the price paid for the exactness of the computations that these devices provide. I have no doubt that physical systems exist that decline to be squeezed into this straitjacket, systems, for example, whose dynamics depend crucially on the continuity, or at least the density, of the time series. For such a system a cellular automaton would not offer a reliable representation.

5.2.4 Computer experiments

In an experiment, as that term is generally used, an experimenter interacts with some part of world in the expectation of learning something about it.

A computer is 'a device or machine capable of accepting information, applying prescribed processes to the information, and supplying the results of those processes'. (I quote Van Nostrand's *Scientific Encyclopedia*, 5th edn, 1968.)

---

[54] Fur further discussion, see Vichniac 1984, sect. 4.5.

[55] It may be relevant that the Introduction to Vichniac's paper, from which the extract above was quoted, was 'adapted in part from grant proposals' (Vichniac 1984, 114). Elsewhere (1984, 113) he is more guarded: 'The validity of the simulation by cellular automata of space-discrete systems (lattices) is however conditional on the taming of the effects of the two other discretenesses (time and variables, see section 4 [which deals with the Ising model]).'

So, when physicists talk of 'running experiments on the computer', what do they mean? Not, presumably, that these experiments are to be performed in order to learn something about computers. But, if not, then what else? I will approach the question in four steps, which the reader may well regard as a slippery slope. Each step will involve an example.

Observe, first of all, that not all of the models used in physics are abstract mathematical models. Material models, like the scale models used in wind tunnels, are also used. Joseph Smeaton's work on water-wheels in the eighteenth century provides a historical example. In the 1750s Smeaton compared the advantages of overshot and undershot wheels by experimenting with a model water-wheel, in which the relevant parameters could be readily adjusted. As Baird remarks (1995 446),

> Smeaton's water-wheel is not, however, a scaled down version of some particular water-wheel he hoped to build. His model serves the more abstract purpose of allowing him to better understand how water-wheels in general extract power from water in motion . . . In quite specific ways it functioned just as theory functions.

The dynamic of the model – in both the mechanical sense and the epistemic – was supplied by a physical process, the same process, in fact, that is at work in full-sized water-wheels. By using this dynamic Smeaton drew conclusions about a water-wheel's behaviour. More precisely, he demonstrated conclusions for the model wheel, which he assumed would apply to a full-sized version. In this enterprise Smeaton was clearly conducting experiments, albeit experiments on a *model* water-wheel.

The second example comes from one of Richard Feynman's *Lectures on Physics* (1963, v.1, 25.8), in which these two equations are displayed.

$$m(d^2x/dt^2) + \gamma m(dx/dt) + kx = F$$

$$L(d^2q/dt^2) + R(dq/dt) + q/C = V$$

Both equations describe damped forced oscillations. In the upper equation, where x and t denote distance and time respectively, the oscillations are those of a mechanical system like a pendulum when an external force is applied; m is the mass of the system, and F the external force, which is assumed to vary sinusoidally with time. The oscillations are damped by a frictional force that varies directly with the speed of the mass: $F_f = \gamma m(dx/dt)$. The factor k determines the natural frequency n of oscillation of the system according to the formula $n = \sqrt{k}/2\pi$.

The lower equation also describes oscillations, but they are the electrical oscillations of a circuit under a sinusoidally varying voltage V. Here L is the self-inductance of the circuit, R its electrical resistance, and C its capacitance. Since q denotes electric charge, dq/dt denotes electric current, and d^2q/dt^2 the rate of change of current.

The two equations are in exact correspondence with each other (but notice that R corresponds to γm, and $1/C$ to k). It follows that we can investigate the behaviour of any mechanical system subject to damped forced oscillations by looking at the corresponding electrical circuit – not by solving its equations, which will be the same for both systems – but by building the electrical circuit and establishing its behaviour by experiment. Feynman adds:

This [the electrical circuit] is called an *analog computer*. It is a device which imitates the problem that we want to solve by making another problem, which has the same equation, but in another circumstance of nature, and which is easier to build, to measure, to adjust, and to destroy!

As in the first example, this procedure clearly involves experimentation – experimentation on a computer, to boot. The added factor in this example is that the results of these experiments are interpreted in terms of a totally different type of system. The physical processes at work in the electrical circuit, and which supply the epistemic dynamic of the representation, are not the same as those at work in the mechanical system.

The third example involves the cellular automaton. As I pointed out, a realisation of a cellular automaton (a *cellular automaton machine* or *CA machine*) is itself a computer, and can be used to provide simulations of a physical system's behaviour. But, unlike an ordinary computer, a cellular automaton is tailored towards a specific system; the structure of the automaton is designed to match crucial structural properties of the system being modelled, like its topology and its symmetries. Since a CA machine is a physical system; its behaviour can be the subject of experiments. Some of these experiments – those that explore the properties that the CA system has in common with the modelled system – will yield conclusions that can be interpreted in terms of that system.

In the first two examples the idea of performing experiments on one system to obtain information about another is unexceptionable, and in the third it does not seem far-fetched. In the fourth and final example, however, it may seem problematic. For this example I return to the computer simulation of atomic interactions between a pin and a plate (section 5.2.2). Would we say that Landman and his collaborators were *conducting experiments?* If not, I suggest, the reason is this. As in the three earlier examples, their simulation made use of a material object, in this instance a high powered computer. And in order to make the tip of the image of the pin descend towards the image of the plate at a certain speed, various processes were set in train within the computer. But, whereas in the cases of the model water-wheel, the electrical circuit, and the cellular automaton there was an obvious correspondence between the processes at work in the simulating device and those at work in the system being modelled, in this case no obvious correspondence exists. Indeed, that is the reason why the all-purpose, high-speed computer is so versatile; its design enables it to execute a great variety of abstract pro-grammes. The internal processes within such a computer are governed by physical laws, but different computers will apply different laws, or the same laws in different ways, to perform the same tasks. To use a cliché, the same software can be implemented in many different kinds of hard-ware. Thus Landman's methodology differed from Smeaton's in this respect, that Smeaton investigated the properties of a full-sized water-wheel by examining the properties of a model water-wheel, but Landman did not investigate the dynamics of pin-and-plate interactions by examining the dynamics of a computer.

Nonetheless, like the systems examined in the three earlier examples, the computer is a material system, and in using it to provide answers to our questions we are relying on the physical processes at work within it. If we ask what gives rise to, or what licenses expressions like 'running experiments on the computer', the analogies between the four examples are part of the answer.

Another is the degree to which, in the late twentieth century, the lived world of physicists who perform orthodox experiments conforms to the world of those who do their experiments on a computer. The orthodox experimentalist works surrounded by electronic devices for enhancing signal at the expense of noise, for sharpening images or smoothing out irregularities, for performing statistical analyses of data collected, etcetera, etcetera.

Consider, as an example, the displaying of results. On the desk in front of me as I write is a book, open to reveal a striking but enigmatic picture. Its central panel could be a reproduction of an abstract paint-ing. It is a vividly coloured rectangle, five inches wide and half as high. No fewer than sixteen different colours are present in it, ranging from black and dark blue at the lower left hand side to bright red and white near its upper right hand corner. Separating these areas, a broad river of intermediate colours, blues and greens below, yellows and oranges above, meanders across the page, as though tubes of non-miscible inks had been squeezed out to produce it. The colours nowhere blend into one another, and the sharpness with which they are defined suggests that the picture has been prepared on a computer. That impression is confirmed by the shape of the broad black border that surrounds the panel, and gives the complete picture the proportions of a television screen.

A heading 'SPACE RESEARCH INSTITUTE AS USSR' appears in the upper section of the border, and in the lower section there is a horizontal strip in which the spectrum of colours from the panel are displayed, with the black end labelled '-3.5', a light green segment in the centre, '.5', and the white end, '4.5 MK'. From these clues we can infer that the panel presents the results of experiment or observation. But, if the former, are they the computerised results of an experiment, or the results of an experiment on a computer? Even the legend 'RADIO MAP' below the panel, and the coordinates running along its bottom edge and up one side are not conclusive. There are well-known maps of Treasure Island and Middle Earth.

In fact, the picture presents observations made by the Russian spacecraft RELIKT 1.[56] The coloured panel is a projection onto the plane of the Galactic Sphere, a celestial sphere in which coordinates are measured from the galactic plane rather than the plane of the Earth's equator. Recorded on it are the intensities of the Microwave Background Radiation (the echo of the Big Bang) from different regions of the sky at a wavelength of 8 mm; hence the phrase, 'Radio Map'. But what we see on the page are not the 'raw data' of these observations; they have been carefully processed for our consumption:

The image is a 'differential map' in the sense that all the intensities are measured relative to the intensity of the Microwave Background Radiation at a fixed point in the sky which has brightness temperature 2.75 K. The colour coding of the intensity corresponds to a temperature difference between the maximum and minimum of $+8$ mK and -4.5 mK respectively. (Longair 1989, 96)

The image has also been 'smoothed to a beam width of 20°' (Longair 1989, 185).

In section 5.2.2, I commented on the sense of realism that visual

---

[56] It appears on p. 185 of Longair (1989).

presentations impart to computer simulations. Ironically, this sense of realism is reinforced, if not induced, by the manifest artificiality with which the results of orthodox experiments are presented. Both confirm us in our use of the expression, 'experiments on a computer'.

Nonetheless, however selective the sensors on the Russian spacecraft were, however vigorously the signals from them were massaged, and however arbitrary was the colour-coding of the final picture, the distribution of colours on the page before me is due in large part to the interactions between the sensors and the external world. Indeed, the function of the commentary that accompanies the picture, and tells us what procedures were used in its production, is precisely to help us understand what kind of information it provides.

This prompts the question: what kind of information can be obtained from computer experiments, investigations in which the only interaction is between the experimentalist and the computer? Well, it can be information about the actual world, about possible worlds, or about impossible worlds. For the first, paradoxical case, let us return, for the final time, to the two-dimensional Ising model. Amongst the computer experiments performed by Bruce and Wallace (see section 5.1.7), one compared the behaviour at the critical point of two versions of the model. In the first version the site variable took one of two values (-1 or $+1$); in the second it took one of three values (-1, 0, or $+1$). That experiment showed that, whereas the local configurations for the two versions were quite different, their configuration spectra both approached the same limit under coarse-graining. This result neatly illustrates the fact that critical behaviour at large length scales may be independent of seemingly important features of the atomic processes that give rise to it. It acquires significance through the importance of the Ising model in our understanding of critical phenomena. Because the Ising model acts as a representative, as well as a representation, of a whole universality class of systems, the experiment tells us something about the actual world.

Examples of computer experiments that provide information about a possible world are not far to seek. One such would be a simulation to determine the effects on the Earth's orbit, were a heavy asteroid to come close to it. Conceptually more interesting, however, are experiments to investigate the behaviour of physical entities that may or may not exist. It is, for example, possible that cosmic 'defects' – persistent inhomogeneities of the Higgs field, analogous to magnetic domain walls – existed in the early universe. If so, they may give rise to fluctuations in the Microwave Background Radiation that is our legacy from that epoch. Alternatively, such fluctuations may be due to an early period of cosmic inflation. The effects of defects may be represented by adding additional terms to the equations that, according to our best theory, governed pressure waves in the early universe, and computer experimentation makes it possible to show that the differences between the predictions of the two models are large enough to be registered by the new generation of satellite-based instruments. They 'translate into observable differences in the microwave sky' (Albrecht *et al.* 1996, 1413).[57] Thus these computer experiments tell us about two possible worlds, one of which may happen to be our own.

A computer experiment that gives us information about impossible worlds is one that investigate the effects of changes in the fundamental constants of physics or of its fundamental laws, since those constants and those laws define the notion of physical possibility. Such experiments have been performed by cosmologists, and seized on by proponents of the anthropic principle, for whom a draught of Hume (1776), twice daily, is recommended.

The fact that computer experimentation can give us information about many kinds of worlds, however, tells us something that should have been obvious all along: that, lacking other data, we can never evaluate the information that these experiments provide. We know that the computer simulation of the Ising model gives us information about the actual world, because we have independent evidence of the model's significance; we will know whether or not the theory of cosmic defects is adequate, not via computer experiments, but through the use of satellite-based instruments. In other words, though the use of computer simulation may usher in a new methodology of theorising, computer experimentation is in a crucial respect on a par with all other kinds of theoretical speculation; the difference between them lies in the richness and variety of the information that the computer provides about the theoretical consequences of those speculations. Of course, doing 'imaginary physics', charting the physics of hypothetical worlds, may bring its own rewards. As Giuseppe Verdi said, it is a fine thing to copy reality, but a better thing to invent it.

---

[57] The final footnote is an appropriate place for me to acknowledge a debt to Andy Albrecht. A lengthy conversation with him, at a time when I had barely begun to prepare this paper, taught me a lot about the use of computers in physics. My thanks also to the two editors of the volume, whose patience, though finite, was remarkable.

REFERENCES

Ahlers, Guenter (1980). 'Critical Phenomena at Low Temperature', *Reviews of Modern Physics*, 52, 489–503.
Albrecht, Andreas, David Coulson, Pedro Ferreira, and Joao Magueijo (1996). 'Causality, Randomness, and the Microwave Background', *Physical Review Letters*, 76, 1413–16.
Amit, Daniel J. (1984). *Field Theory, the Renormalization Group, and Critical Phenomena*, 2nd edn. Singapore: World Scientific.
Andrews, Thomas (1869). 'On the Continuity of the Gaseous and Liquid States of Matter', *Philosophical Transactions of the Royal Society*, 159, 57.
Baird, Davis (1989). 'Scientific Instruments as Elements of Scientific Knowledge: Smeaton's Model Waterwheel and Franklin's Pulse Glass', Unpublished paper.
  (1995). 'Modelling in a Material Medium', pp. 441–51, in Hull *et al.* (1995).
Balzarini, D. and K. Ohra (1972). 'Co-Existence Curve of Sulfur Hexafluoride', *Physical Review Letters*, 29, 840–5.
Binder, K. (1984a). 'Monte Carlo Investigation of Phase Transitions and Critical Phenomena', pp. 1–105, in Domb and Green (1984).
  (1984b). *Applications of the Monte Carlo Method in Statistical Physics*. Berlin: Springer-Verlag.
Binder, K. and D. Stauffer (1984). 'A Simple Introduction to Monte Carlo Simulation and Some Specialized Topics', pp. 1–36, in Binder (1984b).
Bragg, W. L. and E. J. Williams (1934). 'The Effect of Thermal Agitation on Atomic Arrangement in Alloy', *Proceedings of the Royal Society*, A145, 699–730.
Bruce, Alastair, and David Wallace (1989). 'Critical Point Phenomena, Universal Physics at Large Length Scales', pp. 236–67, in Davies (1989).
Brush, Stephen G. (1967). 'History of the Lenz–Ising Model', *Reviews of Modern Physics*, 30, 883–93.
Burke, Arthur W. (1970). *Essays on Cellular Automata*. Urbana: University of Illinois Press.
Cartwright, Nancy (1983). *How the Laws of Physics Lie*. Oxford: Clarendon Press.
Darden, Lindley (ed.) (1997). *PSA 1996*. Vol. 2 (Philosophy of Science Association).
Davies, Paul (ed.) (1989). *The New Physics*. Cambridge: Cambridge University Press.
Domb, Cyril (1960). 'On the Theory of Cooperative Phenomena', *Adv. Physics Phil. Mag. Suppl.*, 9, 149–61.
  (1996). *The Critical Point*. London: Taylor and Francis.
Domb, Cyril, and Martin F. Sykes (1961). 'Use of Series Expansions for the Ising Model Susceptibility and Excluded Volume Problem', *Journal of Mathematical Physics*, 2, 63–7.
Domb, C. and M. S. Green (eds.) (1976). *Phase Transitions and Critical Phenomena*. Vol. 5b. London: Academic Press.

Fernandez, Roberto, Jürg Fröhlich, and Alan D. Sokal (1992). *Random Walks, Critical Phenomena, and Triviality in Quantum Field Theory*. Berlin: Springer Verlag.

Feynman, Richard P., Robert B. Leighton, and Matthew Sands (1963). *The Feynman Lectures on Physics*. Vol. 1. Reading, MA: Addison Wesley.

Fine, Arthur, Mickey Forbes, and Linda Wessels (eds.) (1991). *PSA 1990*. Vol. 2. East Lansing, MI: Philosophy of Science Association.

Fisher, Michael E. (1981). 'Simple Ising Models still Thrive', *Physica*, 106A, 28–47.

(1983). 'Scaling, Universality, and Renormalization Group Theory,' pp. 1–139, in Hahne (1983).

Ford, Joseph (1989). 'What is Chaos that We Should be Mindful of It?', pp. 348–71, in Davies (1989).

French, Peter A., Theodore E. Uehling, and Howard K. Wettstein (eds.) (1993). *Midwest Studies in Philosophy XVIII: Philosophy of Science*. Notre Dame, IN: University of Notre Dame Press.

Gardner, Martin (1983). *Wheels, Life, and Other Mathematical Amusements*. New York: Freeman.

Goodman, Nelson (1968). *Languages of Art*. Indianapolis: Bobbs Merrill.

Griffiths, Robert B. (1964). 'Peierls' Proof of Spontaneous Magnetization of a Two-Dimensional Ising Model', *Physical Review*, A136, 437–9.

(1970). 'Dependence of Critical Indices upon a Parameter', *Physical Review Letters*, 24, 1479–89.

Hahne, F. J. W. (ed.) (1983). *Critical Phenomena (Lecture Notes in Physics 186)*. Berlin: Springer Verlag.

Hartmann, Stephen (1996). 'The World as a Process', pp. 77–100, in Hegselmann *et al.* (1996).

Hegselmann, R. *et al.* (eds.) (1996). *Modeling and Simulation in the Social Sciences from the Philosophy of Science Point of View*. Dordrecht: Kluwer.

Heisenberg, Werner (1928). 'Theory of Ferromagnetism', *Zeitschrift für Physik*, 49, 619–36.

Hoogland, A., J. Spaa, B. Selman, and A. Compagner (1983). 'A Special Purpose Processor for the Monte Carlo Simulation of Ising Spin Systems', *Journal of Computational Physics*, 51, 250–60.

Hughes, R. I. G. (1993). 'Theoretical Explanation', pp. 132–53, in French *et al.* (1993).

(1997). 'Models and Representation', in Darden (1997, pp. S325–336).

Hull, David, Mickey Forbes, and Richard M. Burian (eds.) (1995). *PSA 1994*. Vol. 2. East Lansing, MI: Philosophy of Science Association.

Hume, David [1776] (1947). *Dialogues Concerning Natural Religion*. Ed. Norman Kemp Smith. Indianapolis: Bobbs Merrill.

Humphreys, Paul (1991). 'Computer Simulation', pp. 497–506, in Fine *et al.* (1991).

Ising, Ernst (1925). 'A Contribution to the Theory of Ferromagnetism', *Zeitschrift für Physik*, 31, 253–8.

Jammer, Max (1966). *The Conceptual Development of Quantum Mechanics.* New York: McGraw-Hill.

Lakatos, Imre (1978). *Philosophical Papers, Vol. 1: The Methodology of Scientific Research Programmes.* Ed. John Worrall and Gregory Currie. Cambridge: Cambridge University Press.

Landau, L. D. (1937). 'On the Theory of Phase Transitions,' Tr. and repr. in Landau (1965), 193–216.

— (1965). *Collected Papers.* Ed. D. ter Haar. Dordrecht: Kluwer.

Landau, L. D., and E. M. Lifshitz (1958). *Statistical Physics.* London: Pergamon Press.

Landman, U. *et al.* (1990). 'Atomistic Mechanics and Dynamics of Adhesion, Micro-Indentation, and Fracture', *Science,* 248, 454–61.

Lee, E. W. (1963). *Magnetism.* Harmondsworth, Middlesex: Penguin Books.

Longair, Malcolm (1989). 'The New Astrophysics', pp. 94–208, in Davies (1989).

Niemeijer, Th., and J. M. J. van Leeuven (1974). 'Wilson Theory for 2-dimensional Spin Systems', *Physica,* 71, 17–40.

Pearson, Robert B., John L. Richardson, and Doug Toussaint (1983). 'A Fast Processor for Monte-Carlo Simulation', *Journal of Computational Science,* 51, 241–9.

Peierls, Rudolf (1936). 'Ising's Model of Ferromagnetism'. *Proceedings of the Cambridge Philosophical Society,* 32, 477–81.

— (1985). *Bird of Passage: Recollections of a Physicist.* Princeton, NJ: Princeton University Press.

Pfeuty, Pierre, and Gerard Toulouse (1977). *Introduction to the Renormalization Group and to Critical Phenomena.* Trans. G. Barton. New York: John Wiley.

Poincaré, Henri (1908). *Science et Méthode.* Paris: Flammarion.

Rohrlich, Fritz (1991). 'Computer Simulation in the Physical Sciences', pp. 507–18, in Fine *et al.* (1991).

Ruelle, David (1991). *Chance and Chaos.* Princeton, NJ: Princeton University Press.

Stanley, H. Eugene (1971). *Introduction to Phase Transitions and Critical Phenomena.* Oxford: Clarendon Press.

Toffoli, Tommaso (1984). 'Cellular Automata as an Alternative to (Rather than an Approximation of) Differential Equations in Modeling Physics', *Physica,* 10D, 117–27.

Thouless, David J. (1989). 'Condensed Matter Physics in less than Three Dimensions', pp. 209–35, in Davies (1989).

Vichniac, Gérard Y. (1984). 'Simulating Physics with Cellular Automata'. *Physica,* 10D, 96–116.

Wilson, Kenneth G. (1975). 'The Renormalization Group: Critical Phenomena and the Kondo Problem', *Reviews of Modern Physics,* 47, 773–840.

Wolfram, Stephen (1984). Preface to *Physica* 10D (special issue on cellular automata), vii–xii.

CHAPTER 6

Techniques of modelling and paper-tools in classical chemistry

Ursula Klein

6.1 INTRODUCTION

Among the many different meanings of the category of a model currently applied by historians and philosophers of science, one refers to the fact that scientists often have to invest a considerable amount of work in order to match an accepted fundamental theory, abstract theoretical principles, a general conceptual scheme etc., with new empirical objects. If a smooth assimilation of the particular and concrete object to the general and abstract intellectual framework is impossible scientists build 'models' linking the particular to the general.[1]

In my paper I study a paradigmatic achievement of model building in nineteenth-century chemistry when chemists attempted to extend an accepted conceptual framework to a new experimental domain. In the late seventeenth century and the early eighteenth, chemists created a conceptual framework that encompassed concepts like chemical compound, constitution, affinity, chemical decomposition and recomposition, chemical reaction etc. This scheme shaped the identification and classification of substances as well as experiments investigating their transformations. In the eighteenth century, chemists applied this network of concepts nearly exclusively to inorganic substances. Ordering a particular inorganic substance such as an acid, an alkali or a salt into this network required a series of experiments but was normally seen as unproblematic. However, chemists faced many problems when they began to apply this conceptual network to experiments performed with organic substances. This chapter studies some of these problems, and the way chemists solved them by creating models of their new epistemic objects which structured the experimental phenomena in accordance with the general conceptual

---

[1] Referring to models in modern physics Cartwright, for example, argues that the objects in the model fit the experimental phenomena into a fundamental theory since they match both some aspects of the phenomena and the mathematical needs of the theory. See Cartwright (1983).

scheme. In order to build the models of organic compounds and reactions chemists applied chemical formulas introduced by Jöns Jacob Berzelius in 1813. I argue that there was a paradox involved in the chemists' early application of Berzelian formulas. Berzelius introduced chemical formulas in order to represent the composition of chemical compounds according to 'chemical atomism'. Yet, the intended models built by applying chemical formulas fitted a non-atomistic conceptual framework. This paradox is resolved by considering chemical formulas as 'paper-tools'. By paper-tools I mean stabilized external representations which are applied as material resources for constructing new representations or models. The chemists widely accepted chemical formulas in the 1830s not because they intended to explain or predict experiments in an atomistic way but because chemical formulas could be used as paper-tools for the construction of models fitting their traditional non-atomistic conceptual framework. Applied as paper-tools for building the traditional kind of models, the original representational function of chemical formulas was irrelevant for the historical actors.

In the first part of this chapter I demonstrate how chemical formulas functioned as paper-tools for building traditional models of the constitution of organic compounds and their reactions. I analyse a series of experiments and models built by the German chemist Justus Liebig in 1832 and two years later by the French chemist Jean Dumas. This part starts with a brief analysis of Berzelian chemical formulas. I then describe how eighteenth-century chemists applied the general conceptual framework of chemical compound and reaction to particular experiments with inorganic substances. Both steps are preconditions for my analysis of the extension of this conceptual framework to experiments in organic chemistry and one of the first successes of modelling them.

In the second part of this chapter I demonstrate how the manipulation of chemical formulas displayed new possibilities of explaining reactions transcending the original goals of the chemists. Following the 'suggestions' arising from tinkering with chemical formulas, the French chemist Jean Dumas went a step further than was necessary for his initial purposes of model building. He introduced a new model of the chemical reaction of an organic substance which became a paradigmatic object of the new concept of substitution incorporating chemical atomism. This additional step was not only done *ad hoc*, it also ironically undermined a central aspect of Dumas' goals. I argue from my case study that it was the chemists' pragmatic application of chemical formulas as paper-tools for traditional modelling which paved the way for the introduction of a new atomistic model of chemical reactions.

6.2 CHEMICAL FORMULAS AS PAPER-TOOLS FOR TRADITIONAL MODEL BUILDING

6.2.1 The theory of proportions and Berzelian formulas

When Berzelius launched his formulaic notation in 1813, he intended a short and simple symbolic representation of the qualitative and quantitative composition of compounds according to what he called 'theory of proportions'.[2] This theory postulated that all chemical elements and compounds can be divided into portions which have a determined and characteristic relative weight which can be 'derived' from stoichiometry. Stoichiometry was a combination of experimentation, namely quantitative analysis, and a method of comparing the experimental data registered in tables. In two aspects the theory of proportions was underdetermined by experimental knowledge. First, what the comparison of data coming from the quantitative analysis of many inorganic compounds showed was that elements combine with each other in determined proportions of weight or in small integer multiples or submultiples of it. The statement that a smallest proportion of weight can be ascribed to an isolated element was a theoretical conclusion that gave a simple and plausible explanation of the data. However, alternative explanations, e.g. dynamical ones, were in principle possible. Second, and historically more relevant, the majority of data coming from the quantitative analysis of organic compounds did not confirm the universality of the 'stoichiometric laws' underlying this assumption. Nevertheless, chemists found ways to explain these anomalies within the framework of their theory of proportions.

The theory of proportions had much in common with John Dalton's atomistic theory (see Dalton 1808). Besides the shared metaphysical commitments, both of them assumed that the compound atoms (later 'molecules') of chemical compounds consisted of very few elemental atoms distinguishing them from Newtonian atomism.[3] The most important

---

[2] See Berzelius (1813 and 1814).

[3] This assumption was crucial for explaining the stoichiometric laws. Only if one assumed a small number of combining elemental atoms, like 1 atom a + 1 atom b or 2 atoms a + 1 atom b, were the observed regularities on the macroscopic level (particularly the law of multiple proportions) probable. They were not probable if one assumed that, e. g., 12 atoms a could combine with 17 atoms b or 105 with 124. In such cases there would be no remarkable discontinuity on the macroscopic level, the law of multiple proportions would be improbable. This point was much stressed

difference between Dalton's theory and the theory of proportions was that the latter did not ascribe any additional mechanical property to its postulated entities besides their weight. All questions like those about their size, shape, arrangement in space etc., and even of the dimension of weight, remained open within the theory of proportions. Many chemists refrained from any precise definition of the ontological and explanatory status of their proportions of weight. Alan Rocke has coined the term 'chemical atomism' for this specific theory that has much in common with our familiar expectations of nineteenth-century physical and philosophical atomism but bears, on the other hand, vexing differences.[4]

The crucial point in understanding the details of the theory of proportions or 'chemical atomism' compared with Dalton's atomism or any other form of physical and philosophical atomism is its mode of representation. The theory of proportions was represented by the Berzelian chemical formulas and inseparably tied to them.[5] Berzelian formulas in their still familiar form consisted of letters and numbers, like H^2O for water or $C^8H^{12}O^2$ for alcohol.[6] The letters symbolised the portion of an element and its determined and dimensionless weight registered in tables.[7] The numbers behind each letter indicated the number of these entities contained in the smallest portion of the compound substance. The algebraic form of this notation was exactly what Berzelius needed in order to leave the ontological status of the symbolised entities as open as possible. Berzelian formulas left enough space for various physical interpretations and it was this ontological vagueness which was of great help in their quick acceptance and uncontroversial status, even in those times when forms of physical atomism were under attack. Moreover, the chemical meaning of these signs depended on the context of their application. Since the dimension of the weight of the portions of substances was not determined it was possible to interpret them both as

---

by Berzelius, see Berzelius 1820. It was also the basis for explaining the anomalies coming from the quantitative analysis of organic compounds. The argument here was that the compound atoms of these compounds consisted of many atoms and that therefore the stoichiometric laws could not be observed. See Berzelius (1820, 44 f).

[4] See Rocke (1984).

[5] It is beyond the scope of this chapter to argue this matter in detail. The historical analysis of the formation of the Berzelian formulas demonstrates that Berzelius intended to strip Daltonian atomism of all its statements which went too far beyond experimentation. See Berzelius (1813 and 1814).

[6] In 1814 Berzelius wrote, e. g., the formula for water as $2H + O$. By the late 1820s it was written as H^2O (from the mid-1830s also as H_2O).

[7] There were different systems of relative proportions of weight in the first half of the nineteenth century. In Berzelius' system the proportion of oxygen was 100, that of hydrogen 6,636 and that of water 113,272. For these different systems see Rocke (1984, 82).

macroscopic and as microscopic entities depending on the problem the chemists wanted to solve.

In the first decade after their invention chemical formulas were not applied in the dominant areas of the experimental chemical practice. Atomists like Dalton rejected them since their algebraic form did not sufficiently determine their atomistic meaning in the sense of his own ontologically much more specific atomism. Taking advantage of the chemists' unfamiliarity with mathematical representations, he called them 'horrifying'. 'A young student of chemistry', he remarked, 'might as soon learn Hebrew as make himself acquainted with them'.[8] On the other hand, scientists trained in mathematics, particularly the British, objected that the Berzelian formulas were a caricature of algebra. In 1837, however, the British Association for the Advancement of Science recommended using the Berzelian formulas, arguing that they were extremely helpful in organic chemistry.[9] How did it happen that precisely in that domain of chemistry where chemical atomism was not empirically supported, leading scientists recommended the application of chemical formulas? In order to answer this question we need to follow the way chemical formulas were applied in organic chemistry. The analysis of this practice reveals that chemical formulas turned out to be extremely successful paper-tools for modelling the constitution and the reactions of organic compounds according to a general conceptual framework, which could up to then only be applied in inorganic chemistry.[10] The next part of my paper deals with this general conceptual framework and with the general reasons why organic compounds could not be assimilated to it without applying chemical formulas.

6.2.2 The conceptual framework of chemical compound and reaction and its models

From the early eighteenth century on, chemical experiments with such inorganic substances as metals, metal oxides and salts were shaped by a

---

[8] Quoted in Brock (1992, 139).   [9] For this debate see Alborn (1989), and Brock (1986).
[10] It should be noted that there is a widespread opinion among chemists, philosophers and historians of science that chemical formulas were only convenient means of representing existing knowledge, mere abbreviations and means of communication. Neither historians nor philosophers of science have carefully studied the techniques of manipulating formulas, and the knowledge producing power of chemical formulas when applied as paper-tools in chemical practice. For example, among historians, Maurice Crosland deals with formulas as 'abbreviations' in his 'Historical Studies in the Language of Chemistry' (1963), the French philosopher François Dagognet understands formulas as a sort of 'stenography' in his 'Tableaux et Language de la Chimie' (1969).

general explanatory framework that I call 'concept of chemical compound and reaction'.[11] It encompassed terms such as chemical compound, constitution, chemical composition and decomposition, chemical analysis, affinity and even homogeneous or pure chemical substance, i.e. laboratory substances which turned out to be comparatively stable entities in cycles of chemical decompositions and recompositions. The concept of chemical compound and reaction referred to the unobservable components of chemical substances, and to the hidden processes between chemical substances that were assumed to underlie the phenomenology of chemical operations. It translated a chemical laboratory language referring to instruments, vessels, manipulations and skills into a language referring to definite homogeneous substances and their hidden mutual interactions. The most important assumptions made in this conceptual framework were, first, that apparently homogeneous substances which behaved like relatively stable building blocks in chemical transformations consisted of heterogeneous components into which they could be decomposed and from which they could be recomposed. Secondly, it assumed that all chemical transformations were in fact recombinations of the components of the transformed substances into the products of the transformation, where the components preserved their integrity like building blocks. This recombination was conceived of as a symmetrical interaction between the components directed by internal relations between pairs of substances. It is this building block image of chemical substances and of chemical transformations enriched by the assumption of mutual relations between the recombining chemical substances that I call 'concept of chemical compound and chemical reaction'.[12] Referring to this network of concepts as a 'theory' would miss its epistemic function. Unlike theories it was stabilised to such a degree that it was never questioned, not to say 'tested', by chemists before the twentieth century; it was enduring, and it functioned like a transcendental condition for both the performance and explanation of experiments.[13] Its details become particularly clear by comparing it with analogous

---

[11] See Klein (1994a, 1994b and 1996).

[12] The term chemical reaction was coined in the late eighteenth century. It was only from the early nineteenth century that the term can be frequently found in publications. It should also be noted that this conceptual framework is still present in twentieth-century chemical laboratory language and in the chemical nomenclature.

[13] Closely related to its fundamental epistemic function is the fact that it was rarely explicitly articulated. It was implicit in all eighteenth- and nineteenth-century publications in inorganic chemistry which were not merely technological descriptions of preparations of substances. Its most comprehensive explicit representation in the eighteenth century were the chemical affinity tables published from 1718 onward. For a detailed argument, see Klein (1994a, 1994b and 1996).

alchemical explanations, in particular with the Paracelsian version of alchemy.[14] Paracelsianism explained chemical substances as completely homogeneous bodies, the same in all their parts, and their transformations either as extractions of an inner 'essence' or as an ennobling of their qualities.[15]

Two aspects of this conceptual framework must be emphasised. First, it was not atomistic. Its entities were not atoms but chemical substances. Secondly, it was compatible with different physical interpretations of the nature of the internal relations between the chemical substances which directed the reaction and which caused the coherence of the compound. The most prominent physical explanation was provided by the Newtonian theory of matter.[16] Although the Newtonian theory of chemical attraction was accepted by many eighteenth-century chemists and still played a role in nineteenth-century chemistry, it was neither uncontroversial nor considered relevant for the dominant areas of the experimental practice of chemistry.

How was the general concept of chemical compound and reaction applied to experiments investigating particular inorganic substances and their transformations? The traditional method which chemists wanted to extend to the experimental investigation of the constitution of organic substances and their reactions in the 1830s was done as follows. First, the particular chemical substances applied in a chemical experiment had to be identified. Within the framework of chemical compound and reaction the most important aspect of the identification of a chemical substance was knowledge of its composition.[17] Hence, the chemists did experiments which they interpreted as chemical analysis of the substances. Secondly, the conceptual scheme required that all of the products of the chemical transformation had to be isolated and identified. Only if all reaction products as well as their composition were known was it possible to reconstruct how the constituents of the basic substances recombine

---

[14] For this comparison, see Klein (1994a and 1994b).

[15] Besides that there are many epistemic objects and concepts in Paracelsianism which do not allow such a comparison. In other words, the Paracelsian understanding of bodies and their transformations was incompatible with the concept of chemical compound and reaction.

[16] See Thackray (1970).

[17] In the eighteenth and nineteenth century chemical substances were defined by a broad variety of aspects that included observable qualities like taste, smell or colour, measurable physical magnitudes like specific weight and the boiling or melting point, their so-called chemical qualities, i.e. their chemical behaviour towards reagents, and their composition and constitution. Although the definition and distinction of a chemical substance involved a plurality of aspects and methods, chemists saw their 'composition' and 'constitution' as the most relevant criteria both for chemical classification and for the investigation of chemical reactions.

into the products of the reaction. The criterion of a complete isolation of each of the products was the conservation of mass, i.e. that the sum of the mass of the products of the reaction equalled that of the basic substances.[18] Lavoiser introduced schemes of balancing the masses of the basic substances and of the products of the reaction. It is important to note that in order to formulate these schemes he measured the masses of all basic substances and of all reaction products in each particular case of a chemical transformation. As we shall see below, this was no longer necessary if chemical formulas were applied since they represented reacting masses in a general form. The next step was the identification of each reaction product. Since the knowledge of the particular composition of the reaction products was the most important step of identification on which both the reconstruction of the reaction and the classification of the substance were based, chemists analysed the reaction products. They saw a chemical analysis as confirmed if the substance could be recomposed from the products of analysis. If a chemical analysis could not be completed by the resynthesis of a substance in one step, which was often the case, additional transformations were done which eventually resulted in the resynthesis of a basic substance. The criteria for the distinction between chemical analysis and resynthesis often remained unarticulated in a particular case. The chemical affinity tables gave the involved affinities and provided important background knowledge which exemplified possible recombinations. In all controversial cases it was again quantitative experimentation that chemists saw as decisive. A product of chemical transformation whose mass was smaller with regard to the mass of one of the basic substances, and which contained some of the elements of this basic substance, was viewed as a product of its decomposition. This kind of reasoning eventually combined in the construction of the scheme of the particular reaction which defined the recombining building blocks of the basic substances and how they recombined into the reaction products.

I now come to the problems of the application of this conceptual framework to organic substances and to the chemists' model building.[19] The conception of the chemical compound and reaction was developed

---

[18] In many cases in inorganic chemistry the assumption that all of the reaction products were isolated was based on skill. However, when controversies occurred chemists did the measurement. Quantitative measurement was particularly improved in the last decades of the eighteenth century by Lavoisier as a means of arguing for his theory of oxidation and against the theory of phlogiston.

[19] These problems were first of all discussed in attempts at formulating a theory of organic radicals in the 1830s. See, e. g. Dumas and Liebig (1837).

in inorganic chemistry, chemical transformations of salts being the paradigmatic referents. Whereas experiments performed with salts normally yielded a few, often only two, reaction products that were comparatively easy to isolate, the situation was different in experiments done in organic chemistry. In the chemical transformation of organic substances many reaction products were the norm. Since most of them were liquids or gases and since they decomposed easily when heated, their isolation from each other was technically difficult. In many of the nineteenth-century experiments studying the transformations of organic substances it was impossible to isolate and identify all of the products of a reaction. In nearly all the cases it was impossible to do that in any quantitatively complete way. The problem of measuring the quantities of reacting substances was even more difficult, since it also existed for the basic substances. Chemists who had been investigating chemical transformations of organic substances in the early nineteenth century and had attempted to reconstruct the creation of the cascade of organic reaction products were convinced that the basic organic substances underwent, either simultaneously or in the course of time, different reactions.[20] It was, however, impossible to measure how much of the originally measured quantity of a basic organic substance was transformed in each pathway of reaction. In short, it was normally impossible to measure the actual masses of the substances involved in a concrete transformation process of organic substances. Before chemical formulas were available, these problems made model building of the chemical reactions and the constitution of organic compounds nearly impossible. There were indeed only very few attempts to do that before the 1820s. One of the exceptions was the different models which aimed at explaining the formation of ether from alcohol and sulphuric acid. These remained, however, highly controversial, since they all rested on incomplete knowledge about the quantitative relations. In the next section I analyse one of the first attempts of the chemists to apply chemical formulas as paper-tools for modelling chemical reactions in organic chemistry.

6.2.3 The experiment and the formation of a model of reaction in organic chemistry

In 1832 the German chemist Justus Liebig published the results of a series of experiments centred on investigating and eventually modelling

---

[20] These experiments were mostly done with alcohol and acids. See. e.g., Fourcroy and Vauquelin (1797), Thenard (1807 and 1809), Saussure (1814), Gay-Lussac (1815), Dumas and Boullay (1827 and 1828).

the chemical reaction that occurs when chlorine is introduced into alcohol.[21] Two years later these experiments were repeated by Jean-Baptiste Dumas in France.[22] Liebig had begun to model the chemical reaction by applying formulas, but it was his French colleague who further elaborated this kind of model building. Within the span of a few years, his results convinced both the French chemical community and most foreign chemists. The aim and the problems of the experiment with alcohol and chlorine were described by Liebig as follows:

The erroneous and contradictory assumptions that have been made about the nature of these bodies [of the products of the transformation of alcohol; UK] and the decomposition of alcohol by chlorine stimulated me several times to study these problems. However, the variety of the products and compounds that I saw being created, as well as the false way that I took then, were perhaps the reasons why I did not continue these studies. Intending to gain precise knowledge of all these processes, I tried to learn above all about the bodies which might be created in the complex action of chlorine on alcohol. (Liebig 1832, 184)

Liebig was clear about his main goal. It was the explanation of the transformation which occurs when alcohol was mixed with chlorine. Within the conceptual framework of chemical compound and reaction explaining a transformation meant creating the particular model of the recombination of the basic substances into the products of the reaction. According to Liebig the explanation of this transformation was extremely difficult since many different products were created which were not yet all identified. Dumas mentioned another difficulty that was a hindrance to the identification of all products of the reaction, namely that several of these products could not be isolated: 'the complicated reaction of chlorine with alcohol creates a variety of very different substances difficult to isolate' (Dumas 1835, 611). The first products that could be observed during the transformation of alcohol and chlorine were hydrochloric acid gas and an oily substance. It was this oily substance in particular which caused many experimental problems, and whose identity remained controversial. Liebig noted that chemists who had earlier experimented with alcohol and chlorine had always directed their attention to this problematic substance, and had believed that the transformation process was completed when it was produced.[23] He varied the experiment and eventually found that after the oily substance another reaction product was created. Since

---

[21] See Liebig (1832). For a more detailed analysis of this experiment see Klein (1998).
[22] See Dumas (1834).   [23] See Liebig (1832, 182).

this product was crystalline, it could easily be isolated. After testing its
properties by applying the usual set of reagents, and after quantitative
analysis, Liebig was convinced that he had found a new substance. He
called it chloral to indicate that it was produced from chlorine and
alcohol.

Because Liebig was unable to identify the oily product that was at first
created in the experiment, he gave up his original goal of building a
model of the chemical reaction between alcohol and chlorine and
turned to the investigation of the newly produced chemical substance.
Two years later when Dumas repeated all of Liebig's experiments, he,
too, did not succeed in identifying the oily substance. Yet, Dumas did not
postpone the goal of model building. Whereas Liebig still pursued the
traditional method of modelling a chemical reaction based on the identification of all products of a chemical reaction, and hence refrained
from modelling the reaction between alcohol and chlorine, his French
colleague Dumas understood that chemical formulas could be applied
as paper-tools for modelling the reaction even without knowing all reaction products. Based on the quantitative analysis of chloral he constructed its formula as $C^8O^2H^2Ch^6$ (Ch was the symbol for chlorine).
Having already determined the formula of alcohol as $C^8H^{12}O^2$ or C^8H^8
$+ H^4O^2$ and those of numerous inorganic compounds, including the
formula of hydrochloric acid (ChH), he constructed a model of the reaction in which the masses of the basic substances and of the products of
the reaction were balanced, as can be easily seen by comparing the
letters and numbers of the formulas. The model of the reaction was
complete without necessarily considering all products observed in the
operation, in particular the unidentified oily product produced before
chloral:

The reaction demands:
4 vol. alcohol = 4 vol. hydrogenated carbon = C^8H^8
                 4 vol. vapour of water = H^4O^2
16 vol. of chlorine   [Ch^{16}; UK].
It eventually produces
20 vol. of hydrochloric acid = $Ch^{10}H^{10}$
4 vol. of chloral   = $C^8O^2H^2Ch^6$ (Dumas 1834, 140)

This scheme of balancing the basic substances and the products of a
chemical transformation is a model of the reaction between alcohol and
chlorine that specifies the general conceptual framework of chemical
compound and reaction since it shows how the constituent elements
of the basic substances, alcohol and chlorine, recombine to form the

reaction products hydrochloric acid and chloral. The quantitative comparison of the masses of the reacting substances or the volumes of them, respectively, indicated by the letters and numbers of the formulas of the basic substances and the reaction products, proves that nothing was lost or gained and that the reaction was complete.[24]

My thesis that the modelling of a chemical reaction done here was traditional, i.e. a non-atomistic, may, at first glance, seem paradoxical to the contemporary reader. The point that I want to make is closely related to my concept of paper-tools. The explicit goal of Liebig and Dumas was to build traditional models of the constitution and reactions of alcohol. What the chemists wanted to know was how the elements of alcohol are arranged among each other and how these compound constituents of alcohol recombined in the reaction between alcohol and chlorine. Chemical formulas were applied as tools to reach this goal. In this context they functioned as surrogates for macroscopic masses of reacting substances that Liebig and Dumas could not measure directly in concrete experiments, and as elements of schemes balancing the basic substances and the reaction products. Although the visual image of the models of the reaction displayed a discontinuous constitution of the chemical compounds, this was not the relevant aspect for the historical actors. The finished model of the reaction constructed by means of formulas could as well have been formulated in the following way:

a alcohol + b chlorine = c chloral + d hydrochloric acid; where a, b, c, d, are the reacting masses and a + b = c + d.

In other words, the atomistic meaning of the signs does not constitute any difference in the model.

The example demonstrates that chemists applied chemical formulas as paper-tools for building a traditional model of recombining chemical substances. Several aspects of paper-tools should be emphasised here. First, representations which function as paper-tools must be stabilised or taken for granted in order to be applied as prerequisites for the production of new representations. Secondly, unlike concepts, ideas, theories, etc., which often function as unquestioned prerequisites for the production of knowledge, the extra-mental representation is crucial for paper-tools. Paper-tools are both intellectual and material. They embody knowledge – in the case of chemical formulas it is the theory

---

[24] The French chemists mostly reconstructed the relations of the volumes of chemical substances in chemical reactions following Gay-Lussac. See, e.g., Gay-Lussac (1815). Volumes were related to proportions of weight via density measurements.

of proportions – and exist in an external medium, are visible and can be manipulated according to their 'syntax'. The signs and the syntax of chemical formulas were not sufficiently determined by the theory of proportions. There were many ways that chemical atomism might have been extra-mentally represented. Besides Berzelian formulas, graphical models (e.g., those of Dalton) were possible as well. Compared with the syntax of graphical models, Berzelian formulas were particularly well suited for the traditional modelling of chemical reactions. They combined letters and numbers in a way similar to algebraic notations and had all their advantages. If you compare the numbers of portions of the basic substances and of the products of the reaction you can immediately see that the mass is conserved. Moreover, these systems of signs made calculations of reacting masses in balancing schemes possible without simultaneously invoking any specific physical or chemical meanings and their accompanying problems. It was possible to apply them as components in balancing schemes which modelled the ingredients and products of a chemical reaction in isolation from all other explanatory problems connected with a physical theory of atoms. The modelling of the reaction could be done without, for example, considering problems of spatial arrangement and movement that would have been raised by graphical atomistic models.

6.2.4 The variation of the model and the assimilation of additional experimental knowledge

When Liebig performed his experiment with alcohol and chlorine, French chemists were concerned with problems of classifying organic substances like alcohol, ethers, and other reaction products of alcohol. In 1834, when Dumas repeated Liebig's experiments, he had been working for about seven years on this subject.[25] The problem can be simply described as follows. Alcohol and most of its reaction products did not match the traditional classification in vegetable and animal chemistry.[26] For example, ethers produced from alcohol and different acids were chemical artifacts not found in nature. Vegetable and animal chemistry only studied natural components of plants and animals and classified them according to their natural origin and their observable properties.

[25] See Dumas and Boulley (1827 and 1828).
[26] In order to make a distinction between the early form of organic chemistry before c. 1830 which mostly referred to natural substances and the later form, I call the former vegetable and animal chemistry.

The members of these classes were viewed as untouched natural entities.[27] Together with his colleague Boulley, Dumas tried to find an adequate classification for artificial 'organic' substances like ethers using analogies with taxonomic principles in inorganic chemistry.[28]

After Lavoiser, chemical compounds were viewed in inorganic chemistry as having a hierarchically constructed binary constitution. The paradigmatic cases were salts considered as consisting of two immediate substances, an acid and a base, which themselves consisted of two substantial parts, namely oxygen and a metal or a non-metal, respectively. This concept of binarity was based on the interpretation of chemical transformations of salts as their decomposition into the two immediate and stable components acid and base and their recomposition into a new salt.[29] It is important to note that the general concept of binarity that was relevant for the dominant areas of the chemical practice was not an atomistic concept of the constitution of compounds. Binarity was linked to the concept of the chemical substance and it designated the arrangement of substances within a chemical compound.

In the 1830s, the leading European chemists, among them both Dumas and Liebig, assumed a binary constitution of organic compounds, in analogy with the binary constitution of inorganic compounds.[30] Based on this shared assumption often called 'theory of radicals' chemists tried to construct binary models of the particular organic compounds in which one of the two constituting substances, the 'organic radical', was viewed as the part which was characteristic of organic compounds and on which their classification could be based. Now, on the epistemological level of the particular models of binarity many differences between chemists occurred which often caused passionate debates.[31] In 1834, when Dumas modelled the reaction between alcohol and chlorine there was a fierce debate between Dumas and Jacob Berzelius on the binary constitution of alcohol and its reaction products. Dumas' binary model of alcohol $C^8H^{12}O^2$ was $C^8H^8 + H^4O^2$.

---

[27] For this traditional taxonomy, see, e. g., Fourcroy (1801–1802, vols. 7 and 8).

[28] This was based on earlier works of Saussure and Gay-Lussac. See Saussure (1814) and Gay-Lussac (1815).

[29] A rather comprehensive presentation of this conception of binarity is given in Dumas' criticism in 1840. See Dumas (1840).

[30] See the programmatic paper written by Dumas and Liebig in 1837 (Dumas and Liebig 1837).

[31] Although the analysis of the causes of the differences would certainly be revealing for the general subject matter of my paper, for reasons of space I cannot do this here. I only want to mention that the differences in the model building that underly the debates were not primarily caused by differences of experimental knowledge but by different concepts of the 'organic'.

According to this model[32] alcohol consisted of water and the compound organic radical C^8H^8. Dumas' model had been attacked by Berzelius who by this time had a high international reputation.[33] In this situation Dumas saw an opportunity to argue for his binary model of alcohol by means of the solution of an actual problem, namely the explanation of the reaction between alcohol and chlorine. This was possible since models of the constitution of chemical compounds had to be based on the investigation of chemical reactions. In other words, models of reactions and models of constitution were only different sides of the same coin. In both cases the manipulation of chemical formulas was crucial.

Dumas argued for his model of the binarity of alcohol, which involved modelling the reaction between alcohol and chlorine in the following way. The first model for the chemical reaction between alcohol and chlorine noted above included the statement that 10 atoms or portions of the hydrogen of alcohol were replaced by 6 portions of chlorine. This can be clearly seen in a simplification of the first model. In this simplified model Dumas only compared the formulas of alcohol and chloral:

The formula of alcohol is: $C^8H^{12}O^2$
that of chloral is $C^8H^2O^2Ch^6$
alcohol has lost H^{10}
and gained Ch^6
for producing one atom of chloral from each atom of alcohol [. . .] (Dumas 1835, 70)

Now, without giving detailed explanations Dumas argued *ad hoc*, that the equivalence of 10 portions of hydrogen with 6 of chlorine was an 'anomaly'.[34] In order to resolve it he split the formula equation into two parts:

$$C^8H^8 + H^4O^2 + Ch^4 = C^8H^8O^2 + Ch^4H^4$$

$$C^8H^8O^2 + Ch^{12} = C^8H^2Ch^6O^2 + Ch^6H^6 \text{ (1834, 143)}.$$

By virtue of this modified formula equation Dumas explained how the action of chlorine on the hydrogen of alcohol differed according to the location of the hydrogen in the binary constitution of alcohol. First, the hydrogen of alcohol contained in water is removed by chlorine without a chlorine substitution. Then, in a second step, exactly 6 portions of

---

[32] It is not accidental that Dumas formulated this model as $C^8H^8 + H^4O^2$, and not as $C^8H^8 + 2$ H^2O since it was substances and not atoms which were relevant for any particular model of binarity.

[33] See, e. g., Berzelius (1829, 286ff.), and Berzelius (1833). Berzelius modelled alcohol as an oxide.

[34] See Dumas (1834, 140f).

hydrogen contained in C^8H^8 were removed and substituted by 6 portions of chlorine.

Dumas envisioned his new model of the reaction supporting his own model of the binary constitution of alcohol. His binary model of alcohol included, however, a statement that was to some degree incompatible with the general concept of binarity and the general theory of organic radicals. In that general conceptual framework, modelling the reaction of a binary compound by means of chemical formulas would have been successful if one could show that the two constituents of the basic substance, in our case C^8H^8 and H^4O^2 of alcohol, behaved like integer building blocks. Yet, this was not quite what Dumas' model showed nor what could be concluded from it. Dumas argued that the formula equations demonstrated that chlorine acted differently on the hydrogen contained in the two constituents and that this action depended on the different 'state' of hydrogen (Dumas 1834, 115). This argument gave some plausibility to his model of the binary constitution of alcohol but it neither excluded the fact that the different chemical behaviour of hydrogen could be explained by different models nor did it assume that the two components of alcohol were stable building blocks.[35] Specifically, the statement that hydrogen could be removed and substituted by chlorine in the organic radical was an *ad-hoc* modification of the general theory of radicals which paralleled organic radicals with chemical elements.[36] I come back to this problem and demonstrate additional consequences of it in the second part of my paper.

Dumas' modified model of the reaction between alcohol and chlorine turned out to be extremely successful as an explanation of another urgent experimental problem, namely that of the nature or composition of the oily substance that was produced in the experiment before chloral. Dumas' partition of the formula equation and the rearrangement of the signs of the formulas in the two-part equation structured the transformation of alcohol and chlorine as a two-step reaction. The important point here was that this model yielded the following new formula of a substance: $C^8H^8O^2$. The model showed that the substance represented by this formula was produced before chloral, and that it was consumed in the next step when chloral was created. This intermediary substance in the model explained the unresolved experimental problem

---

[35] Indeed, two years later Liebig explained the different behaviour of hydrogen by a different model of the reaction that applied a different formula of alcohol. See Liebig (1836, 159).

[36] This aspect of Dumas' paper was also emphasised by a commentary of Poggendorff. See Poggendorff (1836, 98).

of the identity of the oily substance. The oily substance was either a chemical compound with the formula $C^8H^8O^2$, or it was a mixture of this compound with side products of the reaction not shown in the model.[37] One year later Liebig happened to identify the substance with the formula $C^8H^8O^2$ as a new chemical compound, called aldehyde.[38] It then became clear that the oily substance was indeed a mixture containing aldehyde and additional decomposition products. In addition, the divided formula equation incorporated another particular phenomenon of the experiment, namely the fact that it was considerably time-consuming (even several days) and that the crystalline substance identified in a series of experiments as chloral could only be observed at the end of the process. In sum, the modified model integrated additional experimental phenomena which were characteristic of organic reactions. With regard to Dumas' main intentions for modifying the first model of the reaction, this was an unpredicted side-effect. It resulted from the manipulation of chemical formulas as paper-tools in accordance with Dumas' model of the binary constitution of alcohol.

6.3 THE INTRODUCTION OF A NEW MODEL: THE PARADIGMATIC OBJECT OF THE CONCEPT OF SUBSTITUTION

In their two functions described above, chemical formulas were applied as paper-tools in order to resolve problems of model building in organic chemistry within the existing and unquestioned conceptual framework of chemical compound, binary constitution and reaction. Particularly the second modified model of the chemical reaction between alcohol and chlorine implied *ad hoc* alterations to the general concept of binarity; but it was still a traditional model of the constitution of a chemical compound focusing on chemical substances as the components of compounds. In both cases the models incorporated additional observations such as the time-consuming technology of the experiment and the production of intermediary substances. However, the representational function of the chemical formulas intended by Berzelius (i.e., that they symbolised chemical atoms) was irrelevant for the created models. We now have to complete our analysis by demonstrating how the tools applied by Dumas for building non-atomistic models displayed their own 'logic' and enforced a new and atomistic model of the chemical reaction.

---

[37] See Dumas (1835, vol. 5, 610ff).   [38] See Liebig (1835, 143).

A year after Dumas' first publication of the second model of the reaction between alcohol and chlorine:

$$C^8H^8 + H^4O^2 + Ch^4 = C^8H^8O^2 + Ch^4H^4$$

$$C^8H^8O^2 + Ch^{12} = C^8H^2Ch^6O^2 + Ch^6H^6 \text{ (1834, p. 143)}$$

he added the following reflections:

If alcohol has the formula C^8H^8, H^4O^2 chlorine can withdraw H^4 without replacing it, transforming alcohol into acetic ether $C^8H^8O^2$, which is what really happens. From this point of time, each atom of hydrogen that has been withdrawn will be replaced by one atom of chlorine, and *without being here concerned with the intermediary compounds*, we state that chloral $C^8H^2O^2Ch^6$ is formed [. . .] (Dumas 1835, 101; emphasis added)

What Dumas had in mind by saying 'without being here concerned with the intermediary compounds' and what he formulated more explicitly a few years later is the assumption that one might experimentally detect a series of chlorinated products if the atoms of hydrogen in $C^8H^8O^2$ are withdrawn and replaced stepwise by the atoms of chlorine, one after the other. Instead of Dumas' verbal statement, the stepwise substitution of one atom of hydrogen by one atom of chlorine and the production of a series of intermediary products can be modelled as follows:[39]

$$C^8H^8O^2 + Ch^2 = C^8H^7ChO^2 + HCh$$

$$C^8H^7ChO^2 + Ch^2 = C^8H^6Ch^2O^2 + HCh$$

$$\vdots$$

$$C^8H^3Ch^5O2 + Ch^2 = C^8H^2Ch^6O^2 + HCh.$$

In this new model of the reaction between alcohol and chlorine chemical formulas have a new function. They are still visual things which are manipulated on paper like building blocks but this time their atomistic significance also becomes relevant. In this new application, an atomistic model of the reaction was constructed which postulates observable differences on the experimental level compared with the previous traditional models. The new model predicts a series of intermediary substances produced from the intermediary substance that is represented by the formula $C^8H^8O^2$ before chloral can be observed. It embodies a difference not implied in the theory of proportions, namely that there are chemical reactions in which the chemical atoms of the basic substances do not recombine simultaneously but stepwise.

---

[39] Dumas did this kind of modelling in 1840. See Dumas (1840, 154 and 160ff).

The new model of substitution showed that there was no such thing as a compound radical that was as stable as an element. It challenged not only Dumas' model of the binary constitution of alcohol but also the shared general goal of the leading European chemists who tried to extend the concept of binarity from inorganic to organic chemistry. The application of chemical formulas as paper-tools for modelling the reactions and the binary constitution of organic substances ironically undermined the goals of the historical actors. In 1835, Dumas introduced the notion of substitution for his new model and extended it to some other chemical reactions of organic compounds.[40] For this extension and generalisation of 'substitution' the particular model of the reaction between alcohol and chlorine became paradigmatic. Dumas further elaborated his new concept of substitution between 1834 and 1840.[41] Only in 1840, after he had performed additional experiments and been challenged by attacks from Berzelius, did he fully realise the theoretical consequences of his concept of substitution, and explicitly dismissed the concept of binarity replacing it with a new theory of the unitary atomistic structure of organic compounds.[42] 'Substitution' became a key concept of modern organic chemistry after 1840. The new concept of chemical reactions was, however, not the consequence of an atomistic research programme or of any other form of deliberate testing of a possible atomistic hypothesis, but of the socially available tools of representation that had been applied originally for quite different ends.

6.4 CONCLUSION

Chemical formulas were introduced by Berzelius in 1813 for representing the composition of chemical compounds according to chemical atomism. However, when chemical formulas became relevant in chemical practice they functioned at first as paper-tools for traditional non-atomistic goals. Chemists applied them like building blocks for modelling the binary constitution of organic compounds. They also applied them as elements for constructing schemes of balancing the basic substances and reaction products in organic chemistry. These balancing schemes

---

[40] See Dumas (1835, 69).

[41] For reasons of space I cannot tell the story of the construction of this new concept which was accompanied by controversies. These controversies did, however, not concern the basic assumptions of the concept of substitution which is still valid today and the atomistic modelling tied to it. For the social aspects of this controversy see Kim (1996); for methodological issues in the controversy see Brooke (1973).   [42] See Dumas (1840).

were the models which fitted into the conceptual scheme of chemical compound and reaction and which structured the complex transformations of organic substances in accordance with specific experimental phenomena. The fact that Berzelian formulas consisted of a sequence of letters and numbers which could be rearranged like building blocks made them particularly apt for both applications. However, Berzelian formulas also displayed a discontinuous constitution which 'showed' possible consequences of chemical atomism. Dumas' new model of the reaction between alcohol and chlorine as a 'substitution' represented and explained the experimentally studied chemical reaction as one in which chemical atoms of hydrogen are withdrawn stepwise and substituted by chemical atoms of chlorine. Dumas was not looking for a new explanation of organic reactions when he started experimenting and modelling, rather it was 'suggested' by the manipulation of chemical formulas applied as paper-tools for the traditional model building. The irony of the story is that his concept of substitution undermined his original goal of extending the concept of binarity from inorganic to organic chemistry. Moreover, during the 1840s the concept of substitution became a key concept of modern organic chemistry thereby reflecting its transformed experimental culture. For the subsequent applications of the concept of substitution, the particular reaction between alcohol and chlorine and its *ad hoc* modelling became a paradigmatic object.

REFERENCES

Alborn, Timothy L. (1989). 'Negotiating Notation: Chemical Symbols and British Society, 1831–1835', *Annals of Science*, 46, 437–60.

Berzelius, Jöns Jacob (1813). 'Experiments on the Nature of Azote, of Hydrogen, and of Ammonia, and upon the Degrees of Oxidation of Which Azote is Susceptible', *Annals of Philosophy*, 2, 276–84, 357–68.

(1814). 'Essay on the Cause of Chemical Proportions, and on Some Circumstances Related to Them; together with a Short and Easy Method of Expressing Them', *Annals of Philosophy*, 3, 51–62, 93–106, 244–57 and 353–64.

(1820). *Versuch über die Theorie der chemischen Proportionen und über die chemischen Wirkungen der Electricität; nebst Tabellen über die Atomgewichte der meisten unorganischen Stoffe und deren Zusammensetzungen.* Dresden: Arnoldische Buchhandlung.

(1829). *Jahres-Bericht über die Fortschritte der Physischen Wissenschaften*, vol. 8.

(1833). 'Betrachtungen über die Zusammensetzung der organischen Atome', *Annalen der Physik und Chemie*, 28, 617–30.

Brock, William H. (1992). *The Fontana History of Chemistry*. London: Harper Collins.

(1986). 'The British Association Committee on Chemical Symbols 1834: Edward Turner's Letter to British Chemists and a Reply by William Prout', *Ambix*, 33 (1), 33–42.

Brooke, John Hedley (1973). 'Chlorine Substitution and the Future of Organic Chemistry. Methodological Issues in the Laurent-Berzelius Correspondence (1843–1844)', *Studies in History and Philosophy of Science*, 4, 47–94.

Cartwright, Nancy (1983). *How the Laws of Physics Lie*. Oxford: Clarendon Press.

Crosland, Maurice (1962). *Historical Studies in the Language of Chemistry*. London: Heinemann Educational Books Ltd.

Dagognet, François (1969). *Tableaux et Language de la Chimie*. Paris: Editions du Seuil.

Dalton, John (1808). *A New System of Chemical Philosophy*. Reprinted with an introduction by Alexander Joseph, 1964, New York: Philosophical Library.

Dumas, Jean B. (1834). 'Recherches de Chimie Organique', *Annales de Chimie et de Physique*, 56, 113–50.

(1835). *Traité de Chimie Appliquée aux Arts*. 8 vols. (1828–46). Paris: Béchet Jeune, vol. 5.

(1840). 'Mémoire sur la loi des substitutions et la théorie des types', *Comptes Rendus*, 10, 149–78.

Dumas, Jean B. and P. Boullay (1827). 'Mémoire sur la Formation de l'Éther sulfurique', *Annales de Chimie et de Physique*, 36, 294–310.

(1828). 'Mémoire sur les Ethers composés', *Annales de Chimie et de Physique*, 37, 15–53.

Dumas, Jean B. and Justus Liebig (1837). 'Note sur l'état actuel de la Chimie organique', *Comptes Rendus*, 5, 567–72.

Fourcroy, Antoine F. de. (1801–1802). *Systême des Connaissances Chimiques et de leurs Applications aux Phénomènes de la Nature et de l'Art*, 11 vols. Paris.

Fourcroy, Antoine F. de and Nicolas L. Vauquelin (1797). 'De l'Action de l'Acide sulfurique sur l'alcool, et de la formation de l'Ether', *Annales de Chimie*, 23, 20315.

Gay-Lussac, Joseph L. (1815). 'Lettre de M. Gay-Lussac à M. Clément, sur l'analyse de l'alcool et de l'éther sulfurique, et sur les produits de la fermentation', *Annales de Chimie*, 95, 311–18.

Heidelberger, Michael and Friedrich Steinle (eds.) (1988). *Experimental Essays – Versuche zum Experiment*. Baden-Baden: Nomos Verlagsgesellschaft.

Kim, Mi Gyung (1996). 'Constructing Symbolic Spaces: Chemical Molecules in the Académie des Sciences', *Ambix*, 43 (1), 1–31.

Klein, Ursula (1994a). *Verbindung und Affinität. Die Grundlegung der Neuzeitlichen Chemie an der Wende vom 17. zum 18. Jahrundert*. Basel: Birkhäuser.

(1994b). 'Origin of the Concept of Chemical Compound', *Science in Context*, 7 (2), 163–204.

(1996). 'The Chemical Workshop Tradition and the Experimental Practice – Discontinuities within Continuities', *Science in Context*, 9 (3), 251–87.

(1998). 'Paving a Way through the Jungle of Organic Chemistry – Experimenting within Changing Systems of Order.' In Heidelberger and Steinle (1998).

Liebig, Justus (1832). 'Ueber die Verbindungen, welche durch die Einwirkung des Chlors auf Alkohol, Aether, ölbildendes Gas und Essiggeist entstehen', *Annalen der Pharmacie*, 1, 182–230.

(1835). 'Ueber die Producte der Oxydation des Alkohols', *Annalen der Pharmacie*, 14, 133–67.

(1836). 'Sur un nouvel éther qui procure aux vins leurs odeur particulière', *Annales de Chimie et de Physique*, 63, 113–63.

Poggendorff, J. C. (1836). 'Elementar-Zusammensetzung der bisher zerlegten Substanzen organischen Ursprungs, nach den zuverlässigeren Angaben zusammengestellt vom Herausgeber', *Annalen der Physik und Chemie*, 37, 1–162.

Rocke, Alan J. (1984). *Chemical Atomism in the Nineteenth Century: From Dalton to Cannizzaro*. Columbus: Ohio State University Press.

Saussure, Théodore (1814). Nouvelles Observations sur la composition de l'alcool et de l'éther sulfurique. *Ann. Chim.*, 89, 273–305.

Thenard, Louis Jacques (1807). 'Troisième mémoire sur les éthers; Des produits qu'on obtient en traitant l'alcool par les muriates métalliques, l'acide muriatique oxigéné et l'acide acétique', *Mémoires de Physique et de Chimie de la Société d' Arceuil*,1, 140–60.

(1809). 'De l'action des acides végétaux sur l'alcool, sans l'intermède et avec l'intermède des acides minéraux', *Mémoires de Physique et de Chimie de la Société d'Arcueil*, 2, 5–22.

Thackray, Arnold (1970). *Atoms and Powers: An Essay on Newtonian Matter-Theory and the Development of Chemistry*. London: Oxford University Press.

CHAPTER 7

The role of models in the application of scientific theories: epistemological implications

Mauricio Suárez

7.1 INTRODUCTION

The theme of this book is reflected in the slogan 'scientific models *mediate* between theory and the real world'. It is a theme with, at least, two aspects. One aspect is methodological. Model building is a pervasive feature of the methodology (or methodologies) employed by scientists to arrive at theoretical representations of real systems, and to manipulate reality. Many of the contributors to this book engage with the methodological issues, and they all agree that the activity of model building is central to scientific practice. The methodological implications of the *slogan* are clear: much of scientific practice, perhaps the totality of it, would be impossible without models.

Another aspect of the theme relates to issues such as the nature of explanation, the form of scientific confirmation and the debate over scientific realism. These are traditional philosophical issues, and in this paper I concentrate on one of them: models provide theories with genuine empirical content, by 'filling in' the abstract descriptions afforded by theory, hence making it possible to apply theories to natural phenomena. How do models perform this role? What are the consequences for the realism issue? The focus of this paper is on the implications of models for the epistemology of scientific knowledge.

7.2 MODELS AS MEDIATORS

There are many kinds of models in science. In this paper I focus on one of them: *mediating models*. First, in this section, I introduce the notion of a mediating model, and I briefly outline some of its main features. In the

Research for this paper was partly carried out at the Centre for the Philosophy of Natural and Social Sciences, London School of Economics. I thank everyone involved in the Research Group in Models in Physics and Economics for their encouragement.

remaining sections I make the notion more precise by considering the key role that mediating models play in the application of scientific theories, and the implications of mediating models for the epistemology of science.

Mediating models have been recently discussed by a number of authors. Adam Morton[1] has referred to them as the providers of physical insight; Margaret Morrison[2] has studied and discussed some of their properties carefully; and among historians of science, Norton Wise[3] has unearthed some of the mediating models and instruments that operated in Enlightenment France.

7.2.1 Features of mediating models

Mediating models always stand between theory and the physical world. Their main function is to enable us to apply scientific theory to natural phenomena. A mediating model often involves a novel conception of a particular physical phenomenon that facilitates the application of some established physical theory to such phenomenon. Morrison has identified three main features. First, mediating models are not derivable from theory. In a very specific sense the construction of these models is not *theory-driven*; I will emphasise this feature later on in this paper. Secondly, these models are not necessitated by the empirical data either (although they may be consistent with the data and they can be suggested by the phenomena). In contrast to a data-model which is determined by the data together with established statistical techniques, a mediating model *'is more than simply a phenomenological classification constructed as a convenient way of representing [data]'* (Morrison forthcoming b). In other words, mediating models typically involve substantial theoretical and conceptual assumptions. Finally mediating models have a very significant property: they can replace physical systems as the central objects of scientific inquiry. Morrison (forthcoming b) writes:

Not only do models function in their own right by providing solutions to and explanations of particular problems and processes, but in some cases they even supplant the physical system they were designed to represent and become the primary object of inquiry. In other words, investigation proceeds on the basis of the model and its structural constraints rather than the model being developed piecemeal in response to empirical data or phenomena.

---

[1] Morton (1993); also in conversation, Bristol, May 1997.
[2] Morrison (1998) and (forthcoming a and b).   [3] Wise (1993).

This is an essential feature of mediating models; it distinguishes this type of model from other closely related types, such as for instance Heinz Post's *floating models*. As reported by Redhead (1980), floating models may also satisfy the first two features ascribed to mediating models. Redhead (1980, 158) describes a floating model as

a model which is disconnected from a fundamental theory T by a computation gap in the sense that we cannot justify mathematically the validity of the approximations being made but which also fails to match experiment with its own (model) predictions. So it is disconnected from the fundamental theory and the empirical facts. In Post's graphic terminology the model 'floats' at both ends. It has, in this sense, no theoretical or empirical support.

Post's parody of a floating model was an example he called the Farm Gate Contraction. Redhead reports this example as follows:

A farmer investigates the relation between the length of the diagonal strut and the length of the rails and stiles of a farm gate. Although he is familiar with Euclid the derivation of Pythagoras's theorem is utterly beyond his deductive powers. So he invents a model theory, a linear one, in which the lengths are related by $l = x + y$ instead of $l = \sqrt{x^2 + y^2}$. Now [the model] has many properties analogous to [the theory] for $x = 0$ or $y = 0$ it gives correct values for l and l increases monotonically with x or y in the model as in the correct theory. But detailed measurement shows that [the model] is false. So the farmer now introduces a new effect, the Farm Gate Contraction, to explain the mismatch between the predictions of the model and the experimental results.

The Farm Gate Contraction is a correction to a floating model. The model, even when corrected in this way, is certainly not *required* by the data, as is shown by the fact that there are alternative models that fit the data just as well (the 'correct' theory is one of them); and it is not supported by any fundamental theory as it is only an inspired (although ultimately mistaken) initial guess. Floating models are not derivable from either theory or empirical data. In that sense a mediating model is a kind of floating model.

However a mediating model has a further essential feature, one that is not necessary for a floating model. While a floating model may convey no new knowledge at all, a mediating model mediates between high level theory and the world by conveying some *particular* or *local* knowledge specific to the effect or phenomenon that is being modelled. This is why the model itself becomes the active focus of scientific research. While a floating model is typically only a computational tool, a mediating model is a carrier of specific, or 'local' knowledge. Morrison (forthcoming b) writes:

It is exactly in these kinds of cases, where the model takes on a life of its own, that its true role as a mediator becomes apparent. Because investigation centres on the model rather than nature itself its representative role is enhanced to the point where the model serves as a source of mediated knowledge rather than as simply a mediator between high level theory and the world.

Hence this third feature, the capacity a model may have to replace the phenomenon itself as the focus of scientific research, is an essential feature of mediating models. It distinguishes mediating models from the far larger class of floating models. In this chapter I develop a further feature of mediating models, which is essential for a full understanding of the role that these models play in the application of scientific theories. Mediating models will often fix the criteria that we use to refine our theoretical descriptions of a phenomenon. These criteria are required to apply theory successfully to the world. Before discussing this fourth feature of mediating models it may be worth emphasising the differences with some of the types of model that are commonly discussed in the literature.

7.2.2 Mediating models in the philosophy of science

A very distinguished, although perhaps languishing, philosophical tradition equates models with interpretations of theory. This tradition assimilates the distinction between scientific theories and scientific models to the syntax/semantics distinction in linguistics. The theory is a purely syntactical entity, while the models provide us with the semantics of the scientific discourse. The relation between the models and the theory is one of satisfaction: the model must make the theory's axioms true.

It is difficult to see how models are to literally 'mediate between theory and the world' if the view of models as providing the semantics of theories is correct. If models are interpretations, or partial interpretations, of theories they are in a sense supererogatory on theory. A theory will define an elementary class of models; hence it will greatly restrict the class of permitted models. An inconsistent theory, for instance, restricts the class of permitted models to the empty set. However, it is a presupposition of the notion of models as mediators that there are three distinct objects (theories, models, and the world) and that they are ordered with the theory at the most abstract end, the world at the opposite end, and the model as the interface between the two. Moreover the model conveys specific physical knowledge. The view of models as interpretations of theories allows for a trichotomy between theory, model and world but it seems to order these objects the wrong way around, with models at the most abstract end, and theories at the interface (as model/theory/world rather than as theory/model/world). Moreover, it implies that models do not convey any significant novel physical information that is not already encoded in theories. Surely this is partly the reason why proponents of this view have so often attempted to construe the relation of confirmation as a purely syntactical connection between a theory, on the one hand, and evidence, on the other.

It is possible on the syntactic view to see the world itself as a possible model of a theory. The theory is a set of axioms in some formal system, and it implicitly defines an elementary class of models. We may then say that a theory is true if it has the world as one of its models, and false if the world is not among its models. In so far as the world itself is to be a model, the distinction between model and the world collapses, and we are left with a dichotomy theory/world. So on this view, models mediate between the theory and the world only in the sense that the set of permitted models of a theory can be said to include the world itself. The activity of model building reduces, on this account, to investigating ways the world would have to be if some specific scientific theory was true. This assumes, once more, that the totality of scientific knowledge about the world is encoded in theories.

There is also, of course, the semantic conception of theories advocated by Suppes, van Fraassen and others. Here the distinction between theory and model collapses as, according to the semantic view, theories *are* models – they are really nothing but collections of models. On this view there is a hierarchical structure of models, from low-level data-models to high-level theoretical models. So the contrast between theories and models disappears. Besides, on the semantic view of theories the domain of application of a scientific theory is assimilated to its domain of empirical adequacy.[4] But mediating models play a key role in the application of theories, precisely in cases in which the theory's domain of application does not coincide with its domain of empirical adequacy. Hence the semantic view lacks the resources to provide us with an understanding of how, in practice, models mediate between theory and the world.

---

[4] This is argued in detail in my PhD thesis (Suárez 1997).

7.3 THEORY-APPLICATION: THE ROLE OF MODELS

In this section I describe a specific proposal for theory-application that involves models as idealisations. This proposal, essentially due to Ernan McMullin, is intended to go further than the traditional accounts of scientific theorising, by placing the activity of model-building at the very core of scientific practice. I argue that, despite its intention, McMullin's proposal effectively dispenses with the need for models as mediators because it invariably construes models as approximations to theories. In section 7.4 I try to illuminate and explicate this practical role of models as mediators by using an example from the history of superconductivity. In section 7.5 I discuss the epistemological implications.

7.3.1 Forms of idealisation

How does scientific theory get applied to the world? Ernan McMullin (1985) has proposed a realist account of theory-application. Theoretical descriptions, argues McMullin, are always idealised; they apply only under very special circumstances, often not realisable in practice. But the idealisation inherent in theory is not epistemologically problematic. Although theoretical descriptions are often not *absolutely* true or false, they are *approximately* true or false.

McMullin finds support for this view in Galileo's idealisation techniques. In *The New Sciences* Salviati, Galileo's alter ego, argues against the Aristotelian views of some of Galileo's contemporaries, personified mainly in the character of Simplicio. The discussion centres around the techniques of approximation required to apply theory to concrete problem situations and to validate the theoretical claims of Galilean mechanics. Two examples are repeatedly used: parabolic trajectories of projectiles, and motion of rolling objects on inclined planes. Consider the latter. Galileo's claim is of course that the motion of a perfectly symmetrical sphere under the earth's gravitational pull on a frictionless plane in a vacuum follows a very strict mechanical law. But any real plane will exhibit friction, any real object is bound to be only imperfectly spherical, and in any actual experiment there is bound to be dampening due to the presence of air. To establish his mechanical conclusions on the basis of actual experiments, Galileo has to claim that the imperfections can be accounted for, and that there is a well established and unique method of introducing corrections into theory to account for 'impediments', the imperfections of nature.

In order to show that there is indeed such a method, Galileo (and McMullin) need to appeal to the notion of approximation. There are, broadly speaking, two methods for approximating theory to the world. One is the approximation of the theory to the problem situation brought about by introducing corrections into the theoretical description – the theory is refined to bring it closer to the problem-situation. The other is the approximation of the problem-situation to the theory by means of simplifications of the problem-situation itself. In the latter case the theory is left untouched, while the problem-situation is altered; in the former case the converse is true: the problem-situation is left untouched, while the theoretical description is corrected.

Let me first consider the former kind of approximation whereby the theoretical description is refined to bring it closer to the problem-situation. This is a form of approximation towards the real case: the corrections introduced into the theoretical description are intended to account for the imperfections that occur in the problem-situation. The same method can be reversed (this is not yet the second method of approximation) by *subtracting*, rather than adding, the required corrections. We may call this an *idealisation*; for the result of such subtraction is of course a more, rather than less, idealised description of the problem-situation. The important feature of this idealisation is that the subtraction of corrections is performed on the theoretical construction, while the description of the problem-situation is left entirely unaffected. For this reason McMullin (1985, 256) calls the first form of approximation *construct* idealisation.

The second method of approximation brings the problem-situation closer to theory. We idealise the description of the problem-situation, while leaving the theoretical construction unaffected. McMullin calls this *causal idealisation* because the description of the causes present in the problem-situation is altered to bring the description into the domain of the theory. In the practice of physics this process can come in either of two forms. It can come first in the form of conceptual redescriptions of the problem-situation, performed only in thought, and not in reality. In such 'thought-experiments' interfering causes are idealised away and the result is a simplified description of the problem-situation. Secondly, there is also the possibility of physical 'shielding' of the experimental apparatus, which will involve changes in the actual experimental set-up.

Such changes are designed to minimise the influence of interfering causes, or to block such influences out altogether. It is perhaps instructive to quote Galileo in full:

> We are trying to investigate what would happen to moveables very diverse in weight, in a medium quite devoid of resistance, so that the whole difference of speed existing between these moveables would have to be referred to inequality of weight alone. Hence just one space entirely void of air – and of every other body, however thin and yielding – would be suitable for showing us sensibly that which we seek. Since we lack such a space, let us (instead) observe what happens in the thinnest and least resistant media, comparing this with what happens in others less thin and more resistant. If we find in fact that moveables of different weight differ less and less in speed as they are situated in more and more yielding media, and that finally, despite extreme difference of weight, their diversity of speed in the most tenuous medium of all (though not void) is found to be very small and almost unobservable, then it seems to me that we may believe, by a highly probable guess, that in the void all speeds would be entirely equal. (quoted in McMullin 1985, 267)

It is uncertain whether Galileo actually performed any of these experiments. If he did, he would certainly have needed to use a technique of 'shielding' to minimise the influence of interfering causes. If, on the other hand, he did not actually perform the experiments then in this passage he is describing a series of *thought-experiments* that gradually minimise the effects of interfering causes – in the mind, of course, not in reality. The dynamics of moveables in the void that he concludes will exhibit equal speeds is in either case a *causal* idealisation. Starting with a concrete problem-situation (i.e. the motion of an object in the earth's atmosphere) Galileo constructs a set of simpler problem-situations. If relations between quantities measurable in these gradually simpler thought experiments converge to a law we can then enunciate the law for the ideal (simplest) case. The resulting law is a *causal* idealisation, because the simplifications correspond to missing causes in the problem-situation.

McMullin summarises the main features of each form of idealisation concisely as follows:

> We have seen that idealization in this context takes on two main forms. In construct idealization, the models on which theoretical understanding is built are deliberately fashioned so as to leave aside part of the complexity of the concrete order. In causal idealization the physical world itself is consciously simplified; an artificial ('experimental') context is constructed within which questions about law-like correlations between physical variables can be unambiguously answered. Causal idealization, instead of being carried out

experimentally, can also be performed in thought, when we focus on the single causal line in abstraction from others and ask 'what would happen if'. (1985, 273)

In this chapter I focus only on *construct idealisation*, the kind of idealisation whereby simplifications are worked out on the theoretical description, rather than on the problem-situation. This is because I believe that every case of theory-application will involve, in practice, at least some degree of construct idealisation. *Construct* idealisation requires no thought-experiments, nor does it require tampering with the real experimental situation. Only one problem-situation, namely the real case, is entertained. It is the theoretical description that gets modified by introducing correction factors that represent 'impediments', the special circumstances that make up the particular problem-situation. In other words, in construct idealisation, the theoretical description is refined gradually to make it applicable to the problemsituation.

In actual practice we look for approximations to the theory that can be applied to a particular problem-situation. Redhead (1980) refers to these approximations as *impoverishment* models. The theoretical description may be very complicated: there may be no analytic solutions to the theoretical equations. How then can we derive the correct impoverishment model? How can we choose among all possible approximations the very one that accurately represents the behaviour of the system? The important point, that I shall now stress, is that the theory itself must contain the information required to select the correct approximation if the approximation in question is to count as a *de-idealisation* of theory.

7.3.2 Idealisation and scientific realism

A theory can be applied by finding a simplifying approximation to it that is adequate for the description of a phenomenon. Not all approximations, however, guarantee that the theory is confirmed by its applications. It is essential to McMullin's realism that the corrections introduced into the theoretical description should not be *ad hoc*. The corrections have to be well motivated *from the point of view of theory*. If the theory is to receive confirmation boosts from its applications, the corrections need to be not only consistent with the theory, but also if not dictated by, at least *suggested by*, the theory. If in a particular application the necessary corrections turned out to be inconsistent with the theory, the theory could be said to be disconfirmed; if the corrections were consistent with the theory, but not suggested by it, the theory would neither receive a confirmatory boost nor a disconfirmatory one. McMullin explicitly acknowledges this important point: according to the (*construct*) idealisation picture of theory application, the manipulations exerted on the theoretical description must be 'theory-driven' because the theory itself is to be truth-apt (a 'candidate for truth' in Hacking's (1982) terminology) and is to gain confirmation through its applications. If the corrections were not suggested by the theory then the resulting description would be *ad hoc* and, from the point of view of a realist epistemology, it would be unable to provide any evidence for the truth of the theory. Thus McMullin writes:

> The implications of construct idealization, both formal and material, are thus truth-bearing in a very strong sense. Theoretical laws [. . .] give an approximate fit with empirical laws reporting on observation. It is precisely this lack of perfect fit that sets in motion the processes of self-correction and imaginative extension described above [i.e. *deidealisation*]. If the model is a good one, these processes are not *ad hoc*; they are suggested by the model itself. Where the processes *are* of an *ad hoc* sort, the implication is that the model is not a good one; the uncorrect laws derived from it could then be described as 'false' or defective, even if they do give an approximate fit with empirical laws. The reason is that the model from which they derive lacks the means for self-correction which is the best testimony of its truth. (1985, 264)

In this passage McMullin is not using the term 'model' to describe a mediating model, as I do in this paper. I have taken 'mediating models' to be distinct from established theory while McMullin is here taking 'model' to stand for a theoretical description, as in the semantic view of theories. McMullin makes it clear that the corrections introduced into a theory to generate predictions in a particular physical problem-situation have to be suggested by the theory itself; otherwise, the corrections would be *ad hoc* and the resulting description, no matter how well it fitted the particular case, would not yield any confirmatory boost for the theory. If the corrections were not suggested by the theory there would be no way to account for the effects that those corrections have upon the final predictions. As McMullin notes (1985, 256) it is essential that there be '*a way of dealing with the fact that construct idealizations "depart from the truth". If this departure is appreciably large, perhaps its effect [. . .] can be estimated and allowed for.*'

By requiring that the corrections into a theoretical model be well motivated from the point of view of theory we make sure that we are always able to estimate their contribution to the final description. In other words, application must be *theory-driven* in order to provide confirmation for the theory. I shall refer to this sort of theory-driven approximation of the theory to the problem-situation that results in a refinement of the theoretical description as *construct de-idealisation*, or *deidealisation* for short, as an approximation of this kind is nothing but the converse of construct idealisation. In forming construct idealisations we idealise away, by subtracting from the description, those features of the problem-situation that are either (a) irrelevant to the theoretical description, or (b) relevant to the theoretical description, but also known to have effects that are precisely accountable for. (In the latter case construct idealisation is often used for the purpose of improving the mathematical tractability of the problem.) In either (a) or (b) a strict criterion of theoretical relevance is presupposed. It is the theory that tells us the relevant features to be idealised away, and suggests how to account for their effects. The same criterion of theoretical relevance must be in place if the converse process of 'adding back' features is to count as a meaningful *deidealisation*. The requirement that the introduction of corrections into a theoretical model be well motivated from the point of view of theory ensures that this criterion is firmly in place.

The above discussion is perhaps sufficient to make clear why the idealisation account of theory application satisfies the realist's constraints. For applications which follow the idealisation account, the theory receives confirmation boosts from the applications. The corrections that serve to generate successful applications are necessarily consistent with theory, because they are suggested by theory. They are corrections suggested by some strict relevance criterion – a criterion that is wholly and unambiguously theoretically determined. So, an application of a theory that conforms to nature provides a good reason to believe that the theory itself is true.

Let me now briefly address the sense of 'approximate truth' that is involved in the idealisation account. McMullin is not arguing that scientific theories are approximately true or false. The theory, on McMullin's view, contains its own criteria of application; so, indeed, the theory contains all theoretical descriptions of problem-situations in its domain. Hence the theory is either true (if it contains one true description of every problem-situation), or false (if it fails to do so). It is because of this that a successful deidealisation of a scientific theory to a particular problem-situation should always be taken as an indication of the theory's truth: it shows that the theory contains one true description of the problem-situation.

The realist's claim is then rather that *theoretical descriptions* of a particular problem-situation may be approximately true or false. His intuition is roughly as follows: successive approximations of a theory to a problem-situation have a degree of confirmation inversely proportional to their 'distance' from the problem-situation as measured on the 'idealisation scale'; but – for a realist – degree of confirmation is degree of truth; so 'distance in the idealisation scale' measures degree of truth. Given two representations A and B of some concrete problem-situation if A is less idealised than B then, in a very precise sense, A is *truer* than B. To pursue a Galilean example: the representation of a sphere rolling down a frictionless plane is less idealised if described in the actual atmosphere (description *A*) than if described in a vacuum (description *B*). The description in the atmosphere has to involve a measure of the dampening due to air. The realist claims that this description is *truer* than the description of the sphere in a vacuum, in a totally unobjectionable sense of the notion of objective truth. For a scientific realist, such as McMullin, Galilean idealisation provides the *model* for the notion of approximate truth.

7.4 PROBLEMS WITH IDEALISATION

It is always open to the opponent of realism to attack the inference from the past success of a theory to its future success, and from its pervasiveness in practice to its truth. An instrumentalist may after all have no qualms with Galilean idealisation: it is a technique of application, it is often used, and sometimes with some conviction that it carries epistemic weight, but in fact it is only a tool, and it can give no genuine warrant for belief other than the psychological comfort offered by the familiarity of its use. But here I do not attempt a general philosophical rebuttal of the realist view. This would take us one step back, in the direction of the traditional disputes concerning arguments for scientific realism – disputes that have not been settled, possibly because they could never be settled.[5]

On independent grounds the realist view will not work. The realist wants to claim that the idealisation account captures the essential features of the procedure of theory-application. I argue that the idealisation account is seriously flawed and that it can not explain the role of models

---

[5] In the philosophy of science this *quietism*, or perhaps simply 'pessimism', towards the realism/antirealism debate has been most ably defended by Arthur Fine – see chapters 7 and 8 of Fine (1986a).

in scientific practice. The inadequacy of the idealisation account stems from the fact that, in practice, theory-application does not typically follow the pattern of *deidealisation*. But the realist does not rest content with this base-level claim; in addition he claims that the idealisation account also agrees with scientific practice at an *epistemological level*. Scientists' confidence in a scientific theory typically increases on account of its many successful applications. The realist seeks support for the idealisation account also on these epistemological practices of scientists. And, indeed, on the idealisation account a theory gains confirmation through its applications, in the manner described in the previous section.

To sum up, there are two distinct claims that the realist makes on behalf of the idealisation account: first, that this account agrees with the practice of theory-application and second, that it agrees with scientific epistemology. In this chapter I contest the truth of the former claim, and I argue that the latter claim, although true, does not provide ammunition for the realist account of theory-application.

7.4.1 Idealisation and mediating models

I like to illustrate the idealisation account of application with a simple example in mechanics due to Giere (1988, ch. 3). The example brings out very clearly what, in my view, is the major defect in this account. Consider the derivation of the equation of the damped linear oscillator from that of the simple harmonic oscillator. The equation of the simple harmonic oscillator is:

$$m\frac{d^2x}{dt^2} = -\left(\frac{mg}{l}\right)x, \qquad (7.1)$$

while the equation that describes a damped harmonic oscillator is:

$$m\frac{d^2x}{dt^2} = -\left(\frac{mg}{l}\right)x + bv. \qquad (7.2)$$

The process that takes one from the theoretical description of the frictionless harmonic oscillator to the damped harmonic oscillator is a successful deidealisation in the attempt to apply classical mechanics to a real-life pendulum. The extra term bv represents the dampening due to air friction that any real oscillator must be subject to. The introduction of this correction term into the idealised description afforded by the equation of the simple harmonic oscillator is motivated by theoretical considerations: in classical mechanics friction is modelled by a linear function of velocity.[6] By introducing well-motivated corrections into the theoretical description of the simple harmonic oscillator we obtain a less idealised description of a real-life pendulum in ordinary circumstances, namely the description of a damped harmonic oscillator.

Equation (7.2) tends to equation (7.1) in the limit $b \rightarrow 0$, as required for an approximation. Hence the two descriptions agree in the asymptotic limit. There are of course plenty of equations that, just like (7.2), tend to the original equation (7.1) in some mathematical limit. Equation (7.2) is special because it is derived from the equation of the simple harmonic oscillator by a process of deidealisation. The damped harmonic oscillator and the simple harmonic oscillator are objects defined implicitly in the theory by their satisfaction of the corresponding equations; hence it is the theory that determines the relations between them. The correction terms introduced into the equation of the simple harmonic oscillator are justified by the putative relations between the objects themselves. Equation (7.1) is satisfied by a linear oscillator with no friction; equation (7.2) is satisfied by a linear oscillator subject to friction. The theory contains all the necessary techniques to represent this difference formally.

Hence the idealisation account makes superfluous the use of models in theory application. Theories must be seen as entirely self-sufficient in the task of generating genuinely realistic representations of problem-situations.[7] Where the idealisation account is true, or generally true, it follows that models cannot *mediate* between theories and the world: in the application of scientific theories that satisfy the idealisation account, there is essentially no work for mediating models to do.

The idealisation account assumes there is a final representation of every system in the theory's domain of application. In practice we may never be able to write this representation, as it may be hideously complicated; but the representation must exist because it can be approximated to an arbitrary degree by successive deidealisations of the theory. However, even in the simple case of the harmonic oscillator the presumption that such a final theoretical representation exists seems profoundly perplexing. The equation of the damped harmonic oscillator is certainly not a final representation of this kind. It is not a theoretical representation of any concrete real system in the world. Admittedly the equation of the damped harmonic oscillator is a less idealised

---

[6] For a discussion of modelling friction see e.g. Goldstein (1980, 24).

[7] Specifically, and to anticipate the main issue in what is to follow, theories do not (*must not*) rely on independently-standing models in order to fix the corrections required for successful *deidealisations*.

representation than the equation of the simple harmonic oscillator for real-life penduli. But this does guarantee that the theory contains a (true) representation of a real-life pendulum. The theory may be incomplete; there may well be some aspects of the problem-situation left unaccounted for, even after all the relevant corrections suggested by the theory have been added in.

But now the promised sense in which models were to mediate between theory and the world is definitely lost: models mediate only between theory and further models. On the idealisation account the theory does all the work required for its own application by determining, in stages, sets of increasingly *less idealised* representations. These representations, however, may never truly represent anything real at all.

7.5 HOW MODELS MEDIATE: THE CASE OF SUPERCONDUCTIVITY

The problem becomes acute when it is noticed that in practice the criteria of theoretical relevance presupposed by the idealisation account are rarely operative in cases of successful theory-application. On the contrary, it is often the case that scientific representations of effects or phenomena are not arrived at as deidealisations of theory. My case study in superconductivity illustrates one way in which models typically mediate between theory and the world.[8] The first theoretical representation of the Meissner effect was not found by applying a criterion of theoretical relevance for the introduction of corrections into the electromagnetic equations of a superconductor. These correction terms were not given by, and could not have been given by, classical electromagnetic theory but were rather derived from a new *model* of superconductivity. The model was motivated directly by the phenomena, not by theory. The criterion required for the application of electromagnetic theory could only be laid out when the model was in place, and an adequate classical electromagnetic description of superconductivity (the London equations) could then finally be derived.

This is, I want to claim, an important sense in which models *mediate*: they establish the corrections that need to be introduced into a theory in order to generate many of its applications. My case study shows how the derivation of a theoretical representation of a physical effect can result

---

[8] Aspects of this case study have been published in a joint paper with Nancy Cartwright and Towfic Shomar (1995). I want to thank them both for helpful conversations on this section.

from corrections that are suggested by a mediating model, which is independent from theory. The approximation used to generate an appropriate representation is not a deidealisation of theory, because the criterion of relevance that guides the introduction of corrections is not theoretically motivated.

I have chosen the Londons' account of superconductivity for a number of reasons: first, because it is such a well-known episode of successful theory-application; second, because of the high esteem and reputation of the two scientists involved; finally, because it is a case of application that is to a large extent explicitly not a deidealisation. But this case study is not exceptional or isolated; on the contrary, I believe that it is paradigmatic of the activity of theory-application in many branches of physics.

7.5.1 The hallmarks of superconductivity

The electromagnetic treatment that Fritz and Heinz London (1934) proposed for superconductors in 1934 is one of the most celebrated cases of theory-application in the history of twentieth-century physics. It was the first comprehensive electromagnetic theory of superconductivity and it remained the fundamental account of superconductivity for nearly twenty years until the advent of the BCS theory (which was heavily informed by the Londons' account, as were all subsequent theories of superconductivity). Superconductors are materials that exhibit extraordinary conducting behaviour under specific circumstances. The hallmarks of superconducting behaviour are the following two well established phenomenological findings: resistanceless conductivity and the Meissner effect.

In 1911 Kamerlingh Onnes (1913) found that when mercury is cooled below $4.2K°$ its electrical resistance falls to near zero. In 1914 he discovered that the effect does not take place in the presence of an intense magnetic field. This is the first phenomenological trait of superconductivity: under a certain critical transition temperature, and in the absence of strong magnetic fields, a superconductor exhibits almost perfect resistanceless conductivity. Almost perfect resistanceless conductivity is confirmed by the presence of a stationary current through, say, the surface of a superconducting ring. The current flows at virtually the same constant rate and does not die off.

The second, equally important, trait of superconductivity was found in 1933 by Meissner and Ochsenfeld (1933). The *Meissner effect* is the sudden expulsion of magnetic flux from a superconductor when cooled below its transition temperature. The flux in a superconductor is always vanishingly small, regardless of what the flux inside the material was immediately before the phase transition into the domain of super-conductivity took place.[9]

7.5.2 *Applying electromagnetism*

Superconductivity was initially considered an electromagnetic phenom-enon and providing an electromagnetic treatment became the main theoretical task. This was a formidable task in view of the Meissner effect. Maxwell's equations on their own are totally ineffective: for a medium of perfect conductivity (a '*super*conductor') Maxwell's equations are inconsistent with the Meissner effect. Perfect conductivity occurs when the scattering of electrons in a medium of low resistance is so small that the electric current persists even in the absence of a supporting external electric field. For a conductor in a vanishingly small electric field, for which $\mathbf{E} = 0$, Maxwell's second equation $\nabla \times \mathbf{E} = -\dfrac{1}{c}\dfrac{\partial \mathbf{B}}{\partial t}$ pre-dicts that $\dfrac{\partial \mathbf{B}}{\partial t} = 0$ and hence that $\mathbf{B}$, the magnetic field, must remain con-stant in time in the transition to the superconducting state. In other words, Maxwell's equations predict that the flux through a coil sur-rounding the metal must remain unaltered during the phase transition. The experiments of Meissner and Ochsenfeld showed that in fact there is a sudden change in the value of the external magnetic field, consistent with the total expulsion of the magnetic flux density from within the superconductor.[10]

Of course by 1933 there was much more to electromagnetic theory than just Maxwell's equations. In the construction of their theory of perfect conductivity Becker, Sauter and Heller (1933) had to appeal to further assumptions about the media, the shape of the conductor, the forces that propelled the electrons in the absence of electric fields and, crucially, the form of the law that linked the electric current to external

---

[9] A distinction is usually made between Type I and Type II superconductors. In Type I super-conductions *all* magnetic flux is expelled in the phase transition. In Type II superconductors the expulsion is only partial. Type II superconductors only appeared much later, and the distinction played no role in the historical instance that I wish to discuss. In this paper by 'superconductors' I refer to type I superconductors only. These are thin films made out from metals like zinc, aluminium, mercury, lead.

[10] The inconsistency of the Meissner effect, perfect conductivity with $\mathbf{E} = 0$, and Maxwell's equa-tions is often emphasised in textbook discussions (see, for instances, Bleaney and Bleaney (1976, ch. 13) and Hall (1974, ch. 11).

fields. Their 'acceleration' theory accounted for a persistent current in a superconductor, but it was shown by the Londons to be in contradiction with the Meissner effect.

In a normal conductor the current either induces an external electric field or is supported by one, and Ohm's law predicts that the current is directly proportional to the field, $\mathbf{j} = \alpha \mathbf{E}$. With the discovery of resistance-less conductivity, Ohm's law had to be abandoned for superconductivity because the current persists in the absence of an external field. Nevertheless all proposed treatments of superconductivity continued to assume that there existed some relation between the superconducting current and external electric fields – not a proportionality relation obviously, but *some* relation nevertheless. The Londons' fundamental contribution was to make unambiguously clear that superconducting currents are in no way supported by electric fields, but by magnetic fields.

What prompted the Londons' suggestion? Why did previous attempts to understand superconductivity continue to assume that the current was physically linked to electric fields? The answer cannot be found by inspecting the state of electromagnetic theory in 1933. No significant contribution or substantive addition to the theory was made during these years that could help to explain the Londons' breakthrough. The significant event was the proposal, by the Londons, of a new *model*.

Historically, the discovery of the Meissner effect signalled the turning point. This unexpected discovery brought about a change in the *conception* of superconductivity. A superconductor was initially conceived in analogy with ferromagnetism: just as a ferromagnet exhibits a magnetic dipole moment in the absence of any supporting magnetic fields, a superconductor exhibits a permanent current even if unsupported by electric fields. The superconducting current is constant in the absence of an electric field, and what this indicates is that the field is not proportional to the current, as in Ohm's law. As a replacement Becker, Sauter and Heller proposed the following 'acceleration equation', where the field is proportional to the time derivative of the current:

$$\Lambda \frac{d\mathbf{j}}{dt} = \mathbf{E} \qquad (7.3)$$

where $\Lambda = \dfrac{m}{ne^2}$ (a constant that depends upon the mass m, charge e and number density of electrons n). In the absence of an external field ($\mathbf{E} = 0$) the 'acceleration equation' predicts a permanent current: $\dfrac{d\mathbf{j}}{dt} = 0$.

7.5.3 Enter the model

The Londons understood that the Meissner effect pointed to an entirely different model. They modelled a superconductor as one huge diamagnet, and replaced Ohm's law with a new electromagnetic relation between the superconducting current and the *magnetic* field. The Londons went on to attempt a microscopic explanation of the coherence of the 'magnetic dipoles' in terms of a coherent macroscopic quantum superposition.[11]

By modelling a superconductor as a diamagnet the Londons were able to introduce an important correction into the 'acceleration equation' theory of Becker, Sauter and Heller. Diamagnetism is associated with the tendency of electrical charges to shield the interior of a body from an applied magnetic field.[12] Following a proposal by Gorter and Casimir (1934), the Londons began by assuming that a real superconductor is constituted by two different substances: the normal and the superconducting current. They then proposed that Ohm's law be restricted to the normal current in the material, and the description of the superconducting current be supplemented with an equation that determined the relation of the current to the background magnetic flux. The 'London equation' for the superconducting current takes the form:

$$\nabla \times \Lambda\mathbf{j} = -\frac{1}{c}\mathbf{H} \qquad (7.4)$$

where $\mathbf{j}$ is the current, and $\mathbf{H}$ represents the magnetic flux inside the superconductor.

It is important to understand that this equation was not derived from electromagnetic theory, but was suggested by the new model of diamagnetism. Although analogy was certainly involved, this is not just simply a case of reasoning by analogy. The Meissner effect does not just mean that the equations that describe magnetic flux in a superconducting material must be formally analogous to the equations for flux in a diamagnetic material. It rather means that a superconductor *is* a kind of diamagnet. Equation (7.4) was derived from a correction to the solutions of the old 'acceleration equation' theory – a correction

---

[11] Superconductivity is of course ultimately a quantum phenomenon. The definitive quantum treatment was given in 1951 by Bardeen, Cooper and Schrieffer (1957) who explained the emergence of coherence by appealing to the formation of Cooper pairs at low temperatures. The history of the BCS theory is fascinating in its own right, but it is of no relevance to my present argument.   [12] See, for instance, Kittel (1953, ch. 14).

prompted by the conception of the superconductor as a diamagnet. According to this conception the fundamental property of a super-conductor is not nearly perfect conductivity but, of course, the expulsion of the magnetic flux within the material during the transition phase. Superconductivity is no longer characterised as the limit of perfect conductivity, but as the limit of perfect diamagnetism. Hence the phe-nomenon of the expulsion of the magnetic flux cannot, and should not, be explained by the emergence of a superconducting current. Superconductivity is truly characterised by two independent and non-reducible phenomenological hallmarks: perfect conductivity and the Meissner effect.

In the theory of Becker, Sauter and Heller the absence of an electric field entails that the Meissner effect is impossible, as expected from our initial consideration of Maxwell's second equation in the case of perfect conductivity. Indeed the 'acceleration equation' entails the following equation for the magnetic flux inside the superconductor:

$$\Lambda c^2 \nabla^2 \frac{d\mathbf{H}}{dt} = \frac{d\mathbf{H}}{dt}. \tag{7.5}$$

Integrating with respect to time one finds the following nonhomoge-neous equation:

$$\Lambda c^2 \nabla^2 (\mathbf{H} - \mathbf{H}_0) = \mathbf{H} - \mathbf{H}_0 \tag{7.6}$$

$\mathbf{H}_0$ denotes the magnetic field at the time $t = 0$ (i.e. at the time the tran-sition phase occurs). Its value depends entirely on the value of the ambient field because a superconductor behaves exactly like a normal conductor before the phase transition, and the external field penetrates completely. The solutions to this equation are given by $\mathbf{H} = \mathbf{e}^{-\sqrt{\Lambda}cx} + \mathbf{H}_0$, where the exponentials $\mathbf{e}^{-\sqrt{\Lambda}cx}$ decrease very quickly with distance x from the surface of the material. So the 'acceleration equation' predicts that the field inside a superconductor will remain invariant throughout the phase transition. No change in the external flux will be observed and a surrounding coil will experience null induction. As London and London (1934, 72) write of the theory of Becker, Sauter and Heller:

[t]he general solution means, therefore, that practically the original field per-sists for ever in the supraconductor. The field $\vec{H_0}$ is to be regarded as 'frozen in' and represents a permanent memory of the field which existed when the metal was last cooled below the transition temperature [. . .] Until recently the exis-tence of 'frozen in' magnetic fields in supraconductors was believed to be

proved theoretically and experimentally. By Meissner's experiment, however, it has been shown that this point of view cannot be maintained.

On the other hand the Londons' diamagnetic model suggests that the field inside the material once the transition has occurred decreases very quickly with distance x from the surface of the material. So the correct solutions must exclude the value $(\mathbf{H}_0)$ of the initial field, and must contain only the exponentials $\mathbf{e}^{-\sqrt{\Lambda c} x}$. These are solutions to the following homogeneous equation: $\Lambda c^2 \nabla^2 \mathbf{H} = \mathbf{H}$. From this equation, the fundamental equation of superconductivity (7.4) can be derived, since

$$\nabla \times \mathbf{H} = \frac{1}{c} \mathbf{j}.$$

To sum up, the Londons suggested that the superconducting current is maintained by a magnetic field. The relation is of inverse proportionality, so that if the field is greater than a certain threshold value the superconducting current will virtually come to a halt, as predicted by Onnes. This equation was determined, in the manner described above, by a new model of superconductivity; the model was in its own turn suggested by the phenomena. This reconstruction explains why no satisfactory theory of superconductivity was derived before the discovery of the Meissner effect. A novel conception, embodied in the model of the superconductor as one huge diamagnet, was required for a successful electromagnetic treatment of superconductivity, and such conception was not available before the discovery of the Meissner effect.[13]

---

[13] It may be tempting to construe some commentators as suggesting that the Londons' only contribution was to restrict the set of initial conditions in the old 'acceleration equation'. Bardeen, for instance, in his impressive review article (1959) states that 'The Londons added (5.4) to the earlier "acceleration" theory of Becker, Sauter and Heller to account for the Meissner effect.' I do not find the reading of these passages in terms of a restriction of the initial conditions at all possible. And yet, this is precisely how French and Ladyman (1997) have read this episode in their response to Cartwright, Shomar and Suárez (1995), and to previous drafts of this work. But the replacement of the set of solutions that involve the initial field in the superconductor by the family of exponential solutions *is not a restriction* of the old theory to the case where the external field before the transition vanishes, i.e. to the case $\vec{H_0} = 0$. It is true that the 'acceleration' theory and the Londons' theory fully agree in that particular case. Nevertheless, the whole point of the Londons' theory is to show that the flux inside the superconductor is vanishingly small *even if* the initial flux was not zero at the time when the transition took place. Whenever the magnetic field is not vanishingly small outside the material before the transition the theories will yield inconsistent predictions as regards the expulsion of the flux: the 'acceleration' theory predicts no expulsion, while the new theory predicts a brutal change, consistent with the Meissner effect. The Londons of course accept that in the case $\mathbf{B}_0 = 0$ the 'acceleration equation' theory gets it right. But they do not remain silent about those other cases that this theory does not get right. They provide a whole new theory that has the same predictions for the $\mathbf{B}_0 = 0$ case, but gives the correct predictions for the other cases. In general, writing down a new equation for the value of a physical quantity in a theory is not equivalent to restricting the initial conditions on the old equations.

7.6 APPLICATION IN PRACTICE: PROBLEMS FOR REALISM

In providing a macroscopic description of the Meissner effect in electro-magnetic terms, the Londons effectively succeeded in providing a satis-factory application of electromagnetic theory to superconductivity. However, they did not *deidealise* electromagnetic theory. Instead they came up with a model that permitted them to impose a novel constraint upon the original theoretical construction. This case study is not excep-tional; on the contrary, *many* scientific applications are derived in this way. In astrophysics, for example, there are several models of stellar structure. A certain conception of the internal constitution of a star, which determines the form of the convection forces in the stellar plasma, has to be assumed before the quantum theory of radiation can be applied. For each different conception there is a corresponding applica-tion of the theory, a family of models, that could not have been derived from the theory alone. Similar conclusions could be derived from other case studies, some of them contained in this volume, by authors studying the role of mediating models in different areas of science. The idealisa-tion account is then not a universal account of scientific theory-applica-tion. It is far too restrictive. It imposes constraints so severe that they are not always – indeed are rarely – met in practice.

7.6.1 The epistemology of theory-application

What are the epistemological implications of the rejection of the ideal-isation account? I shall focus the discussion closely upon the case study. The Londons built an application of electromagnetic theory to super-conductivity; and yet, on McMullin's account, the theory was in no way confirmed by the phenomenon of superconductivity. Confirmation requires that the theory itself must suggest the introduction of correc-tions into the theoretical description. For, as McMullin points out,[14] a theoretical description is *ad hoc* with respect to a theory that does not suggest or motivate its derivation; and an *ad hoc* description, or hypoth-esis, cannot increase the degree of confirmation of a theory with respect to which it is *ad hoc*.[15]

---

[14] See the discussion in section 7.3.2, and in particular the passage quoted from McMullin (1985).
[15] Hempel too makes this claim (1966, 28–30), although he there ascribes a slightly different meaning to the term *ad hoc*. For Hempel, a hypothesis is *ad hoc*, with respect to some theory, if it has no surplus empirical content over the theory other than the particular phenomenon that it is specifically called to account for.

In constructing their account of superconductivity, the Londons introduced a correction into the previously available theoretical description. The correction was certainly not arbitrary, since it was justified by a new model of superconductivity. However, this model was not suggested by the theory – it was suggested by a newly discovered physical effect. On McMullin's confirmation theory, classical electromagnetism was not in this instance genuinely confirmed at all. Was it *neither* confirmed *nor* disconfirmed, or was it simply disconfirmed? The answer to this question depends on what we take electromagnetic theory to be *circa* 1933.

There are two possible pictures. It is possible to take 'electromagnetic theory' in an extended historical sense, as constituted by all applications to electromagnetic phenomena known to the Londons. The 'acceleration equation' theory is part of electromagnetic theory, when construed in this extended sense. But this theory was seen in light of the Meissner effect, to be highly unrealistic, as it made the false assumption that a superconductor would behave as a ferromagnet; when in fact a superconductor is a diamagnet. And, as we saw, the Londons gave an account that contradicted the acceleration equation theory predictions in a range of cases. Hence, if taken in this 'historical' sense, classical electromagnetism was indeed *disconfirmed* by the Meissner effect.

Alternatively, one may provide an abstract reconstruction of electromagnetic theory. The standard reconstructions normally assume that classical electromagnetism is constituted by the deductive closure of Maxwell's equations. Now, the 'acceleration equation' theory, although not inconsistent with Maxwell's equations, is not a logical consequence of these equations. It can be postulated alongside them, in just the way Ohm's law is often postulated alongside Maxwell's equations, but it cannot be derived from them. Nor is the Londons' account a logical consequence of Maxwell's equations; although it is also consistent with them, and can be postulated alongside them.[16] Thus, neither the 'acceleration equation' theory nor the Londons' account is part of electromagnetic theory, understood in this abstract manner. And it follows that, in this abstract reconstruction, the Londons' account provided neither a confirmatory nor a disconfirmatory boost for classical electromagnetism.

And yet, the Londons' treatment did increase scientists' confidence in electromagnetic theory. Superconductivity had proved difficult to model in classical electromagnetism for a long time, and many were

---

[16] It is perfectly possible for a theory T to be consistent with each of two mutually inconsistent assumptions a and b, – as long as T entails neither a nor b, of course.

beginning to despair that a consistent electromagnetic treatment would ever be found. The diamagnetic conception played a key role in the Londons' explanation of the phenomenon of superconductivity, which reveals the extent to which a mediating model carries genuine physical knowledge. The Londons' theory was generally accepted to account rather accurately for the rate of magnetic flux expulsion from a super-conductor during the phase transition reported by Meissner and Ochsenfeld in their experimental investigations.[17] From this application of electromagnetism we learn that superconductivity is an essentially diamagnetic effect; that a superconductor is not a ferromagnet; and, moreover, as the Londons' account correctly predicts the rates of expul-sion of magnetic flux observed by Meissner and Ochsenfeld, we gain a theoretical understanding of the Meissner effect. The Meissner effect does not appear as a mysterious side-effect of superconductors; instead it takes centre stage, it becomes a fundamental hallmark of super-conductivity.

The Londons' account of superconductivity provided an extra 'boost' of confidence in classical electromagnetism which the old 'acceleration' theory could not provide. But, as we have seen, on McMullin's idealisa-tion account of application, the Meissner effect does not make electro-magnetic theory more likely to be true. It seems that this extra boost of confidence in electromagnetism cannot be captured by the standard realist theory of confirmation, so I shall refer to the kind of support that the Londons' treatment provided for electromagnetism as *degree of confi-dence* rather than degree of confirmation.

The fact that the Londons' equation accounts for the Meissner effect gives grounds to believe that classical electromagnetism is *instrumentally reliable*. But it does not constitute evidence for the truth of classical electromagnetic theory. Here *degree of confidence* and *degree of confirmation* seem to depart. Degree of confidence, unlike degree of confirmation, does not point to the likelihood of the theory to be true; it only points to the reliability of the theory as an instrument in application. The theory is a reliable instrument if it is capable, perhaps when conjoined with good enlightening mediating models, of generating successful applica-tions. And from the fact that the theory is instrumentally successful, the truth of the theory does not follow.

Or does it? Would it not be a miracle if the theory was false, yet instru-mentally successful? Does the instrumental success of scientific theories

[17] Although there was some initial resistance to the Londons' theory on empirical grounds. In par-ticular Von Laue disagreed; for the dispute between Fritz London and Von Laue, see Gavroglu (1995, 123–7).

not argue for scientific realism? Arguments of this kind in favour of realism are, of course, well known in the literature.[18] Typical antirealist responses to this argument are equally well known. For instance, Fine (1986b) responds that the 'no-miracles' argument is riddled with circularity: it assumes that precisely the very sort of inference from explanatory power to truth that realism sanctions and instrumentalism contests for scientific practice, is valid at the 'meta-level' and can be used as part of an argument for realism in general. As a response, scientific realists have turned to the pragmatic virtues of realism, and they have tried to show that no version of antirealism is in any better shape. In particular the debate has focused upon Bas van Fraassen's version of antirealism, known as *constructive empiricism*.[19]

The issues about realism that I am raising are tangential to the recent debate between scientific realists and constructive empiricists. Scientific realism and constructive empiricism share a common core, which is rejected by instrumental reliability. On either view a minimum requirement for the acceptance of a scientific theory is that the theory must be empirically adequate – i.e. that what the theory states is the case about the phenomena must indeed be the case. The constructive empiricist argues that the acceptance of a theory need only involve the belief that it is empirically adequate. Theories may have other virtues besides empirical adequacy – such as simplicity, explanatory power, aesthetic value, or even the virtue of being true . . . – but belief in a theory's empirical adequacy is the only doxastic attitude required for the acceptance of the theory. By contrast, the realist argues that the belief that the theory is true, or likely to be true, and not just empirically adequate, is also required for its acceptance. For the realist a good theory, in addition to being empirically adequate, should also be true, or likely to be true – not only true to the phenomena, but true *tout court*, true to the world.

Thus, the scientific realist and her opponent, the constructive empiricist, agree that only highly confirmed theories should be accepted; we should have confidence in theories that are highly confirmed, and only in those. This is because on either view, confirmation always goes *via* empirical adequacy. A theory is confirmed when its observable predictions are borne out. The theory is empirically adequate if *all* the predictions of the theory – past, present and future –

---

[18] The original 'no-miracle' arguments are due to Putnam (1975), and Boyd (1973 and 1984).
[19] For Van Fraassen's constructive empiricism see Van Fraassen (1976), reprinted with corrections in Van Fraassen (1980). A collection of papers by critics of constructive empiricism, together with responses by Van Fraassen is contained in Churchland and Hooker (1985).

are borne out.[20] The realist takes a high degree of confirmation as a strong indication that the theory is true, or very likely to be true, because on her view empirical adequacy is a guide to truth. So, for the realist a high degree of confirmation is required for acceptance. For the constructive empiricist a high degree of confirmation is only an indication that the theory is empirically adequate, nothing more. But, as the constructive empiricist thinks that belief in the empirical adequacy of a theory is required for its acceptance, he will readily agree with the realist that a high degree of confirmation is a requirement for accepting a theory.

The London model does not raise the degree of confirmation of electromagnetic theory; it raises its *degree of confidence* – that is, it gives us a reason to believe that the theory is instrumentally reliable, i.e. that it will go on to provide successful applications. The instrumental reliability of a theory provides grounds neither to believe that the theory is true, nor that it is empirically adequate – it points neither towards scientific realism, nor towards constructive empiricism.

The contrast between *degree of confidence* and *degree of confirmation* is not captured by the current debate on scientific realism. Degree of confirmation measures either the degree of a theory's empirical adequacy, or of its truth. Degree of confidence, as I would like to define it, is not grounded on an confirmatory relationship of the truth-conferring type between a theory and phenomena. Increased *confidence* in classical electromagnetism need not be accompanied by an increase in one's estimated probability that it correctly describes the world, i.e. that it is true. The success of the London model does not provide warrant for that. Neither does it warrant an increase in one's estimated probability that the theory correctly describes the phenomenal world, i.e. that the theory is empirically adequate. Unlike degree of confirmation, *degree of confidence* is not a function of a theory's empirical adequacy. It is not confirmatory but pragmatic, a function of the success of a theory in generating applications to diverse phenomena whenever conjoined with the appropriate mediating models. I call this feature of theories 'instrumental reliability' in order to distinguish it sharply from empirical adequacy. The instrumental reliability of a theory does not require, nor does it necessarily follow from, its empirical adequacy. This is of course in

---

[20] We may never be in a position to know if a theory is empirically adequate or not. The claim that a theory is empirically adequate carries *precisely the same commitment* to the correctness of a theory's future predictions, as does the claim that the theory is true. In this respect the constructive empiricist sticks his neck out *exactly as much as* the realist.

agreement with my case study: the fact that classical electromagnetic theory can be applied to superconductivity should not be taken as an indication that the theory is *true* to superconductivity phenomena.

7.6.2 Conclusions

Let us grant that scientists do see a theory's applications as providing some degree of confidence in the theory. Does this not argue for the idealisation account, and hence for the realist epistemology that underpins it? Some scientific realists, such as McMullin, think so. The idealisation account, they think, is *required* by scientific epistemology. On the idealisation account the explanatory power of a theory is exhibited through its applications, and the theory is more likely to be true in view of the success of its applications. So, realism is required to make sense of the epistemology of theory-application.

However, the instrumentalist can give a similarly good account of the epistemology of application. Scientists' increased confidence in a theory that has generated many applications is a result of the theory's instrumental success in modelling the phenomena. This gives, at most, confidence that the theory will continue to generate successful applications in the future, i.e. that it is an *instrumentally reliable* theory. And this argues against realism: a successful application of a theory need not constitute evidence that the theory is true. The applications of a scientific theory do not necessarily yield the kind of evidential support for the truth of the theory that scientific realism requires them to.[21]

7.7 FINAL REMARKS

My case study points to the existence of a variety (a 'plurality') of ways in which scientific theories are applied. Some scientific applications are, perhaps, deidealisations of theory. But many successful applications are

---

[21] Why was it felt that a quantum treatment was none the less required? Scientists felt that a more robust explanation of the phenomena could be achieved in quantum mechanics. It was a desire for a more robust explanation that led the search for a quantum treatment, and gave rise to the BCS theory. Whether this subsequent episode constitutes ammunition for scientific realism is beyond the scope of this paper. Antirealists will presumably want to deny the inference to the truth of the BCS explanation, no matter how robust the explanation. Constructive empiricists, for example, could claim that what led the search for the quantum treatment was a desire for an empirically adequate theory to provide the explanation – for classical electromagnetism is not empirically adequate of superconductivity phenomena. We are surely back into the muddy waters of the traditional debate between realism and antirealism.

not; it is in those cases that mediating models play a key dual role. First, they help in the application of theory, by guiding the introduction of corrections into the theory required in order to accurately describe the phenomena. Second, they provide us with physical insight into the nature of the phenomena. Because of that, mediating models are not just *useful fictions*; on the contrary they are carriers of significant and very specific genuine knowledge of the phenomena. However, the role of mediating models in theory-application entails that a realist construal of scientific theory becomes highly problematic. The theory itself is used only as an instrument in application, and no attempt is made to confirm or disconfirm it at all.

REFERENCES

Bardeen, J. (1959). 'Theory of Superconductivity', *Encyclopedia of Physics*, 274–369.

Bardeen, J., L. Cooper and J. Schreiffer (1957). 'Theory of Superconductivity', *Phys. Rev.*, 108, 1175–1204.

Becker, R., F. Sauter and G. Heller (1933). 'Uber die Stromveteilung einer supraleitenden Kugel', *Z. Physik*, 85, 772.

Bleaney, B. and B. Bleaney (1976). *Electricity and Magnetism*, 3rd edn, Oxford, Oxford University Press.

Boyd, R. (1973). 'Realism, Underdetermination and a Casual Theory of Evidence', *Nous*, 7, 1–12.

   (1984). 'The Current Status of Scientific Realism', in J. Leplin (ed.), *Scientific Realism*, Berkeley, CA, University of California Press.

Cartwright, N., T. Shomar and M. Suárez (1995). 'The Tool-Box of Science', pp. 137–49, in *Theories and Models in Scientific Processes*, Poznan Studies in the Philosophy of the Sciences and the Humanities, Rodopi, vol. 44.

Churchland, P. and C. Hooker (eds.) (1985). *Images of Science: Essays in Realism and Empiricism*, Chicago, University of Chicago Press.

Fine, A. (1986a). *The Shaky Game: Einstein, Realism and the Quantum Theory*, Chicago, University of Chicago Press.

   (1986b). 'Unnatural Attitudes: Realist and Instrumentalist Attachments to Science', *Mind*, 95, 149–79.

Frassen, B. van (1976). 'To Save the Phenomena', *Journal of Philosophy*, 73, 623–32.

   (1980). *The Scientific Image*, Oxford, Oxford University Press.

Gavroglu, K. (1995). *Fritz London: A Scientific Biography*, Cambridge, Cambridge University Press.

Giere, R. (1988). *Explaining Science: A Cognitive Approach*, Chicago, University Press of Chicago.

Goldstein, H. (1980). *Classical Mechanics*, Addison-Wesley, 2nd edn.

Gorter, C. and H. Casimir (1934). 'On Supraconductivity I', *Physica*, 1, 306–20.

Hacking, I. (1982). 'Language, Truth and Reason', in M. Hollis and S. Lukes (eds.), *Rationality and Relativism*, Cambridge, MA, MIT Press.

Hall, H. E. (1974). *Solid State Physics*, John Wiley and Sons.

Hempel, C. G. (1966). *Philosophy of Natural Science*, Englewood Cliffs, NJ, Prentice Hall Inc.

Kittel, C. (1953). *Introduction to Solid State Physics*, John Wiley and Sons.

Leplin, J. (ed.) (1984). *Scientific Realism*, Berkeley, CA, University of California Press.

London, E. and H. London (1934). 'The Electromagnetic Equations of the Supraconductor', *Proceedings of the Royal Society*, A149, 71–88.

McMullin, E. (1985). *Galilean Idealization, Studies in History and Philosophy of Science*, 16, 247–73.

Meissner, W. and S. Ochsenfeld (1933). 'Ein Neuer Effect bei Eintritt der Supraleitfahingkeit', *Die Naturwissenschaften*, 787–88.

Morrison, M. (1998). 'Modelling Nature Between Physics and the Physical World', *Philosophia Naturalis*, 38:1, 65–85.

(forthcoming a). 'Approximating the Real: the Role of Idealizations in Physical Theory', in M. Jones and N. Cartwright (eds.), *Idealizations in Physics*, Rodopi.

(forthcoming b). *Mediating Models*.

Morton, A. (1993). 'Mathematical Models: Questions of Trustworthiness', *British Journal for the Philosophy of Science*, 44, 659–74.

Onnes, K. (1913). 'Commun. Kamerlingh onnes Lab', Technical Report 34b, University of Leiden.

Putnam, H. (1975). *Mathematics, Matter and Method*, Cambridge, Cambridge University Press.

Redhead, M. (1980). 'Models in Physics', *British Journal for the Philosophy of Science*, 44, 145–63.

Suárez, M. (1997). 'Models of the World, Data-Models and the Practice of Science: The Semantics of Quantum Theory', PhD thesis, University of London.

Wise, N. (1993). 'Mediations: Enlightenment Balancing Acts, or the Technologies of Rationalism', pp. 207–58, in P. Horwich (ed.), *World Changes: Thomas Kuhn and the Nature of Science*, Cambridge, MA, MIT Press.

CHAPTER 8

Knife-edge caricature modelling: the case of Marx's Reproduction Schema

Geert Reuten

8.1 INTRODUCTION

This paper discusses an early two-sector macroeconomic model of a capitalist economy, that is Marx's so called Schema of Reproduction as set out in Volume II of *Capital* (1885). If we agree with Schumpeter (1954, 15) that progress in economics is very much a matter of the development of gadgets then this particular model would obviously rank high in the bringing forth of gadgets or tools for economic analysis. As one might expect the Schema of Reproduction played an important role in the development of (early) Marxian theory, particularly that of economic cycles (e.g. works of Bauer, Luxemburg, and Grossman). The Schema also influenced a type of non-Marxian economics of the cycle in the early twentieth century – mainly through the work of Tugan-Baranovsky, the first author to take up the Schema in his own work in 1895 (see also Boumans in this volume). Next, in the 1950s and 1960s the model was of influence for the orthodox economics' theories of growth and capital[1] mainly through the work of Kalecki.[2] A major development from the Marxian Reproduction Schema is 'Input–Output Analysis', a technique developed by Leontief (1941, 1953) – for which he was granted the 1973

Earlier versions of this chapter were presented at ISMT 1995, ECHE 1996 and at the workshop Models as Mediators in the Practice of Economic Science 1996. I thank Chris Arthur, Martha Campbell, Mino Carchedi, Paul Mattick Jr, Francisco Louca, Fred Moseley, Patrick Murray and Tony Smith for their comments on earlier versions. I am also grateful to the editors of this volume, Mary Morgan and Margaret Morrison, for their extensive comments. Mary Morgan patiently stimulated me to sharpen the focus of the paper.

[1] More so than in physics, economic science has for over at least two centuries been characterised by coexisting competing paradigms. Throughout this paper I use the term 'orthodox economics' for the dominating mainstream paradigm. Since about 1920 it is 'neoclassical' based – building on the marginalist economics of Jevons, Walras and Marshall – with especially for the period 1950–1975 Keynesian flavours. Marxian economics has always been a minor paradigm – though in several periods somewhat *en vogue* (e.g., the 1920s and 30s and again the 1960s and 70s).

[2] See Kalecki (1935 and 1968); cf. Sawyer (1985, ch. 8).

Nobel Prize in economics – and which is used still today in the National Account Statistics of most OECD countries.[3]

The main aim of the paper is to show the construction of a particular type of economic modelling which one might usefully call 'knife-edge caricature modelling'. I borrow the term 'knife-edge' from Solow (1956, 161), and the term 'caricature' from Gibbard and Varian (1978).

For Gibbard and Varian, caricatures are a particular class of economic models which 'seek to "give an impression" of some aspect of economic reality not by describing it directly, but rather by emphasizing . . . certain selected aspects of the economic situation'. A model applied as a caricature 'involves deliberate distortion . . . in a way that illuminates certain aspects of that reality' (1978, 665, 676).

The term 'knife-edge' in the case of growth theory refers to particular conditions, as set out in the model, that must be fulfilled for there to be steady economic growth; it is hypothesised that the fulfilment of such conditions (the knife-edge) is most unlikely. Running off the knife-edge would take the economy off stable growth with no automatic way to get back on it.[4] Even if the model builder believes – as Marx does for our case – that actual instability is plausible so that the model is in a way counterfactual, the content of the model is nevertheless seen to reveal important characteristics of the economy (as does a caricatural model). It is rather when the model is put *to work* that the knife-edge is revealed. Even so, as we will see, a number of devices for constructing the model lead up to the knife-edge characteristic.

Before going into the case proper (sections 8.3 and 8.4), the next section (8.2) contains some remarks on the case material as well as on the very idea of studying this part of Marx's *Capital* as a case of models and how this might relate to Marx's apparent dialectical method. In an Appendix I provide a recapitulation of the symbols used throughout especially sections 8.3 and 8.4. The paper closes with a methodological analysis of the case (section 8.5).

8.2 INTRODUCTION TO THE CASE MATERIAL: VOLUME II OF *CAPITAL* AND MARX'S METHOD

For some specialists in the field of Marx's theory the whole enterprise of the main part of this paper is suspicious since a modelling approach is

---

[3] As stimulated by the work of Richard Stone. For the link between the Marxian Reproduction Schema and the Leontief input–output analysis, see Lange (1959, ch. 3).

[4] See also Jones (1975, 53–9) for the term 'knife-edge' in the context of the Harrod-Domar growth model.

seen to be incompatible with Marx's method in general. The method adopted by Marx in *Capital* has been the subject of recurring debate. Much of that work seems to point at a systematic-dialectical approach; however, one can certainly also find quite a lot of an apparently 'linear' economic modelling approach.[5] Even if *Capital* might be interpreted in a dialectical fashion (as I think it can), most scholars in the field agree that Marx himself did not reach a full systematic-dialectical presentation. I discuss this issue, particularly for the case of Marx's 'reproduction schemes', in a separate paper (Reuten 1998).[6] In what follows I do not give a thorough treatment of the systematic-dialectics. However, it may be noted that from its viewpoint the model to be discussed below might be considered as what would be called a *moment*, that is 'an element considered in itself, which can be conceptually isolated, and analyzed as such, but which can have no isolated existence'.[7] Thus apparently a dialectical moment is readily comparable to a model of part of the economy/society. However, it is indeed 'a moment' of a general theory, to be connected within the systematic whole.

In relation to this, a few words on my terminology regarding 'model' or 'schema' are appropriate. Marx does not use the term model; he uses the term 'schema'. Without necessarily implying that these notions are the same generally, I will use the terms interchangeably in this chapter. The reason is, as I will argue, that Marx's schema may be conceived of as a model in the modern economic sense. The term model in economics came into fashion only after 1950. It may be noted that one of the

---

[5] The latter interpretation dates back to Sweezy (1942), who particularly interprets the whole of *Capital* as a successive approximation approach. The foundations for the approach of the former interpretation are in Hegel's logic (1812; 1817), which it is well to differentiate from his philosophy of history; the term *systematic*-dialectics aims to delineate this difference. One important characteristic of that method – the one that economists, and perhaps scientists generally, trained in mathematical and formal logical traditions of thought, may find difficult – is that it proceeds via several levels of abstraction that are *conceptually different* from each other. To put it in orthodox language: if chapter 1 of a systematic-dialectical work, seemingly, 'defines' money, the term money will have a *different* meaning – richer, less abstract – some chapters later on. Thus in fact 'definitions' are not fixed in a dialectical method. In contradistinction, conventional models fix concepts by definitions and are in that sense conceptually 'flat' or 'linear'. Anyway this problem will not much hinder us for the discussion of the case of Marx's Schema of Reproduction since that is layered at one particular level of abstraction.

For recent discussions of Marx's method see the papers in Moseley (1993), Bellofiore (1998), Moseley & Campbell (1997) and Arthur and Reuten (1998); see also Likitkijsomboon (1992; 1995), Reuten (1995) and Williams (1998). For the terminology of 'linear' versus dialectical logic, see Arthur (1997).

[6] In that paper I conclude that whilst the case is *compatible* with a systematic-dialectical interpretation, the material favours a modelling interpretation. This, however, allows no further conclusions for the method of other parts and the whole of *Capital*.   [7] Reuten and Williams (1989, 22).

founders of economic model building, Tinbergen, used the term 'schema' in his early work (see Boumans 1992). Although I use the terms interchangeably I prefer the term 'schema' when close to Marx's text and the term 'model' in the methodological argument about it.

Marx's *Capital* is published in three volumes. The case material to be discussed is from Volume II. It was edited by Engels after Marx's death and published in 1885. The work is subtitled 'The Process of Circulation of Capital' and is divided into three parts. The first is on the Circuit of Capital, the second on the Turnover of Capital and the third on the Reproduction and Circulation of the Social Capital. The latter part (chs. 18–21, about 175 pages) makes up the case material for this chapter.

In fact, Volume II is a reordered collection of notebooks – it is unsure to what extent Marx considered them ready for publication. The text for our case was taken from Notebooks II (written in 1870) and VIII (written in 1878).[8] It should be mentioned that this is the last part of *Capital* that Marx worked on before his death. At that time he had already drafted Volume III of the work.[9] As we will see this point may be relevant for the interpretation of the case.

8.3 MARX'S REPRODUCTION MODEL: PRELIMINARY STAGE

8.3.1 Categorical abstractions: generic and determinate

In Part 3 of Volume II, Marx presents a model of the reproduction of the capitalist economy in general. In particular, he constructs a dynamic two-sector macroeconomic model – as far as I know, the first in the history of economics.[10] I first turn to the macroeconomic aspect and

---

[8] All my quotations below are from the English 1978 Fernbach translation and all page references are equally to this edition (Marx 1885). These page references are preceded by a Roman number, indicating the Notebook from which the text was taken. E.g. II, 427 = Marx 1885, from Notebook II, page 427 in the Fernbach translation.

[9] See Engels's Preface to Marx 1885 (pp. 103–4) and Oakley (1983). A discussion of the interconnection of Volumes I and II of *Capital* is in Arthur and Reuten (1998).

[10] 'It is no exaggeration to say that before Kalecki, Frish and Tinbergen no economist except Marx, had obtained a macro-dynamic model rigorously constructed in a scientific way.' (Morishima 1973, 3). 'His theory is probably the origin of macro-economics' (Klein 1947, 154). '. . . the theory adumbrated in Volume II of *Capital* has close affinities with Keynes' (Robinson 1948, 103). See also Jones (1975, 98).

The question of origin is of course a matter of appreciation. Note that the two chapters in which Marx presents his models (chs. 20 and 21) are preceded by an introductory chapter (ch. 18) and a chapter dealing with former presentations of the issue, in particular Smith's and Quesnay's (ch. 19). Here Marx refers with praise to Quesnay's *Tableau Economique* of 1758, suggesting that it greatly inspired him. However, if we compare Quesnay's Tableau and Marx's

next to the division in two sectors; the dynamics will be discussed in section 4.

Whereas Marx's analysis in the previous parts of *Capital* focused on the production of the individual capital and its conditions (Volume I) and the circuits of the individual capital (money capital, production capital, commodity capital) he now considers the interconnection of individual capitals in their social-aggregate context: the material output of the one capital being an input for another, profits or surplus-value being spent on consumer goods or being invested in the enlargement of capital and finally wages being spent on consumer goods. How can all this socially match both materially and in terms of value (money)? In the following quotation we see a number of *categorical abstractions* coming on the scene:[11] we see Marx conceiving the economy as 'social', as a 'movement of the total product' (the individual capitals, being the link in this movement); and we see the categories of consumption as an expenditure of wages and the expenditure of surplus-value (consumption or investment):

For our present purpose . . . [w]e can no longer content ourselves . . . with the value analysis of the product of the individual capital . . . [The] elements of production [of the individual capital – i.e. the inputs], in so far as they are of the objective kind, form as much a component of the social capital as the individual finished product that is exchanged for them and replaced by them. On the other hand, the movement of the part of the social commodity product that is consumed by the worker in spending his wage, and by the capitalist in spending surplus-value, not only forms an integral link in the movement of the total product, but is also interwoven with the movements of the individual capitals . . . (II, 469)[12]

Schema any similarities seem rather remote (or it should be in their aspect of monetary circulation). On Marx's appreciation of and inspiration from Quesnay's *Tableau Economique* for his model of reproduction, see Gehrke and Kurz (1995, esp. 62–9 and 80–4). On Marx's appreciation of Smith in this respect, see Moseley (1998).

[11] Making abstractions is a necessary precondition for modelling: the already abstract entity can then be modelled, e.g. by way of successive approximation.

The term 'abstraction' is often used in several different senses. Most relevant is the difference between what I call 'inclusive' and 'exclusive' abstractions. For example the term 'animal' is an abstraction *including* all beings with certain characteristics (cats, horses, human beings – the latter are again including abstractions by themselves). In this case the term 'abstracting from' means abstracting from difference (or perhaps rather conceiving the unity-in-difference, to phrase it dialectically). On the other hand one might also say: 'in this paper about animals I abstract from all non-mammals'. This would be an *excluding* abstraction; in a way it is a definition *pro temps* of 'animal'. It is equivalent to *assuming* there are no non-mammals. (Evidently excluding abstractions are counterfactual; the point is the relevance of the exclusion.)

In this paper I always use the term abstraction in the first, inclusive, sense – mostly emphasising this by the adjective *categorical* abstraction. Marx, however, sometimes uses the term abstraction in the second, exclusive, sense – for which I prefer the term 'assumption'. [12] See note 8.

These categorical abstractions are further specified as *dual* abstractions revealing, in Marx's view, the real duality of a capitalist economy: material *and* monetary. We saw this in the quotation above ('value analysis of the product' versus 'elements of production . . . of the objective kind'), it is stressed again one page further on:

> As long as we were dealing with capital's value production and the value of its product individually, the natural form of the commodity product was a matter of complete indifference for the analysis, whether it was machines or corn or mirrors. . . . But this purely formal manner of presentation is no longer sufficient once we consider the total social capital and the value of its product. . . . [The latter's movement is] conditioned not just by the mutual relations of the value components of the social product but equally by their use-values, their material shape. (II, 470)

Thus we have the duality of the value dimension (money, which is a homogeneous dimension) and the physical dimension (the 'natural form', 'use-values', the 'material shape' – i.e. heterogeneous, multiple dimensional). The duality of capitalist entities is central to Marx's analysis generally.[13] In the current part of his work it gets a remarkable (macroeconomic) treatment. As we will see later on in some more detail, he assumes prices constant. At the same time all the entries in the model that we are going to discuss are in the monetary dimension ($£$, $\$$): they are entries composed of a material quantity z times their price. But since prices are held constant all variations in those entries are variations in material quantities.

A next categorical abstraction is the division of the total economy into two sectors or 'departments' of production, the one producing means of production the other producing means of consumption. By way of this abstraction some 'order' is achieved already in the apparent mass of social interconnections:

> The society's total product, and thus its total production process, breaks down into two great departments:
> 1. *Means of production*: commodities that possess a [material] form in which they either have to enter . . . [production], or at least can enter this.
> 2. *Means of consumption*: commodities that possess a form in which they enter the individual consumption of the capitalist and working classes.
> In each of these departments, all the various branches of production belonging to it form a single great branch of production . . . The total capital applied in each of these two branches of production forms a separate major department of the social capital. (II, 471)

---

[13] It is often cast in terms of dialectical 'contradictions' (see esp. *Capital I*, ch. 1).

In both of Marx's macroeconomic and his sectoral categories we see a general aspect of his method, that is to make a difference between *generic abstractions* and *determinate abstractions*. The former abstractions are applicable to all kinds of societies, whereas the latter concern their particular form within a particular mode of production – in our case capitalism (see Murray 1988, ch.10). Thus the *determinate* form of the macroeconomic categories is emphasised in their duality of not only being 'materially shaped' but also being determined by the value form of money (commodities, capital). Similarly the categorical abstraction of division into 'means of production' and 'means of consumption' is itself a *generic* functional one. The determinate form of these categories – whilst of course already being introduced in the quotation above by the reference to commodity and capital – is emphasised in the model to be presented below (as we will see the determinate form is in the value constituents of the two departments: constant capital, variable capital and surplus-value).[14]

8.3.2 Assumptions for the model of Simple Reproduction

In setting up his model, or *schema* as it is called, Marx seems to follow a method of successive approximation – the particular way is much similar to the approach we find in the earlier parts of the book. Beginning with a model of *Simple Reproduction*, where the economy reproduces itself each period but there is no growth (capitalists consume all surplus value), he next considers *Expanded Reproduction*, that is, the realistic situation where there is growth in the economy (capitalists accumulate (part of) the surplus-value). In this section I comment on the model of Simple Reproduction.

The schema is presented in numerical form, and an elementary formalisation seems to be derived from it – as we will see later on. Numerical presentation in economics was common in those days, and extended for this field of economics into the 1930s when Kalecki started formalising business-cycle models (see Boumans in this volume).[15] The dimension of the schema is monetary, i.e. £ or $ etc.; given the constant price assumption the entries can be conceived of as material composites (VIII, 469, 473):

---

[14] In fact Marx's text above (p. 471) is followed by definitions of variable and constant capital (pp. 471–2) which emphasise again the dual character of capital: its material constituent and its value constituent.

[15] Kalecki explicitly builds on Marx's reproduction schemes (see especially his 1935 and 1968).

Schema A: Simple Reproduction

$$
\begin{array}{llll}
\text{c} & \text{v} & \text{s} & \text{x}
\end{array}
$$

I. $4000 + 1000 + 1000 = 6000$ (means of production)
II. $2000 + 500 + 500 = 3000$ (means of consumption)

$\overline{6000 + 1500 + 1500 = 9000}$ (social gross product)[16]

where:

I = department I, producing means of production (6000);
II = department II, producing means of consumption (3000);
c = constant capital, the value of the means of production applied;
v = variable capital, the value of the social labour-power applied;
s = surplus-value, the value that is added by labour minus the
    replacement of the variable capital advanced (cf. II, 472).

Thus, whilst the departmental division itself is a generic abstraction, the constituents are determinate for the capitalist mode of production where capital essentially produces capital. In the example above: the capital previously accumulated (7500) and invested in 6000 constant capital plus 1500 variable capital (capital in the form of production capital) generates a capital of 9000 (in the form of commodity capital). Behind the consequently created 'surplus-value' (1500) is the, in terms of Marx's analysis of *Capital I*, physical extraction of surplus-labour. The precondition for such extraction on a social scale *in capitalism* being the availability of a class of free *wage* labourers, that is a class that, since it lacks property in means of production, is enforced to sell its labour power to the owners of capital so as to gain their living. (Note that for 'surplus-value' we may as well – at this stage of Marx's analysis – read 'profits'.)[17]

Although Marx does not comment on the numbers in the schema, they do not seem arbitrary. In an earlier chapter (ch. 17, II, 397–8) Marx quotes an estimate of the ratio of the total capital stock to the total consumption for Britain and Ireland (as reported) by Thompson (1850). This ratio amounts to 3.[18] A similar ratio in the schema above is 2.

[16] Here the 4th column is total gross production (including intermediate production), and the 3rd row is total gross expenditure (including intermediate expenditure). So for the shape of a modern Leontief input-output table, one has to rotate the schema 90 degrees to the west, and move the initial 3rd row to the outer east, with c_1 (4000) and c_2 (2000) remaining in the first quadrant of intermediate expenditure and production.

[17] Surplus value, for Marx, is a composite of: industrial profit, interest and rent. The distribution of surplus-value into those three categories is only made in *Capital III*. Thus at this level of abstraction industrial capital, financial capital and ground capital are still one: capital.

[18] 'Or three times the year's labour of the community . . . 'Tis with the proportions, rather than with the absolute accurate amount of these estimated sums, we are concerned' (William Thompson, *An Inquiry into the Principles of the Distribution of Wealth*, London 1824/1850, quoted by Marx 1884, 398).

However, for the time being it has been assumed that there is no 'fixed' constant capital (see below); and it is provisionally assumed that all surplus value is consumed (dropping these assumptions – as Marx does later on – increases the ratio).

Before discussing what Marx 'does' with this model, I first comment upon the model's underlying 'simplification devices'.

The first assumption has already been mentioned:

a[19] What we are dealing with first of all is reproduction on a simple scale. (II, 469).

This is an assumption for successive approximation (to be dropped in the next chapter – see my section 8.4). However, it is a particular one in this respect; Marx holds that simple reproduction is always an element of expanded reproduction:

> Simple reproduction on the same scale seems to be an abstraction . . . But since, when accumulation takes place, simple reproduction still remains a part of this, and is a real factor in accumulation, this can also be considered by itself. (VIII, 470–1; cf. 487)

As we will see later on there is a *positive heuristic* to this idea for considering 'simple reproduction' an element of 'expanded reproduction' – in the sense that it leads Marx fruitfully to search for certain similar characteristics in the later model; on the other hand there is also a *negative heuristic* to the same idea – in the sense that it seems to block fruitful search for other characteristics.[20]

A second assumption for successive approximation is to be dropped in section 11 of the same chapter (see my section 8.3.4):

b   . . . in dealing with the total social product and its value, it is necessary to abstract at least provisionally from the portions of value transferred to the annual product during the year by the wear and tear of the fixed capital, in as much as this fixed capital is not replaced again in kind in the course of the year. (II, 472–3)

In fact, it is assumed throughout the earlier sections of the chapter that there is no fixed capital (or equally, that all fixed capital is used up during the production period): 'the fixed capital that continues to function in its natural form being excluded by our assumption'. (VIII, 473).

The final four assumptions go to the heart of the model and remain in force for the later model of 'expanded reproduction' (section 8.4).

---

[19] Assumptions are indicated in bold letters throughout this paper.

[20] The reader will notice that I use the terms positive and negative heuristic in a related though not quite the same sense as Lakatos in his famous paper (1970).

The first of these (including two sub-clauses) has been mentioned already:

c   . . . we assume not only that products are exchanged at their values [**c-1**], but also that no revolution in values takes place in the components of the productive capital [**c-2**]. (II, 469)

In effect this assumption boils down to assuming that prices do not change.[21] At first sight this might appear a *ceteris paribus* assumption, but it is not. Marx argues that possible price changes do not affect his model: they can be *neglected* for the problem at stake. This is of course crucial to the whole understanding that the model might provide. In my view therefore the following two quotations are key to the model. On the first part of the assumption (**c-1**) Marx comments:

In as much as prices diverge from values, this circumstance *cannot exert any influence* on the movement of the social capital. *The same mass* of products is exchanged afterwards as before, even though the value relationships in which the individual capitalists are involved are no longer proportionate to their respective advances and to the quantities of surplus-value produced by each of them. (II, 469; emphasis added)[22]

The second quotation (on the second part of the assumption, **c-2**) makes in fact the same point; the punch comes after the italicised 'secondly' at the end of the quote: even unevenly distributed 'revolutions in value' – though affecting the magnitudes of the components of (social) capital – would not change the particular macroeconomic *interconnections* between constant and variable capital (as well as between them and surplus value):

As far as revolutions in value are concerned, they change nothing in the relations between the value components of the total annual product, as long as they are generally and evenly distributed. In so far as they are only partially and unevenly distributed, they represent disturbances which, *firstly*, can be understood only if they are treated as *divergences* from value relations that remain unchanged; *secondly*, however, given proof of the law that one part of the value of the annual product replaces constant capital, and another variable capital, then a revolution . . . would alter only the relative magnitudes of the portions of value that function in one or the other capacity . . . (II, 469–70)

Thus price changes, whatever their source, are neglectable.

---

[21] The first part of the assumption (**c-1**, exchange at values) is not surprising: it fits into the general systematic of *Capital*, and is in fact dropped in Part 2 of Volume III. The second part of the assumption (**c-2**) is remarkable to the extent that in *Capital I* 'revolution in values' has already been shown essential to the capitalist system.

[22] Incidentally, this is relevant for the interpretation of the Volume III Value to Price transformation.

Assumptions **d** and **e** can be readily seen by inspection of the numerical model above: the ratios of $c/(c+v)$ and of s/v are the same for both departments.

d   As to the rate of surplus value (s/v) it is assumed that it is for both departments equal, constant and given (100%). These assumptions are maintained throughout this Part.

Although not commented upon (the rate of surplus-value is treated at length in both Volume I and Volume III of *Capital*), this seems – like assumptions **c** above and **e** and **f** below – a simplifying device of the type *neglectable*, thus possible changes or divergences in the rate are without particular relevance to the problem at hand. (Note that assumptions **a** and **b** are equally simplifying devices, but rather for procedural purposes.)

e   The next assumption concerns the value composition of capital $(c/(c+v))$, which is, for each department, taken as equal, constant and given. These assumptions are maintained throughout chapter 20, but relaxed several times in chapter 21. Marx comments:

What is arbitrarily chosen here, for both departments I and II, is the ratio of variable capital to constant capital; arbitrary also is the identity of this ratio between the departments . . . This identity is assumed here for the sake of simplification, and the assumption of different ratios would not change anything at all in the conditions of the problem or its solution. (VIII, 483)

Thus both simplifications **d** and **e** can be made because their possible (and more complex) departmental divergences do not fundamentally affect the problem and are therefore *neglectable*. This is related to the more severe assumption **c**: the possible divergences at hand would not affect the interconnection between the departments – yet to be developed.[23]

A final assumption, which is maintained throughout the Part, is made explicit much further on in the text:

---

[23] Readers at home with Marx's or Marxian economics will be aware of the interconnection of assumption **c** and the assumptions **d** and **e** as set out by Marx in the famous 'transformation of values into prices of production' of *Capital III*. It may be reminded therefore that this Part of *Capital II* was written after that Part of *Capital III* so it cannot be argued that Marx is neglecting something, the importance of which he had not yet touched upon. Again, the crucial neglectability assumption is the one about prices (**c**). Once that is made the magnitudes of the compositions of capital and the rates of surplus-value do not matter by implication. (This issue will be of special importance for the extension of the model to 'expanded reproduction' (section 3) where Marx sometimes varies these magnitudes.)

From the point of view of Marx's general method all this is most important: the transformations in *Capital* are systematic, not historical (see also Arthur 1997). Thus, e.g., the value to price transformation in Volume III is conceptual and cannot be said to *actually* affect the size of the departments.

f   Capitalist production never exists without foreign trade . . . Bringing
foreign trade into an analysis of the value of the product annually reproduced can . . . only confuse things . . . We therefore completely abstract
from it here . . . (VIII, 546)

This is again an assumption of the type *neglectable* for the current problematic.

   In sum we have two categorical abstractions combining generic and
determinate elements, and six simplifying devices, that is, two assumptions for successive approximation and four neglectability assumptions.
Neglectability assumptions could, but need not be, dropped since dropping them would introduce complication without however affecting the
problematic – thus complication would be irrelevant. On the other
hand, dropping assumptions for successive approximation introduces
relevant complication. Note that a categorical abstraction (such as consumption or departments of production) introduces a complex category.
At the same time it simplifies the interconnection of the entities that the
abstraction refers to (see also note 11).[24]

8.3.3 The model and the departmental interconnections for Simple Reproduction

Generalising his numerical Schema A (above), Marx uses the notation:

$$I_c + I_v + I_s = I$$
$$II_c + II_v + II_s = II$$

Apparently the notation is adopted for shorthand rather than for formal
manipulation. In what follows, I represent this into the notation that has
become conventional in modern Marxian economics:[25]

$$c_1 + v_1 + s_1 = x_1 \tag{8.1}$$

$$c_2 + v_2 + s_2 = x_2 \tag{8.2}$$

$$\overline{c + v + s = x} \tag{8.3}$$

Simple reproduction is defined by the condition:

$$x_1 = c \tag{8.4'}$$

---

[24] From the point of view of a dialectical method it is required to step down to lower (more concrete) levels of abstraction ultimately reaching the level of the everyday phenomena. The ultimate proof that the 'high level' neglectability assumptions are indeed neglectable can only be
sustained at the empirical-phenomenological level.

[25] Although Marx uses his notation throughout the text, e.g. for the derivation of conditions of
reproduction (see below), a full schema, like equations 8.1–8.3, is always cast in numerical terms.

or equally by the condition:

$$x_2 = v + s \qquad (8.4'')$$

Analysing at length the mutual exchange between the departments, which is 'brought about by a money circulation, which both mediates it and makes it harder to comprehend' (VIII, 474), Marx derives from his model the following proportionality condition for simple reproduction (VIII, 478):

$$v_1 + s_1 = c_2 \qquad (8.4)$$

This condition emphasises the interconnection between the two departments as revealed in their mutual exchange.[26] Thus by considering those elements that are, so to say, used up within a department itself (the means of production c_1 for Department I, and for Department II the means of consumption v_2 consumed by its labour and s_2 consumed by the capitalists of this same sector) the interchange between the departments is given by the remaining elements v_1, s_1 and c_2 (see condition 8.4). The point then is that, given the relationship between the two departments, $v_1 + s_1$ and c_2 must not only balance in terms of value but also materially (in their 'natural form'). Thus the amount of means of production necessary as an input for Department II (c_2), as produced in Department I, must balance with the means of consumption for workers and capitalists in Department I (v_1, s_1) as produced in Department II:

The new value product of the year's labour that is created in the natural form of means of production (which can be broken down into $v + s$) is equal to the constant capital value c in the product of the other section of the year's labour, reproduced in the form of means of consumption. If it were smaller than II_c [i.e. c_2], then department II could not completely replace its constant capital; if it were larger, then an unused surplus would be left over. In both cases, the assumption of simple reproduction would be destroyed. (VIII, 483–4)

Here we see simple reproduction's balance on the 'knife-edge' of just this equality as stated in condition (8.4). Marx does not discuss a divergence from the knife-edge; that is postponed till he incorporates fixed capital (see section 8.3.4) and when he introduces expanded reproduction (we come back to this in more detail in section 8.4, when we discuss the latter, and in section 8.5).

---

[26] Condition (4) and the conditions (4') and (4") of course each imply each other: their focus is different. Note that Marx does not use the term equilibrium, but talks of 'proportionate part', and holds that the proportionate part on the left side 'must be equal' to the proportionate part on the right side (VIII, 474, 478).

Thus the complexity of a capitalist exchange economy – with a multitude of similar 'proportionate' exchanges between firms – is reduced to the simplified complexity of just two aggregates. However, these are not just any two aggregates: the particular division is constructed so as to focus the analysis, as we will see in the next section, on problems of accumulation of capital and (cyclical) economic growth. Of course this is one main focus of the whole of Marx's *Capital*. It is remarkable then that the first stage of Marx's 'successive approximation' (simple reproduction) forgoes the accumulation of capital altogether. Apparently he wants to analyse to what extent the very 'skeleton' of merely a stationary economy (that is an economy with a zero growth rate) poses important problems.

8.3.4 Fixed capital

Apart from Marx's analysis on the basis of condition (8.4) we find this aim, of searching to what extent a stationary economy poses problems, especially in his analysis of fixed capital in a state of 'simple reproduction'. In section 11 of chapter 20 (pp. 524–45) Marx drops assumption **b** and considers the effect of the incorporation of fixed capital for his model. Thus in terms of annual reproduction he incorporates constant capital components whose life is longer than a year (cf. VIII, 525).

For his analysis he starts from the same numerical model given above, focusing again on the condition (8.4), i.e. $v_1 + s_1 = c_2$. The general problem is that 'part of the money received from the sale of commodities, which is equal to the wear and tear of the fixed capital, is not transformed back again into . . . productive capital . . . it persists in its money form'; thus we have a 'hoard formation' which is to be expended when the fixed capital components have to be replaced (VIII, 526). So the commodity value 'contains an element for depreciation of . . . fixed capital' (VIII, 528).

Considering the issue from Department II (the right hand side of condition (8.4)), Marx divides up c_2 into an element c_2 proper and an element $c_2(d)$, 'where *d* stands for *déchet* (depreciation)' (VIII, 531). Thus apparently against the hoard formation *d*, part of the Department I commodities (part of $v_1 + s_1$) remain unsold, which seems to 'contradict the assumption of reproduction on the same scale' (VIII, 531–2).

However, as at any point in time fixed capital investments are 'of different ages', 'so each year do quantities of fixed capital reach the end of their life and have to be renewed in kind' as financed out of their depreciation fund (depreciation hoards of money) (VIII, 528–9).

On this basis Marx finds a further specification of the condition for (simple) reproduction, which is that the fixed capital to be renewed each year just equals the sum of the depreciation (in money, e.g. £) of the remaining fixed capital replacing **b** with **b′**:

b′ The precondition here is evidently that the fixed component of department II's constant capital which in any given year has been transformed
back into money to its full value and thus has to be renewed in kind . . .
has to be equal to the annual wear and tear of the other fixed component
of the constant capital in department II which still goes on functioning in
its old natural form . . . (VIII, 540)

Thus by way of reconceptualising the component 'constant capital' fixed capital has been included in the model. With it, however, the condition for simple reproduction has become a more fragile one.

It may be noted that this reconceptualisation of constant capital is in a way a story annexed to the model. In the case of a microeconomic model for one firm the equality of the depreciation (£) and the wear and tear (material) *in part* of the constant capital component would be a dimensional idealisation. For the macroeconomic case at hand, however, it is rather the large numbers that must do an averaging out so that constant capital can consistently be conceived as *in toto* real monetarily and real materially.

Dividing up the departments into sections renewing fixed capital and those still precipitating out fixed capital in money, Marx next analyses the two cases in which the proportionality condition does not hold. In the first case fixed capital has to be renewed, for which there has been insufficient production (underproduction), thus some capitals in some branches are forced to contract production, which has downward spiral effects through the economy. In the second case a renewal of fixed capital anticipated upon by Department I producers is not effected, thus their overproduction forces them to contract production which again spirals through the economy. In both cases, Marx argues, there would develop 'a crisis – a crisis of production – despite reproduction on a constant scale' (VIII, 543–4).[27] Marx emphasises that such 'disproportionate production of fixed and circulating capital . . . can and must arise from the mere *maintenance* of the fixed capital' i.e.

[27] 'Foreign trade could help in both cases . . . [but] only shifts the contradictions to a broader sphere' (p. 544).

with simple reproduction. 'Within capitalist society . . . it is an anarchic element' (VIII, 545).

Evidently Marx here arrives at an important stage of the presentation: even if we assume simple reproduction there is a threat of falling off the knife-edge, and so, in his view, a threat of economic crisis. It is also important for judging Marx's modelling procedure, for indeed he next extends the analysis (see section 8.4) to the more complicated expanded reproduction, by so to say bracketing this problem, or by carrying along the simplifying *condition* (**b′**) of proportionate production of fixed and circulating constant capital such that renewal of fixed capital is equal to depreciation. Evidently for Marx the problem is *not* irrelevant. Rather one might say that, for the sake of completing the analysis, Marx constructs – in the words of Gibbard and Varian (1978) – a helpful 'caricature'.

In this discussion of fixed capital we touched on a particular analytical procedure adopted by Marx for his model: for analysing the problem of fixed capital he divides up the departments into sections. In fact he adopts a similar procedure several times when analysing particular problems. For example in his section 4 when discussing the exchange within Department II, he distinguishes between sub-departments producing means of consumption for workers (expenditure of the variable capital equivalent, i.e. wages) and for capitalists (expenditure of surplus-value).[28]

8.3.5 Final remarks on the model for Simple Reproduction

An important aspect of Marx's Reproduction Schema is that it brings into sharp focus the difference between 'total production value' and 'value-added' (in Marx's terminology, between 'value of the total product' and 'the value product of labour'). This distinction is central to modern macroeconomics since Keynes.[29] Marx, in terms of his schema, introduces the distinction as follows:

On the premise of simple reproduction . . . the total value of the means of consumption annually produced is equal to the annual value product, i.e. equal to

---

[28] Especially the latter division, which one may also see as a three departmental division, has been taken up in later extensions of the model (e.g. Tugan-Baranowski, Luxemburg, Kalecki).

[29] Modern orthodox macroeconomics centres on value-added and its expenditure equivalent and mostly neglects total production value. It is a characteristic of Marxian economics till today to incorporate both items in its models via the 'constant capital' component (intermediate production plus replacement investment in Keynesian terminology).

the total value produced by the labour of the society in the course of the year, and the reason why this must be the case is that with simple reproduction this entire value is consumed . . . [F]or the capitalists in department II, the value of their product breaks down into $c+v+s$ [i.e. $c_2+v_2+s_2$], yet, considered from the social point of view, the value of this product can be broken down into $v+s$. (II, 501–2; cf. p. 506)

He formalises this as:[30]

$$x_2 = (v_1 + s_1) + (v_2 + s_2)$$

which has condition (8.4) at its base.[31] In a later Notebook we find a general statement of the two concepts:

The overall annual reproduction [$c+v+s=x$], the entire product of the current year, is the product of the useful labour of this year [$l^u \rightarrow x$]. But the value of this total product is greater than the portion of its value which embodies the annual labour, i.e. the labour-power spent during this year [$l^v \rightarrow v+s=y$]. The *value product* of the current year, the value newly created during the year in the commodity form [y], is smaller than the *value of the product*, the total value of the mass of commodities produced during the year [x]. (VIII, 513)

Throughout Marx's text much emphasis is on the money circulation within and between the two departments.[32] In this context a final noteworthy aspect of the model of simple reproduction, revealing another similarity with Keynes, is the so-called 'widow's cruse' argument. (It is

---

[30] In his notation (p. 502): $II_{(c+v+s)} = II_{(v+s)} + I_{(v+s)}$.

[31] On the same theme (remember that the numerical schema for Department II runs: $2000_c + 500_v + 500_s = 3000_x$) Marx writes:

'As far as the constant value component of this product of department II is concerned . . . it simply reappears in a new use-value, in a new natural form, the form of means of consumption, whereas it earlier existed in the form of means of production. Its value has been transferred by the labour process from its old natural form to its new one. But the value of this two-thirds of the value of the product, 2000, has not been produced by department II in the current year's valorization process.' (II, 503)

Hence, again, the importance of the formula $x_2 = (v_1 + s_1) + (v_2 + s_2)$. Conversely, for Department I ($4000_c + 1000_v + 1000_s = 6000_x$) the 4000 constant capital

'is equal in value to the means of production consumed in the production of this mass of commodities, a value which reappears in the commodity product of department I. This reappearing value, which was not produced in the production process of department I, but entered it the year before as constant value, as the given value of its means of production, now exists in that entire part of the commodity mass of department I that is not absorbed by department II . . .' (II, 498)

Thus we have $c_1 + v_1 + s_1 = x_1 = c_1 + c_2$.

[32] A recapitulation is on pp. 491–2; cf. chapter 17 on the same issue. See also Campbell (1998) who provides a scholarly discussion of the monetary theory of all of *Capital II*.

derived in Keynes' *Treatise on Money* of 1930 as well as in Kalecki (e.g. 1935); in Kaldor's (1955, 85) well-known phrase it runs: 'capitalists earn what they spend, and workers spend what they earn'.) In the course of outlining money circulation in terms of his model Marx finds the same argument: '*it is the money that department I itself casts into circulation that realizes its own surplus-value*' (VIII, 495 – Marx's emphasis). And in more general terms, 'In relation to the capitalist class as a whole, however, the proposition that it must itself cast into circulation the money needed to realize its surplus-value . . . is not only far from paradoxical, it is in fact a necessary condition of the overall mechanism' (VIII, 497).[33] Note that Kalecki's version of the argument explicitly derives from Marx's schemes.

In sum: the first major achievement of the model of simple reproduction is the construction of a macroeconomics generally, with its particular emphasis on generic and determinate abstractions, specifically concerning the duality of dimensions (monetary and material) of the capitalist mode of production. The second major achievement is to grasp the macroeconomic relations in terms of a two-sector system fitting Marx's approach of generic and determinate abstractions. This leads him to grasp the – now familiar – distinction between 'value of the product' (production value) and 'value product' (value-added) in terms of his model. And the third is the general thread in Marx's analysis: to search for the necessary interconnections of exchange between the two departments of production. Rather than, therefore, the two equations $x_1 = c$, or $x_2 = v + s$, it is the exchange equation $v_1 + s_1 = c_2$ that is central to the analysis. We will see in the next section that a similar equation also provides the guiding thread for Marx's analysis of the macroeconomics of expanded reproduction.

8.4. EXPANDED REPRODUCTION AND THE KNIFE-EDGE OF PROPORTIONALITY

More so than in the previously discussed chapter (ch. 20), the last chapter of Marx's *Capital II* (ch. 21) has the character of an unfinished draft. A main part of the text is a meticulous analysis of how economic growth (twofold) is possible at all. What are the conditions? The import one gets from it is that the two-department abstraction for his model

---

[33] Cf. Marx's chapter 17, II, 409. Thus Kaldor (1955, 85) is wrong when he writes that 'this model' (i.e. 'his' model) 'is the precise opposite of the Ricardian (or Marxian) one', at least as far as Marx is concerned. See also the end of Kaldor's note 1.

(carried on from the previous chapter) is a powerful analytical instrument. For example, in the course of the analysis Marx is able to grasp all kinds of spiral (multiplier) effects – such as on p. 580, where starting from an accumulation in Department I, there results an over-production in Department II, whence a spiral effect influences Department I. At times the two-department division is further differentiated (sub-divisions within departments) so as to get to grips with particular problems. Perhaps most important, his use of the two-department abstraction indeed brings to the fore the problematic of the dual character of capitalist entities, processes and relations. With the exception of this last issue, Marx's end result seems generally not too complicated – as judged from the point of view of the end of twen-tieth-century economic theory on cycles and growth. However, even if that maturation required some eighty years, the real trail-blazing activ-ity was the way in which the problem of this dynamics of the capital-ist economy was posited by Marx, and which I refer to as the 'knife-edge caricature' (see below).

8.4.1 The general frame for the analysis

In Marx's model for simple reproduction, as we have seen, a stationary economy is portrayed: all profits are consumed thus there is no capital accumulation and economic growth. Marx's model for expanded repro-duction outlines an economy in which there is stable economic growth: part of the profits are accumulated in new capital. Any stable economic growth is quite a complicated process, as Marx shows, since some way or the other *all* relevant components of the economy will have to increase at a definite rate. In a capitalist economy a transition from an economy with zero growth (simple reproduction) to an economy with an x% rate of growth – or, which comes to the same problem, from x% to x + y% growth – is not a matter of a conscious plan. In some way the monetary profit incentive has to do the job. It is not evident that it can, Marx submits, rather it is an accident if it could. Nevertheless Marx's aim is to set out the conditions for this accidental situation of balanced growth, which, as we will see, is rather a 'knife-edge':

capitalist production already implies that money plays a role . . . and gives rise to certain conditions for normal exchange that are peculiar to this mode of pro-duction . . . which turn into an equal number of conditions for an abnormal course, possibilities of crisis, since, on the basis of the spontaneous pattern of this production, this balance itself is an accident. (VIII, 570–1)

The process of moving from one path of economic growth to another, Marx stresses over and again in his text, is likely to be interrupted by economic crises of overproduction and underproduction. Nevertheless, he forgoes these (in section 8.5 I will come back to the methodological reason for it). A main part of his model for expanded reproduction therefore consists of what Gibbard and Varian (1978) have called a *caricature*, a 'deliberate distortion' in order to analyse (in Marx's case) the conditions for a balanced economic growth.

Prior to setting out the model for expanded reproduction proper Marx outlines the main requirement for either a transition from simple to expanded reproduction, or a transition to further expansion, that is to a higher growth path. This requirement he apparently learns from his model of simple reproduction and the effort to reconstruct it for expanded reproduction: 'in order to make the transition from simple reproduction to expanded reproduction, production in department I must be in a position to produce fewer elements of constant capital for department II, but all the more for department I' (VIII, 572). In effect, then, Department I would substitute part of surplus-value (s_1) from the means of consumption (some equivalent part of c_2), to spending it on additional means of production (which are now equivalent to that available in commodity form from Department I). Here we see the main problem for a transition since Department II would thus be stuck with an unsold commodity stock equivalent to that: 'There would thus be an overproduction in department II, corresponding in value precisely to the expansion of production that took place in department I' (VIII, 580). Now the 'normal' reaction to this overproduction in Department II would be for Department II to cut back production – which would be fine if it were to the extent of the means of production they could not get from Department I anyway. However, given their overproduction, they might want to cut back production more than that – and thus buy even less means of production: 'The over-production in department II might in fact react so strongly on department I . . . [that the] latter would thus be inhibited even in their reproduction on the same scale, and inhibited, moreover, by the very attempt to expand it' (VIII, 580).

A real paradox. Marx brings up the problem and refers back to it several times, but does not analyze it any further. In fact, as we will see in section 8.4.3, by additional caricatural assumptions he makes his model work in the way that effective overproduction does not arise.

8.4.2 Assumptions for the model of Expanded Reproduction

From the model of simple reproduction (section 8.3.2) Marx carries over to that for expanded reproduction the assumptions **c** to **f** (assumption **a** was the one of simple reproduction). However, assumption **e**, about the composition of capital, is sometimes relaxed so as to allow for divergent compositions as between the departments; nevertheless, within a department it remains constant.

The earlier assumption **b'**, on fixed constant capital (section 8.3.4), needs adapting since we now will have an addition to fixed capital. Apart from that Marx's analysis of it (VIII, 565–70) is much in line with its discussion for simple reproduction: from the side of individual capitals it runs in gradual lumps of hoarding (depreciation allowances) and discrete dishoarding (investment); within a department and its branches, one section of capitalists will be engaged in stages of the former ('one-sided sale'), while another section actually buys additional elements of constant capital ('one-sided purchase'). Apparently Marx does not want to focus on this problem and thus assumes, even for the case of expanded reproduction, that in the aggregate the hoarding for investment in fixed capital equals dishoarding so that a

b″ balance exists . . . the values of the one-sided purchases and the one-sided sales cover each other. (VIII, 570)

Since he over and again stresses that there is *not* such a balance this assumption is obviously a caricature.

g In the same vein Marx assumes a sufficient monetary accommodation for expanded reproduction (VIII, 576).[34]

h A further delimitation of the problematic is revealed in the assumption of a sufficient labour force, i.e. that 'labour-power is always on hand' (VIII, 577).

This assumption, however, is not an analytical one, as Marx for its explanation refers back to *Capital I*. Thus it is rather an approximation for which 'no argument within the model' is given (Gibbard and Varian 1978, 670).

For completeness I add that Marx does not aim to set out the transition from simple to expanded reproduction but rather the transition from some growth rate to a higher one (VIII, 566). This is, as I have already indicated, not very important as both kinds of transition pose the same problems.

---

[34] These and other monetary aspects of *Capital II* are dealt with in detail in a competent paper by Campbell (1998). Especially she sets out why Marx purposely abstracts here from credit money.

8.4.3 The schemes for Expanded Reproduction

For the analysis of expanded reproduction, Marx uses three numerical schemes, which I refer to as Schemas B, C and D.[35] Marx treats Schema B very briefly (see my note 43), and its analysis is apparently a preliminary one. Below I present an outline of Schema C, which is also the best worked out case in Marx's text. Towards the end of this section I make some remarks on Schema D.

Once again these schemas are in numerical form; each with different starting values. For all schemas, it is at first sight unclear why particularly these specific starting values have been chosen – only towards the end of the chapter does it become clear that they are meant to be representative cases for three particular circumstances. (Quite apart from this it is also obvious from the text that Marx tried to employ 'easy numbers' for his calculations.)

Each schema (B, C, D) is presented for a sequence of periods, each representing the *production* in that period. At the end of each period capitalists in each department plan to accumulate capital for an expanded production in the next period (= *intended exchange arrangement*). Thus they aim to use more means of production (c) and labour-power (v) than they did in the current period. However, these plans may not match e.g. the means of production that have actually been produced in the current period, thus there might be over- or underproduction in comparison with these plans. Thus especially for the case of underproduction there may be bottlenecks preventing steady growth. At the end of each period then the confrontation of the realised *production* and the *intended exchange arrangement* gives rise to some *actual exchange arrangement* which is the basis for the next round of production.

Once we are in a situation that the *intended* exchange arrangements match the *actual* arrangements (and therefore also production), and no new changes in parameters occur, we are on a steady growth path. I will call a situation of a fixed set of parameters a 'regime'. Marx then analyses the transition from one regime to another by varying just one parameter, which is 'the rate of accumulation out of surplus-value' for Department I (α_1).[36]

In the way Marx makes his model work (at least for Schema C, as we

---

[35] In the text these are mentioned as follows: Schema B = 'schema a' (pp. 581–5); Schema C = 'first example' (pp. 586–9); Schema D = 'second example' (pp. 589–95).

[36] In particular, he assumes that in Department I half of surplus-value is being accumulated; the rate for the other department stays, as intended, initially at the old rate (in the proportions of the existing compositions of capital in each department). See pp. 586 and 590. Note that for the preliminary Schema B, Marx assumes an intended rate of accumulation of 50% for *both* departments (p. 582). As we will see, that has no effect on the actual rate of accumulation for Department II.

will see) there is only one period of *transition* from the old regime to the new one. Thus starting from a steady state regime in period 1, and changing the regime at the end of that period (intended), a new steady state will already be reached in period 3.

Thus schematically we have the following sequence:

a. **period 1:** **production old regime – steady state**
b. end period 1: intended arrangement for old regime (would old regime have continued; matches a)
c. end period 1: intended arrangement for new regime (would have to match a)
d. end period 1: actual arrangement for new regime (= basis for production period 2)
e. **period 2:** **production new regime – transition**
f. end period 2: intended arrangement for new regime (would have to match e)
g. end period 2: actual arrangement for new regime (= basis for production period 3)
h. **period 3:** **production new regime – steady state**

Although I interpret the starting situation (period 1) of each schema as one of proportionality for a specific steady state growth path, Marx does not say this explicitly. Nor does he calculate the steady state parameters for the starting situation (as I will do below). (And as we see later on, his omission to do this may have put him on the wrong track for his conclusions from the model.)

The schemes of production (a, e, h) that I present below are identical to the ones that Marx gives. The other schemes (b, c, f, g) are presented by Marx in different and varying formats.

I use the following notation:

g = rate of growth

u = surplus-value consumed by or via capitalists ('unproductive consumption')

Δc = surplus-value accumulated in constant capital

Δv = surplus-value accumulated in variable capital

Thus we have for surplus-value (s):

$$s = u + \Delta c + \Delta v$$

The actual rate of accumulation out of surplus-value (α) is defined as:

$$\alpha = (\Delta c + \Delta v) : s$$

(α' = rate for the old regime; α = rate for the new regime; the *intended*, or planned, rate of accumulation is indicated by α^p).

The parameters for Marx's Schema C (old regime) can only be made explicit by his numbers. These are for the composition of capital:

$$c_1:(c_1 + v_1) = \gamma_1 \qquad = 0.80 \qquad\qquad (8.5)$$

$$c_2:(c_2 + v_2) = \gamma_2 \qquad = 0.67 \qquad\qquad (8.6)$$

For the rate of surplus-value:

$$s_1:v_1 = \epsilon \qquad = 1 \qquad\qquad (8.7)$$

$$s_2:v_2 = \epsilon \qquad = 1 \qquad\qquad (8.8)$$

For the rate of accumulation out of surplus-value:

$$(\Delta c_1 + \Delta v_1):s_1 = \alpha_1 \qquad = 0.45 \qquad\qquad (8.9)$$

$$(\Delta c_2 + \Delta v_2):s_2 = \alpha_2 \qquad = 0.27 \qquad\qquad (8.10)$$

Where Δc and Δv have the same proportions as in (8.5) and (8.6):

$$\Delta c_1:(\Delta c_1 + \Delta v_1) = \gamma_1 \qquad = 0.80 \qquad\qquad (8.11)$$

$$\Delta c_2:(\Delta c_2 + \Delta v_2) = \gamma_2 \qquad = 0.67 \qquad\qquad (8.12)$$

Thus there is no technical change – at least no change in the value composition of capital (assumption **c-2**).

The remainder of (potential) surplus-value is the 'unproductive consumption' (u) by or via capitalists:

$$u_1 = (1 - \alpha_1)s_1 \qquad\qquad (8.13)$$

$$u_2 = (1 - \alpha_2)s_2 \qquad\qquad (8.14)$$

Thus 'hoarding' is set aside, that is all incomes are expended – at least in the aggregate. (In his text, however, Marx devotes considerable attention to hoarding, for example in the opening section of chapter 21 (cf. section 8.3.1). Indeed he conceives of hoarding as crucial to the circulation and reproduction process – see Campbell 1998.)

Schema C: Expanded Reproduction
I reiterate that for the model below the ratios $c/c+v$ and s/v are given and constant. Thus once we have a starting value for e.g. c the numerical values for the other variables follow.

Schema C-a. Period 1: Production old regime – steady state (VIII, 586)

	c	v	s	x
I.	4000 +	1000 +	1000 =	6000
II.	1500 +	750 +	750 =	3000

$$5500 + 1750 + 1750 = 9000$$

Since $(x_1 - c)/c = (6000–5500)/5500 = 9.1\%$, this might be a schema of proportionality for a steady growth path of $g = 9.1\%$, *iff* $\alpha_1' = [(4000 + 1000) \times 9.1\%]{:}1000 = 45.5\%$; $\alpha_2' = [(1500 + 750) \times 9.1\%]{:}750 = 27.3\%$. (Marx does not calculate these ratio's).[37]

Accordingly, had the old regime continued, we would have had the following intended exchange arrangement at the *end* of period 1 (Marx does not mention this):

Schema C-b. End period 1: Intended exchange arrangement for old regime (would old regime have continued; matches schema C-a)

	c	v	u	Δv	Δc	x		
I.	4000 +	1000 +	545 +	91 +	364 =	6000	$(\alpha_1^p{}' = 45.5\%)$	$[= \alpha_1']$
II.	1500 +	750 +	545 +	68 +	137 =	3000	$(\alpha_2^p{}' = 27.3\%)$	$[= \alpha_2']$

$$5500 + 1750 + 1091 + 159 + 500 = 9000$$

(Throughout rounded off to whole numbers.)

Here u, Δv and Δc are the (intended) destination of the total of profits s.

This schema *b* matches schema *a* so the intended exchange arrangement can also be the actual exchange arrangement ($x_1 = 6000 = c + \Delta c$ and $x_2 = 3000 = v + u + \Delta v$).

The part of the surplus product that is accumulated (Δv and Δc) seems to have a different status from the other components (c, v, u). Although particularly Δv is materially produced within the period under consideration, this part of (potential) surplus-value is only realised within the next, when the extra labour-power is hired (VIII, 580–1). The realisation of Δc can be conceived of in the same way (VIII, 575). Thus the sale and purchase of these components of scale increase, in a way lag behind. Of course it applies to all components, and not just the last-mentioned, that their production and circulation – even within a period under consideration – involves complex intertemporal processes:

---

[37] Equivalently: since $\Delta c = 6000–5500 = 500$ and since for steady state growth the ratio c_1/c_2 is fixed, we can find $\Delta c = \Delta c_1 + \Delta c_2$. Next, given the ratio $c/(c+v)$ we also find $\Delta v_1 + \Delta v_2 = \Delta v$. From these values then we derive the necessary rates of accumulation $\alpha_1' = (\Delta c_1 + \Delta v_1)/s_1 = 45.5\%$ and $\alpha_2' = (\Delta c_2 + \Delta v_2)/s_2 = 27.3\%$.

The continuous supply of labour-power on the part of the working class in department I, the transformation of one part of departments I's commodity capital back into the money form of variable capital, the replacement of a part of department II's commodity capital by natural elements of constant capital II_c [i.e. c_2] – these necessary preconditions all mutually require one another, but they are mediated by a very complicated process which involves three processes of circulation that proceed independently, even if they are intertwined with one another. The very complexity of the processes provides many occasions for it to take an abnormal course. (VIII, 571)

Nevertheless the lagging behind of realisation, Marx concludes, is not the vital point of difference between simple and expanded reproduction:

Just as the current year concludes . . . with a commodity stock for the next, so it began with a commodity stock on the same side left over from the previous year. In analysing the annual reproduction – reduced to its most abstract expression – we must thus cancel out the stock on both sides . . . and thus we have the total product of an average year as the object of our analysis. (VIII, 581)

Thus we have yet another reduction of the problem.

Now instead of carrying on at the old regime (schema b) at the end of period 1, Department I – instead of planning exchange according to the old arrangement (schema b) – decides to increase the rate of accumulation (Department II intends to maintain the old rate). Thus Marx fixes $\alpha_1 = 50\%$ and then analyses the transition numerically. For this he takes as starting point the condition (8.4) for simple reproduction ($v_1 + s_1 = c_2$), *gradually* developing this in the course of the examples into a condition for expanded reproduction:

It is self-evident that, on the assumption of accumulation, $I_{(v + s)}$ [i.e. $v_1 + s_1$] is greater than II_c [i.e. c_2] . . . since (1) department I incorporates a part of its surplus product into its own capital and transforms . . . [Δc_1] of this into constant capital, so that it cannot simultaneously exchange this . . . for means of consumption; and (2) department I has to supply the material for the constant capital needed for accumulation within department II [Δc_2] out of its surplus product . . . (VIII, 590)

Thus we have:

$$(v_1 + s_1) - \Delta c_1 = c_2 + \Delta c_2$$

or

$$(v_1 + u_1) + \Delta v_1 = c_2 + \Delta c_2{}^{38} \qquad (8.15)$$

---

[38] This also derives from the balance equation: $x_1 = (c_1 + \Delta c_1) + (c + \Delta c_2)$
or from: $x_2 = (v_1 + u_1 + \Delta v_1) + (v_2 + u_2 + \Delta v_2)$.

In further presenting the numerical schemes, I will indicate for each schema whether it satisfies this condition. Marx does not do this. Again, he derives generalisations *from* his numerical schemes. Thus they are not illustrations, but rather analytical tools.

So, for Schema C-b we have condition (8.15) satisfied, as

$$1000 + 545 + 91 = 1500 + 137 \qquad \text{(rounded off)}$$

Following on from the change in the rate of accumulation ($\alpha_1 = 50\%$) we get, instead of Schema C-b, the following intended arrangement at the end of period 1:

Schema C-c. End of period 1: Intended exchange arrangement for new regime (would have to match C-a)

c	v	u	Δv	Δc	x	
I.	4000 +	1000 +	500 +	100 +	400 = 6000	($\alpha_1^P = 50\%$)
II.	1500 +	750 +	545 +	68 +	137 = 3000	($\alpha_2^P = \alpha_2^{P\prime} = 27.3\%$)

$$5500 + 1750 + 1045 + 168 + 537 = 9000$$

With these plans there is imbalance, the intended arrangement does not match production (C-a):

$$v_1 + u_1 + \Delta v_1 < c_2 + \Delta c_2 \qquad (1600 < 1637)$$

This situation cannot be. There are fewer means of production on offer (6000) than there is intended demand for (5500 + 537). Conversely there are more means of consumption on offer (3000) than the intended demand (1750 + 1045 + 168). So what happens? In fact Marx lets the course of development be dictated by department I as they hold the means of production.[39] (Note that it is assumed there are no price changes.) Thus Department I fulfils its plans and Department II is stuck with a shortage of means of production (37), plus an equivalent unsold stock of commodities for consumption. However, proportionate to the shortage in means of production (of 37) it will then hire

---

[39] Several commentators have complained about this 'dictate' of department I producers *vis-à-vis* department II producers. For example Robinson (1951, 19): 'On the face of it, this is obviously absurd'; and Morishima (1973, 118): Marx's 'very peculiar investment function'. From Marx's point of view, however, the assumption may not seem too unreasonable. Given that for Marx the property of means of production in general (by the capitalist class) is a cornerstone to his analysis, it does not seem odd that when considering that capitalist class, the vantage point is the *production* of means of production (Department I). If producers of means of production could profitably accumulate capital within their own branch they would be foolish not to, and instead sell those means of production. (In the long run, however, they might do the investment in other branches themselves.)

less extra labour-power (from 68 to 50) giving rise to an extra stock of commodities of 18. (Thus we have the paradox for Department II: eager to expand at over-capacity. If Department II were to react to its over-capacity by decreasing demand for means of production from Department I, then we would have the same paradox for Department I. In sum, a downward spiral would be plausible. Cf. section 8.4.1.) Marx shortcuts the transition – apparently because he wants to make the strongest possible case for 'balance' – by assuming that Department II capitalists absorb the stock of means of consumption $(37 + 18)$ by consuming it unproductively, thus realising their surplus-value to that extent. (We see the 'widow's cruse' in effect – section 8.3.5.) Thus we get the following arrangement (the differences from the previous Schema C-c are in italics):

Schema C-d. End of period 1: Actual exchange arrangement for new regime
(= basis for production period 2)

$$\begin{array}{llllll} & c & v & u & \Delta v & \Delta c & x \\ \text{I.} & 4000 + 1000 + & 500 + 100 + 400 = 6000 & (\alpha_1 = 50\%) \\ \text{II.} & 1500 + \ \ 750 + & 600 + \ \ 50 + 100 = 3000 & (\alpha_2 = 20\%) \end{array}$$

$$\overline{5500 + 1750 + 1100 + 150 + 500 = 9000}$$

(where condition 8.15 is met: $1000 + 500 + 100 = 1500 + 100$).

This is the 'rational' reaction for Department II to have, $\alpha_2 = 20\%$ being the result. In effect the plan for Department I to increase the rate of accumulation results in a decreased rate for Department II (and this, according to Marx, is the only way in which an (extra) expansion can come about – VIII, 572). The schema for production in the next period then becomes:

Schema C-e. Period 2: Production new regime – <u>transition</u> (VIII, 587)

$$\begin{array}{lllll} & c & v & s & x \\ \text{I.} & 4400 + 1100 + 1100 = 6600 & (g_1 = 10\%) \\ \text{II.} & 1600 + \ \ 800 + \ \ 800 = 3200 & (g_2 = 6.7\%) \end{array}$$

$$\overline{6000 + 1900 + 1900 = 9800}$$

Consequently the rate of growth for Department I has increased to 10%, and that for II has decreased to 6.7% (both initially at 9.1%).

For the end of period 2, Marx then (implicitly) assumes that Department II intends to reach the old rate of accumulation ($\alpha_2 = 27.3\%$; $\Delta c_2 / s_2 = 18.2\%$, i.e. $\Delta c_2 = 146$) and moreover to catch up with the former level of accumulation (in means of production 36). Thus the intended Δc_2 becomes $146 + 36 = 182$. Department I maintains $\alpha_1 = 50\%$.

Schema C-f. End period 2: Intended exchange arrangement for new regime (would have to match production C-e)

	c	v	u	Δv	Δc	x
I.	4400 + 1100 +	550 + 110 + 440 = 6600				($\alpha_1^P = 50\%$)
II.	1600 + 800 +	527 + 91 + 182 = 3200				($\alpha_2^P = 34\%$)

$$6000 + 1900 + 1077 + 201 + 622 = 9800$$

Again $v_1 + u_1 + \Delta v_1 < c_2 + \Delta c_2$ ($1760 < 1782$), again Department I can dictate the course, and again Department II absorbs the potential over-production (22 plus 11, since labour-power hired decreases proportionally). Accordingly we have for the actual exchange arrangement (differences from schema f in italics):

Schema C-g. End of period 2: Actual exchange arrangement for new regime (= basis for production period 3)

	c	v	u	Δv	Δc	x
I.	4400 + 1100 +	550 + 110 + 440 = 6600				($\alpha_1 = 50\%$)
II.	1600 + 800 +	*560* + *80* + *160* = 3200				($\alpha_2 = 30\%$)

$$6000 + 1900 + 1110 + 190 + 600 = 9800$$

(where condition 8.15 is met: $1100 + 550 + 110 = 1600 + 160$).

Department II has recovered part of the former level of accumulation, but not all. As a result the schema for production in the next period becomes:

Schema C-h. Period 3: Production new regime – new steady state (VIII, 588)

	c	v	s	x
I.	4840 + 1210 + 1210 =	7260 ($g_1 = 10\%$)		
II.	1760 + 880 + 880 =	3520 ($g_2 = 10\%$)		

$$6600 + 2090 + 2090 = 10780$$

With this schema we are finally at the new steady state growth path. From now on all entries can increase at a growth rate of 10% ($g = 10\%$ for both departments). Department II cannot catch up with accumulation any further, so α_2 stays at 30%.[40] Marx calculates the schema for

---

[40] Though for this example it will have caught up in absolute quantity after two more periods, since the growth rate has risen.

three more periods, each period all components of the model of course increase by 10% (VIII, 589).

The transition from the initial growth path (9.1%) to the new one (10%) has been accomplished in two periods because of rather severe assumptions; in fact Marx has shown how fragile an enterprise it is to arrive at a higher growth path. So much for Schema C.[41]

8.4.4 Marx's generalisations concerning over/underproduction

As I have said, Marx's models are not illustrations of results that you already know; they are tools for arriving at a generalisation. In constructing the production schemes he implicitly applies in all his numerical

---

[41] In the literature the object of Marx's reproduction model is variously appreciated, especially the status of its 'equilibrium'. In terms of the interpretation of Marx's text, most commentators seem to give up at some point (early) in Marx's chapter 21 and come up with a reconstruction of their own. Nearest to my own interpretation is that of Desai (see below). A review of that literature is beyond the scope of this paper, therefore I restrict to a few comments on three well-known scholars in the field.

I cannot agree with Foley's (1986, 85) interpretation of what Marx is doing: it is not the case that Marx's *initial* schemes (period 1) were meant to represent reproduction for the *new* rate of accumulation (which they clearly cannot, as Marx indicates). Foley suggests that Marx merely wanted to find an adequate schema for 'period 1' and that the 'discrepancy' between the initial schema and the rate of accumulation 'annoyed Marx', and that he therefore 'devoted several pages of his notes to the attempt to find a schema that would exhibit proportional expanded reproduction.' No, Marx analyses the *process of change* following on from a change in the rate of accumulation. Koshimura (1975, 17–19) equally neglects the transitional process.

Morishima (1973) hardly analyses the properties of Marx's schemes of expanded reproduction or the transitional process (pp. 117–20), concerned as he is to 'replace' Marx's 'special investment function' (see note 39 above) with the 'more reasonable' case for which capitalists of Departments I and II 'have the same propensity to save' (p. 122). Although this is of course a useful exercise it precludes him getting to grips with the schemes and their object themselves. In Morishima's reconstruction the model is one of unstable growth (with, depending on the compositions of capital, either explosive oscillations or monotonic divergence from the balanced growth path – p. 125). The account of Harris (1972) is along similar lines.

Desai (1979, 147–53, 161–71) has a somewhat different view of the periodisation from that outlined above, although he appreciates the important 'ex-ante' versus 'ex-post' character of Marx's schemes. His account de-emphasises the level of abstraction at which the schemes operate, and consequently we differ about the interpretation of the aim of the schemes. Desai also thinks that the dimensions of the schemes are 'labour-values' (so does Mandel, 1978, 38) and that the schemes fail 'to pose the problem of expanded reproduction in the price domain'. On the first point he is wrong (at least, Marx says otherwise – e.g. p. 473) and on the second he neglects Marx's view about its irrelevance for the problem at hand (see my comment on assumption **c**). Finally, and related, he neglects Marx's emphasis on the twofold character of the entities he deals with. Therefore I cannot agree that Marx's problematic is 'entirely confined to the circuit of commodity capital'. (I do not want to disclaim the Marxian theories of these three authors in this field; however, I am concerned here strictly with Marx's reproduction theory.)

examples the formula $v_1 + u_1 + \Delta v_1 = c_2 + \Delta c_2$ (condition 8.15), and explicitly derives the relation from them, albeit not as a formal equation (pp. 590 and 593). Thus on page 593 we read:

With production on an increasing capital basis, $I_{(v + s)}$ [i.e. $v_1 + s_1$] must be equal to II_c [i.e. c_2], plus the part of surplus product that is reincorporated as capital [Δc_1], plus the extra portion of constant capital needed to expand production in department II [Δc_2], and the minimum for this expansion is that without which genuine accumulation, i.e. the actual extension of production in department I, cannot be carried out.[42]

So we have:

$$v_1 + s_1 = c_2 + \Delta c_1 + \Delta c_2 \qquad (8.15')$$

or

$$v_1 + [u_1 + \Delta v_1 + \Delta c_1] = c_2 + \Delta c_1 + \Delta c_2 \qquad (8.15'')$$

Nevertheless, at the very end of the text (pp. 595–7), when Marx is preparing to draw general conclusions from his schemes – especially concerning the point if transition is reached via over- or underproduction – he once again falls back on the simple reproduction condition with which he started:

$$v_1 + s_1 = c_2 \qquad (8.4)$$

modifying it into

$$v_1 + u_1 = c_2 \qquad (8.16)$$

He applies this last formula (8.16) for the parameters of the new regime ($\alpha_1 = 50\%$, thus $u_1 = \frac{1}{2} s_1$) to the initial production (Schema C-a). Indeed for the values of Schema C-a the formula 8.16 holds: $1000 + 500 = 1500$.

Why does he apply this apparently irrelevant or mistaken formula? The easy answer is to refer to the unfinished shape of the text: it was perhaps meant to be followed by a piece indicating the relevant difference between the conditions for simple and expanded reproduction.

However, there is another explanation, which directly relates to Marx's examples. Note that his generalisations (pp. 595–7) follow just after setting out Schema D (pp. 590–5). The problem is not so much that he takes the formula $v_1 + u_1 = c_2$ as a *starting point* of the analysis.

---

[42] The last caveat ('the minimum' etc.) says that we must at least have $v_1 + s_1 \geq c_2 + \Delta c_1$.

Indeed with Schema D, Marx takes an example for which this formula in the way he applies it does *not* hold for the initial situation – as it did for Schemas B and C.[43] The sorry point is that Schema D is an unlucky example: with it he describes the transition to a *decreasing* rate of accumulation and growth, whilst it is apparently meant to describe (further) expansion – taking off with a new rate of accumulation of 50% for Department I as in all his examples. However, since Marx neglects to calculate the relevant *initial* properties of his schemas – especially the rates of accumulation and growth – he seems unaware of this.

Schema D: Expanded reproduction; *a. Period 1: Production, initial situation*

$$
\begin{array}{llll}
\;\; c & v & s & x \\
\text{I.} \;\; 5000 + 1000 & + 1000 & = 7000 \\
\text{II.} \; 1430 + 286^* & + 286^* & = 2002 \\
\hline
 6430 + 1286 & + 1286 & = 9002
\end{array}
$$

* here, Marx has 285

This might be a schema of proportionality for a steady growth path of $g = 8.9\%$, *iff* $\alpha_1' = \alpha_2' = 53.2\%$. (Marx does not calculate these ratio's). The new rate of accumulation thus *decreases* to $\alpha_1 = 50\%$.

For our purposes we do not need to go through this example any further (in the end, the new growth rate will slow down to 8.3%).[44]

Indeed, for the new regime ($\alpha_1 = 50\%$, thus $u_1 = 500$):

$$v_1 + u_1 > c_2 \qquad\qquad \text{(i.e. } 1500 > 1430)$$

---

[43] Schema B has the same relevant properties as Schema C, the one we discussed in the main text above, except that it is somewhat simpler as the compositions of capital are equal. Its initial make-up is:

Schema B: Expanded reproduction; *a. Period 1: Production, initial regime*

$$
\begin{array}{llll}
\;\; c & v & s & x \\
\text{I.} \;\; 4000 + 1000 & + 1000 & = 6000 \\
\text{II.} \; 1500 + 375^* & + 375^* & = 2250 \\
\hline
 5500 + 1375 & + 1375 & = 8250
\end{array}
$$

* Marx has here 376 – apparently to facilitate the calculations.

This might be a schema of proportionality for a steady growth path of $g = 9.1\%$ (6000−5500/5500), iff $\alpha_1' = \alpha_2' = 45.5\%$. (Marx does not mention these proportions). The new rate of accumulation increases to $\alpha_1 = 50\%$.

Note that for the new regime of exchange (end period 1) it just *happens* to be the case that $v_1 + u_1 = c_2$ (1000 + 500 = 1500). But the same applied to Schema C! Apparently Marx is then led to take this formula (much akin to the simple reproduction condition 8.4) as the starting point for his analysis.

[44] Again, Marx is unaware that it goes down since he neglects to calculate the implied initial growth rate of 8.9%.

But this seems irrelevant. What is relevant however, and which is the reason we have potential *overproduction* in Department I, is that:

$$v_1 + u_1 + \Delta v_1 > c_2 + \Delta c_2 \qquad \text{(i.e. } 1000 + 500 + 83 > 1430 + 127$$
$$\text{thus } 1583 > 1557\text{)}$$

The accidental relation between $v_1 + u_1$ and c_2 in his examples lets Marx conclude that $(v_1 + u_1)/c_2$ is an important ratio.

This is an interesting example of how the same heuristic of a model may be positive (fruitful) as well as negative (blocking). Thus Marx's finding from the model of simple reproduction of the exchange condition $v_1 + s_1 = c_2$ apparently makes him search for a similar condition in the extended model (positive heuristic). In the search for it, the first thing he does when presenting his model of expanded reproduction, is come up with $v_1 + u_1 = c_2$. Now it seems that the similarity with $v_1 + s_1 = c_2$ blocks his further search (negative heuristic), which is understandable to the extent that *apparently* $v_1 + u_1 = c_2$ does the job that Marx was looking for, which is to search for the conditions of over/underproduction.

8.4.5 Further analysis of the transition; final remarks

Elsewhere in *Capital* Marx theorises the rate of accumulation out of surplus value (α) as a necessary force in capitalism. In the model for expanded reproduction that we have discussed, however, α_1 is fixed for analytical purposes. α_2, on the other hand, is taken for a semi-*variable*. Its intended value is that of the previous (running) period, in effect however it is a *result*. Unproductive consumption u_2 varies accordingly. In this way, Marx's account short-cuts adaptation after any changes in the system (α; the same might apply for changes in γ or ϵ); it also precludes downward spiral effects: *effective* overproduction is ruled out. Any potential overproduction (given a rate of accumulation α_1) is absorbed via the adaptation in α_2: either by unproductive consumption (for means of consumption) or by accumulation (for means of production).[45]

Expanded reproduction and proportionality, we have seen, are defined by the condition:

$$c_2 + \Delta c_2 = v_1 + u_1 + \Delta v_1 \qquad (8.15)$$

---

[45] The latter happens in Schema D. Whereas Marx lets Department I dictate the course of things (α_1 fixed) – and whilst that may make sense within his line of thought – either or both of α_1 and α_2 might in principle be taken as semi-variables (with 'ex-ante' and 'ex-post' divergences).

which centres the analysis on the interconnecting exchanges between the two departments. In the way Marx has the model work, the possible violation of this condition hinges on the difference between the planned or intended α_2^p and the actually realised α_2.

It is within the logic of Marx's reasoning to start from a given accumulation of capital in each department, from which follow numerical values for the other variables. Thus in the face of the pattern for the parameters α, γ and ϵ (the rate of accumulation out of surplus-value, the composition of capital, and the rate of surplus-value), the starting values c_1 and c_2, or $(c_1 + v_1)$ and $(c_2 + v_2)$, determine the course of things, notably smooth adaptation or *potential* overproduction in Department I or Department II, with their potential downward spiral effects. Each time condition 8.15 may turn out to be an inequality 'at the end' of the period; the resulting accumulation of capital ('ex-post') thus determining the course for the next period. The following three cases can be distinguished by way of reconstruction of Marx's generalisations:[46]

(1) Potential overproduction in department II (cf. Schemas B and C), if:

$$v_1 + u_1 + \Delta v_1 < c_2 + \Delta c_2 \quad \text{(Marx has: } v_1 + u_1 = c_2\text{)}$$

(2) Smooth adaptation, if:

$$v_1 + u_1 + \Delta v_1 = c_2 + \Delta c_2$$

(3) Potential overproduction in department I (cf. Schema D), if:

$$v_1 + u_1 + \Delta v_1 > c_2 + \Delta c_2 \quad \text{(Marx has: } v_1 + u_1 > c_2\text{)}$$

In effect the process of transition that Marx sets out runs as follows. Ensuing upon a (positive) change in the rate of accumulation from a previous α' to a new intended α (requiring a relative increase of Department I), (new) proportionality is established via a re-adaptation of the rates of accumulation α_1 and α_2. In Marx's model the period of transition is short-cut by a pre-emptive re-adaptation for especially α_2, thus absorbing any overproduction and evading downward spirals.

In other words: upon the change of α_1' to α_1, the Δc_1 [that is: $\alpha_1 \gamma_1 s_1$] is a (new) constant fraction of c_1, whence we have a constant rate of

---

[46] As I have indicated (section 8.4.4), Marx implicitly sets out the interconnection in his numerical schemes and also formulates it explicitly; nevertheless for his generalisations he draws back on a modification of his generalisation for simple reproduction.

growth for Department I. However, $v_1 + u_1 + \Delta v_1$ [that is $v_1 + (1-\alpha_1)s_1 + \alpha_1(1-\gamma_1)s_1$] is also a (new) constant fraction of c_1; at the same time it determines $c_2 + \Delta c_2$ [that is: $c_2 + \alpha_2(\gamma_2 s_2)$]: the extra production of means of production in Department I that it does not use up itself and which it sells to Department II – Department II cannot have more, only less; however, given the α_2 planned, it absorbs what is available. Therefore Department II becomes chained to the growth rate of Department I.

In this process of adaptation, Department I thus dictates the course. The ownership of means of production for producing means of production is apparently thought of as crucial: Department II cannot expand unless I does.

In sum: more so than the chapter on simple reproduction, the chapter on expanded reproduction reveals the defects of an unfinished draft and an unfinished analysis. Guiding Marx's generalisations is an adjustment of the condition for simple reproduction. However, the adjustment is not carried through to its full extent; it is nevertheless effected in the numerical schemes. Even if unfinished, the power of the model is revealed very well. Heuristically it also leaves plenty of room for further analysis of dynamic processes. At the core of the model are the same fundamental macroeconomic abstractions, developed into a two-sector approach, as those of simple reproduction. Generally Marx shows that, even setting aside all sorts of complications (his caricatural assumptions), proportionality between the two sectors – or generally: steady state growth – would be like balancing on a 'knife-edge'. In the process of transition from one growth path to another, we saw in effect – as an interesting digression – the 'widow's cruse' mechanism: 'capitalists earn what they spend, and workers spend what they earn' (cf. section 8.3.5).

8.5 GENERAL METHODOLOGICAL ANALYSIS OF THE CASE

At first sight one might characterise Marx's method for developing his model of reproduction as one of successive approximation. (This is the dominating opinion since it was phrased that way by Sweezy in 1942.) This characterisation is appropriate since Marx indeed adopts 'procedural simplifications' to that extent, especially in the moves from Simple Reproduction without fixed capital, to Simple Reproduction with fixed capital and finally to the model of Expanded Reproduction. Nevertheless other aspects of Marx's reproduction model are of much methodological interest. These are taken up in the following six points.[47]

1. Analysis 'from within'. The aim of Marx's reproduction model is to find the conditions for expanded reproduction, especially on a higher scale – transition to a higher growth path. For Marx the driving force of the capitalist system is accumulation of capital. Thus it would be within the system's logic to get on to ever higher growth paths (in £,$!). Elsewhere he shows how this must be a cyclical movement recurrently interrupted by economic crises, either due to labour shortage (*Capital I*) or to technical change (*Capital III*). Although in the Part of *Capital II* that we have discussed Marx reiterates that economic expansion is a crisis prone process, some interpreters of Marx find it difficult to see the combination of emphasis on cycle and crisis together with the apparent steady state growth presentation of the Schemas. However, the approach is consistent if we realise that each time (also in *Capital I* and *III*) Marx's method is directed at analysing capitalism *from within its own logic*.[48] Thus in the Part that we have discussed Marx sets out to make – from his point of view – the strongest case for expansion. In the course of it he again and again finds on his way potentially disturbing factors – even when he discusses the indeed 'simple' reproduction, e.g. monetary circulation (not discussed in this paper) or the replacement of fixed capital. Each time that might in a way have been reason to cut off further analysis since he arrives at a phenomenon that, apparently, can never get into something like steady state growth. The art of his model building is then to set those aside (to 'bracket' them) and to proceed to the 'every day' problem of potential overproduction and underproduction.

2. Strongest case caricatures. On his way to presenting this 'strongest case' Marx seems to construct ever more 'caricatures' of capitalism. What Marx is doing here fits the general description Gibbard and

---

[47] I will emphasise methodological aspects from the point of view of model building. Still other methodological issues would deserve consideration, especially concerning Marx's systematic-dialectics which, apparently, plays a less than subordinate role in the case material examined. This issue is discussed in Reuten (1998).

[48] Thus Marx's aim is to analyse capitalism from its own standards and to assess it by the fulfilment of its own standards. To the extent that these are contradictory it may meet the boundaries of fulfilment. In this respect Marx adopts a method of internal *critique* rather than external *criticism*. In this respect also there is a direct link from Hegel's method to Marx's (see Benhabib 1986, chs. 1–4; and Arthur 1993). Marx applies, by the way, the same method of critique to the assessment of economists before him. This is not to say, of course, that Marx does not have an opinion on his subject matter – he does, and his language often reveals that. His method and his opinion, however, are quite different issues.

Varian (1978) provide for caricatural models in modern economics. For example, to assume that monetary accommodation for capital accumulation or fixed capital replacement is *not* a problem in capitalism is a caricature; it is evidently – at least for Marx – not an approximation of reality. However, if I interpret Marx correctly, this is a caricature that lives up to the system's self-image and therefore this, rather than another possible caricature, is appropriate. In the end then he arrives at only a few strokes (to refer to the drawing metaphor from which the term caricature is taken). The end caricature then is to imagine that even here a balance exists. What has been shown then is not that balanced growth is impossible. No, it is shown that it is possible, although it is balancing on a knife-edge. All the same all the caricatural assumptions made under way are even so many potential points for setting into motion economic crisis.

3. Neglectability: mastery of the object of inquiry. Besides procedural assumptions for successive approximation, and next to these 'strongest case caricatures', the construction, or make up, of the model is predominantly determined by 'neglectability assumptions' (for our case the assumptions **c** through **f**: constant prices, constant compositions of capital and a constant rate of surplus-value).[49] This type of assumption shows in my view the model builder's true theoretical mastery of the object of inquiry that the model deals with: the ability to omit so as to show important aspects or even the essence of the entity – affixing reality by omitting part of reality. For our case: prices etc. can be taken constant because their variation would not matter. Although these assumptions may look like, or even have the form of, *ceteris paribus* assumptions, they are not. For it is hypothesised that the violation of these assumptions do not affect the subject matter.

Together with the caricatural assumptions, those of neglectability show the virtue or vice of a model and therefore also the real art of the model building. For our case especially, the thesis that price changes may be neglected is quite a claim (for one might argue, as orthodox economists would, that price changes are central to equilibrating mechanisms in a capitalist economy, thus also in preventing falling too far from the knife-edge, or putting the economy on or towards the knife-edge). The model's vice? Indeed, since prices are held constant and as each entry in Marx's schema is a price times quantity entity (measured in £, $ etc.), all changes in the entries are changes in material quantities. Bringing in

---

[49] In the make-up of Marx's model we also saw one assumption of 'approximation', that is the one on a sufficient labour force to accommodate the accumulation of capital.

price changes from the start would certainly have given a different focus to the model. However, the central point seems to be that, rather than price adaptations, quantity adaptations primarily matter, or at least that quantity adaptations bring home the issue that matters for the 'social reproduction of capital'. The model's virtue?

4. Make-up and room for experiment: heuristic potential.   The make-up of the model is not only determined by its caricatures and neglected points but also by what I have called throughout this paper its particular *categorical abstractions* (for our case this concerns primarily the macroeconomic categories, their two-departmental division as well as their generic and determinate aspects). Although these are indeed characteristics of the model and come *into* model building, they are rather the result of theoretical activity and have to be argued for theoretically. Nevertheless, they determine the scope of the model and particularly its *heuristic potential*.

For example: (a) Marx's treatment of fixed capital would have to be different – perhaps an idealisation – if the model were at a micro level of abstraction: in that case the condition of wear and tear being equal to depreciation would have come down to assuming away fixed capital; (b) the two-department division leaves scope for a three or more department division – as Marx himself sometimes did and as many economists working in the Marxian tradition did after him (e.g. Tugan-Baranovsky, Luxemburg, Kalecki); (c) as to the generic/determinate aspect: surprisingly Marx's model for a capitalist economy was adopted as a gadget for Soviet planning in the early 1920s, apparently its generic aspect being lifted out;[50] (d) the last two items together apparently were of inspiration for Leontief to develop the Schemes into input–output tables and their analysis – now again applied to a capitalist economy.

The heuristic potential of the model is also determined by the particular way in which it can be manipulated or the particular way in which one can (mentally) experiment with it. This may be a matter of the extent to which the model allows for sensibly playing with its variables or its parameters so as to increase our insight. Even if Marx, as we have seen, considers changes in the composition of capital $(c/(c + v))$ or in the rate of surplus-value (s/v) neglectable for the problem of reproduction, such changes are central to his theory of the development of the rate of profit and the cyclical development of capitalism (cf. *Capital III*, Part Three). In fact several chapters relating to the composition of capital and the rate of profit in *Capital III* contain examples and models with

---

[50] See e.g. Lange (1959, ch. 3), Jasny (1962) and Desai (1979).

similar elements as the ones we have seen above. Not much imagination is required to combine those – as many Marxian economists have – and to analyse cyclical development within the context of a reproduction schema (early authors doing this have been e.g. Bauer, Grossman, Luxemburg and Pannekoek). It is also a short step to build into the model a government sector or e.g. monopoly pricing. All this indeed reveals the heuristic potential of the model. Marx himself, as I have indicated, did not reach this since the reproduction model discussed above was the last part of *Capital* he worked on. Nevertheless it has generated quite a family of models.

Having discussed the aim for Marx's model as well as its make-up and the heuristic potential thereof we may now turn to the working of the model. I first make an observation on the internal working of the model, turning finally to the working of the model in its theoretical context.

5. Fruitful versus blocking working heuristics. We have seen that Marx's model for Simple Reproduction (Schema A above) works through the interdepartmental exchange condition (8.4): $v_1 + s_1 = c_2$. This condition (rather than any other such as conditions (8.4') or (8.4'') which would formally do as well) is the focus for his analysis of the working of the model.

When starting to extend this model to Expanded Reproduction (Schema C above) the exchange condition (8.4) provides the apparently *fruitful* heuristic for *preemptively* finding the working of the Expanded Reproduction model. A modification of (8.4) provides him explicitly with $v_1 + u_1 = c_2$, where u_1 stands for the expenditure of Department I capitalists on means of consumption, as similarly s_1 in the simple model stood for expenditure of Department I capitalists on means of consumption. This would have been fine for a *first* modification; indeed it requires further modification – as it stands it is incorrect.

Nevertheless when making the expanded model actually work in its various phases, Marx implicitly applies the (correct) exchange condition (8.15): $v_1 + u_1 + \Delta v_1 = c_2 + \Delta c_2$ which is a further modification of (8.4). Indeed his models are correct in this respect. It is remarkable that he also finds condition (8.15) from his models explicitly – albeit not in terms of a formal equation. So far (8.15) seems to provide him with a fruitful heuristic.

However, when analysing the particular quantitative properties of the three varieties of his expanded reproduction model, the initial *preemptive* finding for the working of the Expanded Reproduction model provides a 'heuristic blocking'. Why? Apparently no reason whatsoever lies in the construction of the model – as I have shown in section 8.4.3. The reasons for the insufficient modification may be related to the draft stage of the text (notebooks), or as I have suggested, to bad luck in choosing the model examples (especially Schema D which, contrary to Marx's intention, sets out contraction instead of growth). The particular model examples happen to be consistent with the conclusions that Marx draws from them.

Generally it seems that this aspect of the carrying over of the working of a model to its extension is similar to the way in which an analogy model can generate a positive or a negative heuristic (see chapter 12 by Morgan in this volume).

This point about the working of the model in relation to particular quantitative properties is rather a detail. More important is the general context for the working of the model to which I now turn.

6. Immediate story and wider project.   If my story so far is correct then we may conclude that from the point of view of *model building* there is not a gulf between Marx and modern orthodox economics. Nevertheless a model, like a painting, shows differently depending on the exhibition.

In part this is a matter of telling a story around the model proper – in our case that story is told in the original text immediately surrounding the model (see also chapter 4 by Boumans in this volume). For example, if in a formal model we arrive at an equation such as $v_1 + u_1 + \Delta v_1 = c_2 + \Delta c_2$ (or $v_1 + u_1 = c_2$ for that matter) nothing in the *technical make-up* of the model tells us if 'behind' this is a movement to or away from equilibrium so that it represents (tendentially) the 'normal' case or the 'abnormal' case. For Marx the knife-edge is no more than an indeterminate possibility, 'founded' upon caricatures representing the self-image of capitalism (cf. points 1 and 2 above). In his immediate story however, he reveals no more than the conditions for falling off the knife-edge leading up to economic crisis, without telling us about the course of such crises and their aftermath (leading back to a new balance in an economic upturn?).

It might be tempting to argue that Marx just takes returning economic crises for granted (the nineteenth-century empirical regularity). Such an argument (in our case) would neglect that model building, even if we add its immediate surrounding story, does not operate in a methodological and theoretical void. Models are part of explicit (cf. Lakatos) and implicit research programmes/projects. For Marx it is certainly the case that the model we have discussed is no more than 'a moment' of a wider research project (in the systematic-dialectical sense indicated in section 8.2). Economic crises due to over- and underproduction are on the one hand banal and on the other erratic: they are not law-like processes but rather due to the general 'anarchy of the market'. Marx is therefore prepared to build on the knife-edge, that is on the balance, so as to move on in Volume III of *Capital* to the centrepiece of the capitalist self image: technical progress. Indeed he moves on to what – in Marx's view – is the strongest case for capitalism. This is also the domain for which he does detect law-like processes. For Marx therefore the analysis of the cyclical course of capitalist development within the frame of the reproduction model would be superfluous or rather besides the point of the model.

However, as with the artist's painting, once alienated, the model builder cannot choose the exhibition. Unlike the painting, a model can be both conserved and adapted at the same time so figuring at several galleries.

APPENDIX: RECAPITULATION OF SYMBOLS USED

I   = Department I, producing means of production
II = Department II, producing means of consumption
c   = constant capital, the value of the means of production applied
v   = variable capital, the value of the social labour-power applied
s   = surplus-value, the value that is added by labour minus the replacement of the variable capital advanced (v)
g   = rate of growth
Δc = surplus-value accumulated in constant capital
Δv = surplus-value accumulated in variable capital
u   = surplus-value consumed by or via capitalists ('unproductive consumption')
α = actual rate of accumulation out of surplus-value (new regime), defined as: $\alpha = (\Delta c + \Delta v) : s$
α' = actual rate of accumulation out of surplus-value (old regime)
α^P = *intended*, or planned, rate of accumulation out of surplus-value
γ = composition of capital, defined as $\gamma = c : (c + v)$
ϵ = rate of surplus-value, defined as $\epsilon = s : v$

REFERENCES

Note: All years in brackets are the original dates of publication as referred to in the text – if appropriate several editions are indicated by superscripts. If the *printing* quoted from may differ it is provided where relevant.

Arthur, Christopher J. (1993). 'Hegel's *Logic* and Marx's *Capital*', in Moseley (ed.), 1993: 63–88.

    (1997). 'Against the logical–historical method: dialectical derivation versus linear logic', in Moseley and Campbell, 1997.

Arthur, Christopher and Geert Reuten (1998). 'Marx's Capital II, The Circulation of Capital', in Arthur and Reuten (eds.) 1–16.

Arthur, Christopher and Geert Reuten (eds.) (1998). *The Circulation of Capital: Essays on Book II of Marx's 'Capital'*, London/New York: Macmillan/St. Martin's.

Bellofiore, Riccardo (ed.) (1998). *Marxian Economics – A Reappraisal*, Vols. I and II, London/New York, Macmillan/St. Martin's.

Benhabib, Seyla (1986). *Critique, Norm, and Utopia*, New York: Columbia University Press.

Boumans, Marcel (1992). *A Case of Limited Physics Transfer: Jan Tinbergen's Resources for Re-shaping Economics*, Amsterdam: Thesis Publishers.

Campbell, Martha (1998). 'Money in The Circulation of Capital', in Arthur and Reuten (eds.), 1998: 129–57.

Desai, Meghnad (1979). *Marxian Economics*, Oxford: Basil Blackwell.

Foley, Duncan K. (1986). *Understanding Capital; Marx's Economic Theory*, Cambridge MA/London: Harvard University Press.

Gehrke, Christian and Heinz D. Kurz (1995). 'Karl Marx on physiocracy', *The European Journal of the History of Economic Thought*, 2/1, 1995: 53–90.

Gibbard, Allan and Hal R. Varian (1978). 'Economic Models', *The Journal of Philosophy*, Vol LXXV, no 11, November: 664–77.

Harris, Donald J. (1972). 'On Marx's scheme of reproduction and accumulation', *Journal of Political Economy*, 80/3–I: 503–22.

Hegel, G. W. F. (1812[1], 1831[2]). *Wissenschaft der Logik*, Engl. transl. (1969) of the 1923 Lasson edn, A. V. Miller, *Hegel's Science of Logic*, Humanities Press 1989.

    (1817[1], 1830[3]). *Enzyklopädie der Philosophischen Wissenschaften im Grundrisse I, Die Wissenschaft der Logik*, Engl. transl. of the third edition (1873[1]), T. F. Geraets, W. A. Suchting and H. S. Harris, *The Encyclopaedia Logic*, Indianapolis/Cambridge 1991, Hackett Publishing Company.

Horowitz, David (ed.)(1968). *Marx and Modern Economics*, New York/London: Monthly Review Press.

Jasny, N. (1962). 'The Soviet balance of national income and the American input-output analysis', *L'industria*, no.1: 51–7.

Jones, Hywel (1975). *An Introduction to Modern Theories of Economic Growth*, Nairobi/Ontario: Nelson.

Kaldor, N. (1955/56). 'Alternative theories of distribution', *Review of Economic Studies*, vol. 23, 1955–6: 94–100, reprinted in A. Sen (ed.), *Growth Economics*, Harmondsworth, Penguin Books 1970: 81–91.

Kalecki, Michal (1935, 1952). 'A macroeconomic theory of business cycles', *Econometrica*, July 1935; revised as 'The determinants of profits' in *Theory of Economic Dynamics*, Kelley, New York 1969.

(1968). 'The Marxian equations of reproduction and modern economics', *Social Science Information*, Vol. 7.

Klein, Lawrence R. (1947). 'Theories of effective demand and employment', *The Journal of Political Economy*, April 1947, reprinted in Horowitz 1968: 138–75.

Koshimura, Shinzaburo (1975; 1st Japanese edn 1956 or 1957?). *Theory of Capital Reproduction and Accumulation*; ed. J. G. Schwartz, Engl. transl. Toshihiro Ataka, Dumont Press Graphix, Kitchener, Ontario (Canada) 1975.

Lakatos, Imre (1970). Falsification and the Methodology of Scientific Research Programmes, in I. Lakatos and A. Musgrave (eds.), *Criticisms and the Growth of Knowledge*, Cambridge: Cambridge University Press 1974: 91–196.

Lange, Oscar (1959[1], 1962[2]). *Introduction to Econometrics*, second edition, translated from Polish by E. Lepa. Oxford and Pergamon Press 1966.

Leontief, Wassily (1941). *The Structure of American Economy, 1919–1929; An Empirical Application of Equilibrium Analysis*, Cambridge, MA, Harvard University Press.

Leontief, Wassily (ed.) 1953. *Studies in the Structure of the American Economy; Theoretical and Empirical Explorations in Input-Output Analysis*, New York: Oxford University Press.

Likitkijsomboon, Pichit (1992). 'The Hegelian Dialectic and Marx's *Capital*', *Cambridge Journal of Economics*, Vol. 16/4: 405–19.

(1995). 'Marxian Theories of Value-Form', *Review of Radical Political Economics*, Vol. 27, no 2: 73–105.

Mandel, Ernest (1978). *Introduction to Marx* (1885), English Fernbach translation, pp. 11–79.

Marx, Karl (1867; 1885; 1894). *Capital: A Critique of Political Economy*, Vols. I-III (German originals I:1867[1],1890[4]; II:1885; III:1894), translated by Ben Fowkes (I) and David Fernbach (II + III), Penguin Books, Harmondsworth 1976, 1978, 1981.

(1885[1],1893[2]). ed. F. Engels, *Das Kapital, Kritik der Politischen Okonomie, Band II, Der Zirkulationsprozess des Kapitals*, MEW 24, Dietz Verlag, Berlin 1972, Engl. transl. (Ernest Untermann 1907[1]), David Fernbach, *Capital, A Critique of Political Economy, Vol II*, Penguin Books, Harmondsworth 1978.

Morishima, Michio (1973). *Marx's Economics; A dual theory of value and growth*, Cambridge/New York: Cambridge University Press, 1974.

Moseley, Fred (ed.) (1993). *Marx's Method in 'Capital': A reexamination*, Humanities Press, NJ 1993.

(1998). 'Marx's reproduction schemes and Smith's dogma', in Arthur and Reuten (eds.) 1999: 159–85.

Moseley, Fred and Martha Campbell (eds.) (1997). *New Investigations of Marx's Method*, Humanities Press, New Jersey 1997.

Murray, Patrick (1988). *Marx's Theory of Scientific Knowledge*, Humanities Press, New Jersey/London.

Oakley, Allen (1983). *The Making of Marx's Critical Theory; A Bibliographical Analysis*, Routledge & Kegan Paul, London.

Reuten, Geert (1995). 'Conceptual collapses; a note on value-form theory', *Review of Radical Political Economics*, Vol. 27/3: 104–10.

   (1998). 'The status of Marx's reproduction schemes: conventional or dialectical logic?', in Arthur and Reuten (eds.), 187–229.

Reuten, Geert and Michael Williams (1989). *Value-Form and the State; the tendencies of accumulation and the determination of economic policy in capitalist society*, Routledge, London/New York.

Robinson, Joan (1948). 'Marx and Keynes' (in Italian for *Critica Economica*, 1948); in Robinson, *Collected Economic Papers I*, Oxford 1950, reprinted in Horowitz 1968: 103–16.

   (1951). 'Introduction to Rosa Luxemburg' (1913), in *The Accumulation of Capital*, Routledge and Kegan Paul, London, 1971: 13–28.

Sawyer, Malcolm C. (1985). *The Economics of Michal Kalecki*, London, Macmillan.

Schumpeter, Joseph A. (1954). *History of Economic Analysis* (ed. London E. B. Schumpeter), Allen & Unwin, London, 1972.

Solow, Robert M. (1956). 'A contribution to the theory of economic growth', *Quarterly Journal of Economics*, vol. 70, 1956: 65–94, reprinted in A. Sen (ed.), *Growth Economics*, Penguin Books, Harmondsworth, 1970: 161–92.

Sweezy, Paul A. (1942). *The Theory of Capitalist Development*, Modern Reader Paperbacks, New York/London 1968.

Williams, Michael (1998). 'Money and labour-power: Marx after Hegel, or Smith plus Sraffa?', *Cambridge Journal of Economics*.

CHAPTER 9

Models and the limits of theory: quantum Hamiltonians and the BCS model of superconductivity

Nancy Cartwright

9.1 INTRODUCTION

In the 1960s when studies of theory change were in their heyday, models were no part of theory. Nor did they figure in how we represent what happens in the world. Theory represented the world. Models were there to tell us how to change theory. Their role was heuristic, whether informally, as in Mary Hesse's neutral and negative analogies, or as part of the paraphernalia of a more formally laid out research programme, as with Imré Lakatos. The 1960s were also the heyday of what Fred Suppe (1977) dubbed 'the received view' of theory, the axiomatic view. Theory itself was supposed to be a formal system of internal principles on the one hand – axioms and theorems – and of bridge principles on the other, principles meant to interpret the concepts of the theory, which are only partially defined by the axioms. With the realisation that axiom systems expressed in some one or another formal language are too limited in their expressive power and too bound to the language in which they are formulated, models came to be central to theory – they came to constitute theory. On the semantic view of theories, theories are sets of models. The sets must be precisely delimited in some way or another, but we do not need to confine ourselves to any formal language in specifying exactly what the models are that constitute the theory.

Although doctrines about the relation of models to theory changed from the 1960s to the 1980s, the dominant view of what theories do did not change: theories represent what happens in the world. For the semantic view that means that models represent what happens. One of the working hypotheses of the LSE/Amsterdam/Berlin modelling project has been that this view is mistaken. There are not theories, on

I want to thank Jordi Cat, Sang Wook Yi and Pooja Uchil for contributions to this paper and to its production, and the LSE Models in Physics & Economics Research Project for supporting the research for it.

the one hand, that represent and phenomena, on the other, that get represented, more or less accurately. Rather, as Margaret Morrison (1997) put it in formulating the background to our project, models mediate between theory and the world. The theories I will discuss here are the highly abstract 'fundamental' theories of contemporary physics. I want to defend Morrison's view of models not as constituting these theories but as mediating between them and the world.

Of course there are lots of different kinds of models serving lots of different purposes, from Hesse's and Lakatos' heuristics for theory change to Morrison's own models as contextual tools for explanation and prediction. In this discussion I shall focus on two of these. The first are models that we construct with the aid of theory to represent real arrangements and affairs that take place in the world – or could do so under the right circumstances. I shall call these *representative models*. This is a departure from the terminology I have used before. In *How the Laws of Physics Lie*, (1983) I called these models *phenomenological* to stress the distance between fundamental theory and theory-motivated models that are accurate to the phenomena. But *How the Laws of Physics Lie* supposed, as does the semantic view, that the theory itself in its abstract formulation supplies us with models to represent the world. They just do not represent it all that accurately.

Here I want to argue for a different kind of separation: these theories in physics do not generally represent what happens in the world; only models represent in this way, and the models that do so are not already part of any theory. It is because I want to stress this conclusion that I have changed the label for these models. Following the arguments about capacities initiated in chapter 10 of *Nature's Capacities and their Measurement* (1989) and further developed in *The Dappled World*, (forthcoming) I want to argue that the fundamental principles of theories in physics do not represent what happens; rather, the theory gives purely abstract relations between abstract concepts: it tells us the 'capacities' or 'tendencies' of systems that fall under these concepts. No specific behaviour is fixed until those systems are located in very specific kinds of situations. When we want to represent what happens in these situations we will need to go beyond theory and build a model, a *representative* model.

For a large number of our contemporary theories, such as quantum mechanics, quantum electrodynamics, classical mechanics and classical electromagnetic theory, when we wish to build these representative models in a systematic or principled way, we shall need to use a second kind of model. For all of these theories use abstract concepts: concepts that need fitting out in more concrete form. The models that do this are laid out within the theory itself, in its bridge principles. The received view called these *interpretative* models and I shall retain the name even though it is not an entirely accurate description of the function I think they serve. The second kind of model I focus on then will be the interpretative model.

I begin from the assumption that it is the job of any good science to tell us how to predict what we can of the world as it comes and how to make the world, where we can, predictable in ways we want it to be. The first job of models I shall focus on is that of representing, representing what reliably happens and in what circumstances; and the first job of this chapter will be to distinguish theory from models of this kind. To get models that are true to what happens we must go beyond theory. This is an old thesis of mine. If we want to get things right we shall have to improve on what theories tell us, each time, at the point of application. This is true, so far as I can see, in even the most prized applications that we take to speak most strongly in a theory's favour. This should not surprise us. Physics is hard. Putting it to use – even at the most abstract level of description – is a great creative achievement.

I used to argue this point by explaining how the laws of physics lie. At that time I took for granted the standard account that supposes that what a theory can do stretches exactly as far as its deductive consequences – what I here call the 'vending machine view' of theory. Since then I have spent a lot of time looking at how theories in physics, particularly quantum physics, provide help in making the world predictable, and especially at devices such as lasers and SQUIDs (Superconducting Quantum Interference Devices) whose construction and operation are heavily influenced by quantum mechanics. I have been impressed at the ways we can put together what we know from quantum theory with much else we know to draw conclusions that are no part of the theory in the deductive sense. The knowledge expressed in physics' fundamental principles provides a very powerful tool for building models of phenomena that we have never yet understood and for predicting aspects of their behaviour that we have never foreseen. But the models require a cooperative effort. As Marcel Boumans' chapter in this volume claims for model building in economics, knowledge must be collected from where we can find it, well outside the boundaries of what any single theory says, no matter how fundamental – and universal – we take that theory to be. And not just knowledge but guesses too. When we look at how fundamental theories get applied, it is clear that the *Ansatz* plays a central role.[1]

The Ginsburg-Landau model of superconductivity, which is described by Towfic Shomar (1998) in his PhD dissertation, gives a nice example of both the importance of co-operation and of the role of the *Ansatz*. As Shomar stresses, this model, built upwards from the phenomena themselves, is still for a great many purposes both more useful for prediction and wider in scope than the fundamental model of Bardeen, Cooper and Schrieffer (BCS). The situation is well reflected in the description in a standard text by Orlando and Delin (1990) of the development followed up to the point at which the Ginsburg-Landau model is introduced. As Orlando and Delin report, their text started with electrodynamics as the 'guiding principle' for the study of superconductivity; this led to the first and second London equations.[2] The guiding discipline at the second stage was quantum mechanics, resulting in a 'macroscopic model' in which the superconducting state is described by a quantum wave function. This led to an equation for the supercurrent uniting quantum mechanical concepts with the electrodynamic ones underlying the London equations. The supercurrent equation described flux quantisation and properties of type-II superconductors and led to a description of the Josephson effect. The third stage introduced thermodynamics to get equilibrium properties. Finally, with the introduction of the Ginsburg-Landau model, Orlando and Delin were able to add considerations depending on 'the bi-directional coupling between thermodynamics and electrodynamics in a superconducting system (1990, 508).

This kind of creative and cooperative treatment is not unusual in physics, and the possibility of producing models that go beyond the principles of any of the theories involved in their construction is part of the reason that modern physics is so powerful. So, under the influence of examples like the Ginsburg-Landau model I would no longer make my earlier points by urging that the laws of physics lie, as they inevitably will do when they must speak on their own. Rather, I would put the issue more positively by pointing out how powerful is their voice when put to work in chorus with others.

---

[1] It should be noted that this is not just a matter of the distinction between the logic of discovery and the logic of justification, for my claim is not just about where many of our most useful representative models come from but also about their finished form: these models are not models of any of the theories that contribute to their construction.

[2] See M. Suárez, in this volume, for a discussion of these.

The first point I want to urge in this essay then is one about how far the knowledge contained in the fundamental theories of physics can go towards producing accurate predictive models when they are set to work cooperatively with what else we know or are willing for the occasion to guess. But I shall not go into it in great detail since it is aptly developed and defended throughout this volume. My principal thesis is less optimistic. For I shall also argue that the way our fundamental theories get applied – even when they cooperate – puts serious limits on what we can expect them to do. My chief example will be of the BCS theory of superconductivity, which has been one of the central examples in the LSE modelling project. Readers interested in a short exposition of the core of the argument about the limits of theory in physics can move directly to section 9.6.

9.2 THE 'RECEIVED VIEW'

On the received view good theory already contains all the resources necessary for the representation of the happenings in its prescribed domain. I take this to be a doctrine of the 'received' syntactic view of theories, which takes a theory to be a set of axioms plus their deductive consequences. It is also a doctrine of many standard versions of the semantic view, which takes a theory to be a collection of models.

Consider first the syntactic view. C. G. Hempel and others of his generation taught that the axioms of the theory consist of internal principles, which show the relations among the theoretical concepts, and bridge principles. But Hempel assigned a different role to bridge principles than I do. For Hempel, bridge principles do not provide a way to make abstract terms concrete but rather a way to interpret the terms of theory, whose meanings are constrained but not fixed by the internal principles. Bridge principles, according to Hempel, interpret our theoretical concepts in terms of concepts of which we have an antecedent grasp. On the received view, if we want to see how specific kinds of systems in specific circumstances will behave, we should look to the theorems of the theory, theorems of the form, 'If the situation (e.g., boundary or initial conditions) is X, Y happens.'

Imagine for example that we are interested in a simple well-known case – the motion of a small moving body subject to the gravitational attraction of a larger one. The theorems of classical mechanics will provide us with a description of how this body moves. We may not be able to tell which theorem we want, though, for the properties described in the theory do not match the vocabulary with which our system is presented. That's what the bridge principles are for. 'If the force on a moving body of mass m is GmM/r^2, then the body will move in an elliptical orbit $1/r = 1 + e\cos\varnothing$ (where e is the eccentricity).' To establish the relevance of this theorem to our initial problem we need a bridge principle that tells us that the gravitational force between a large body of mass M and a small mass m is of size GmM/r^2. Otherwise the theory cannot predict an elliptical orbit for a planet.

The bridge principles are crucial; without them the theory cannot be put to use. We may know for example from Schroedinger's equation that a quantum system with an initial state $\varnothing_i = \exp(-i\omega x t)$ and Hamiltonian $H = p^2/2m + V(r) + (e/mc)A(r,t)$ will evolve into the state $\varnothing_f = \int a(\omega,t)\varnothing_i \exp(-i\omega x t)d\omega$, where ω is the frequency of radiation. But this is of no practical consequence till we know that $\varnothing_i$ is one of the excited stationary states for the electrons of an atom, H is the Hamiltonian representing the interaction with the electromagnetic field and from $\varnothing_f$ we can predict an exponentially decaying probability for the atom to remain in its excited state. The usefulness of theory is not the issue here, however. The point is that on the 'received view' the theorems of the theory are supposed to describe what happens in all those situations where the theory matters, whether or not we have the bridge principles to make the predictions about what happens intelligible to us. On this view the only problem we face in applying the theory to a case we are concerned with is to figure out which theoretical description suits the starting conditions of the case.

Essentially the same is true for the conventional version of the semantic view as well. The theory is a set of models. To apply the theory to a given case we have to look through the models to find one where the initial conditions of the model match the initial conditions of the case. Again it helps to have the analogue of bridge principles. When we find a model with an atom in state $\varnothing_i$ subject to Hamiltonian H we may be at a loss to determine if this model fits our excited atom. But if the atoms in the models have additional properties – e.g., they are in states labelled 'ground state', 'first excited state', 'second excited state', and so on – and if the models of the theory are constrained so that no atom has the property labelled 'first excited state' unless it also has a quantum state $\varnothing_i$, then the task of finding a model that matches our atom will be far easier. I stress this matter of bridge principles because I want to make clear that when I urge that the good theory need not contain the resources necessary to represent all the causes of the effects in its prescribed domain,

I am not just pointing out that the representations may not be in a form that is of real use to us unless further information is supplied. Rather I want to deny that the kinds of highly successful theories that we most admire represent what happens, in however usable or unusable a form.

I subscribe neither to the 'received' syntactic of theories nor to this version of the semantic account. For both are cases of the 'vending machine' view. The theory is a vending machine: you feed it input in certain prescribed forms for the desired output; it gurgitates for a while; then it drops out the sought-for representation, plonk, on the tray, fully formed, as Athena from the brain of Zeus. This image of the relation of theory to the models we use to represent the world is hard to fit with what we know of how science works. Producing a model of a new phenomenon such as superconductivity is an incredibly difficult and creative activity. It is how Nobel prizes are won. On the vending machine view you can of course always create a new theory, but there are only two places for any kind of human activity in deploying existing theory to produce representations of what happens, let alone finding a place for genuine creativity. The first: eyeballing the phenomenon, measuring it up, trying to see what can be abstracted from it that has the right form and combination that the vending machine can take as input; secondly – since we cannot actually build the machine that just outputs what the theory should – we do either tedious deduction or clever approximation to get a facsimile of the output the vending machine would produce.

This is not, I think, an unfair caricature of the traditional syntactic/semantic view of theory. For the whole point of the tradition that generates these two views is the elimination of creativity – or whim – in the use of theory to treat the world (Daston and Galison 1992). That was part of the concept of objectivity and warrant that this tradition embraced. On this view of objectivity you get some very good evidence for your theory – a red shift or a Balmer series or a shift in the trend line for the business cycle – and then that evidence can go a very long way for you: it can carry all the way over to some new phenomenon that the theory is supposed to 'predict'.

In *The Scientific Image* Bas van Fraassen (1980) asks: why are we justified in going beyond belief in the empirical content of theory to belief in the theory itself? It is interesting to note that van Fraassen does not restrict belief to the empirical claims we have established by observation or experiment but rather allows belief in the total empirical content. I take it the reason is that he wants to have all the benefits of scientific realism without whatever the cost is supposed to be of a realist commitment.

And for the realist there is a function for belief in theory beyond belief in evidence. For it is the acceptability of the theory that warrants belief in the new phenomena that theory predicts. The question of transfer of warrant from the evidence to the predictions is a short one, since it collapses to the question of transfer of warrant from the evidence to the theory. The collapse is justified because theory is a vending machine: for a given input the predictions are set when the machine is built.

I think that on any reasonable philosophical account of theories of anything like the kind we have reason to believe work in this world, there can be no such simple transfer of warrant. We are in need of a much more textured, and I am afraid much more laborious, account of when and to what degree we might bet on those claims that on the vending machine view are counted as 'the empirical content' or the deductive consequences of theory. The vending machine view is not true to the kind of effort that we know that it takes in physics to get from theories to models that predict what reliably happens; and the hopes that it backs up for a shortcut to warranting a hypothesised model for a given case – just confirm theory and the models will be warranted automatically – is wildly fanciful.[3] For years we insisted theories have the form of vending machines because we wished for a way to mechanise the warrant for our predictions. But that is an argument in favour of reconstructing theories as vending machines only if we have independent reason to believe that this kind of mechanisation is possible. And I have not seen even a good start at showing this.

9.3 CUSTOMISING THE MODELS THAT THEORY PROVIDES

The first step beyond the vending machine view are various accounts that take the deductive consequences of a single theory as the ideal for building representative models but allow for some improvements,[4] usually improvements that *customise* the general model produced by the theory to the special needs of the case at hand. These accounts recognise that a theory may be as good as we have got and yet still need, almost always, to

---

[3] In order to treat warrant more adequately in the face of these kinds of observations, Joel Smith suggests that conclusions carry their warrant with them so that we can survey it at the point of application to make the best informed judgements possible about the chances that the conclusion will obtain in the new circumstances where we envisage applying it. See Mitchell (1997) for a discussion.

[4] It is important to keep in mind that what is suggested are changes to the original models that often are inconsistent with the principles of the theory.

be corrected if it is to provide accurate representations of behaviour in its domain. They nevertheless presuppose that good scientific theories already contain representations of the regular behaviours of systems in their domain even though the predicted behaviours will not for the most part be the behaviours that occur. This is close to my position in *How the Laws of Physics Lie* (1983) and it is the position that Ronald Giere (1988) maintains. A look at Giere's account suggests, however, that his views about the way models relate to real systems differ from mine. Correlatively, Giere is far more optimistic than I am about the limits of theory.

On Giere's (1988) account theories have two parts: models and hypotheses about the models' similarity to real systems. The laws and main equations of a theory are encoded in the definition of the models themselves. A Newtonian particle system is a system that obeys Newton's laws (Giere 1984, 80). Thus, as Giere has put it recently, Newton's principles for mechanics are to be thought of as rules for the construction of models to represent mechanical systems, from comets to pendulums (Giere 1995, 134). In that sense, Giere concludes, as I did in *How the Laws of Physics Lie*, that for models, truth comes cheap (Giere 1995, 131). Newton's laws are necessarily true of the objects in the set of Newtonian models. In his view it is instead the truth of the hypotheses of the theory that should concern us since these indicate the degree of similarity or approximation of models to real systems.

How similar models are to a case will depend on how specifically we characterise the model, and, thereby, on how we fill in or interpret the abstract quantities appearing in the laws that define the set of models. In Newton's second law, $F = ma$, we can specify the value of the force as, for instance, proportional to the displacement of a mechanical body, $F = -kx$, so that the equation of motion is $md^2x/dt^2 = -kx$. The solutions to this more specific equation of motion describe the behaviour of systems we call 'harmonic oscillators'. Two examples are a mass on a spring and a simple pendulum. The first kind of system obeys Hooke's law and the second obeys Galileo's law that the pendulum's period is proportional to the square root of its length and independent of its mass. To derive Galileo's law for the pendulum we need to introduce further specifications into the model beyond the assumption that the force on it is proportional to its displacement. In this case we insist that the force must be of the form $F = -(mg/l)x$, where l is the length of the pendulum and -mg is a uniform gravitational force acting on the pendulum. The model is an *idealisation* of real pendulum in several respects: the two-dimensional representation of oscillations on the x–y plane is reduced to a one-dimensional representation on the x-axis; the swinging bob is a mass-point particle whose motion is restricted to a small angle; and it is only influenced by a gravitational force. The model of the ideal pendulum now obeys Newton's equation of motion, $F = ma$, in the desired more concrete form, $md^2x/dt^2 = -(mg/l)x$.

Giere stresses the importance of this kind of model in treating the world. On his view about theoretical hypotheses, the models of a theory, and thus indirectly the laws that define them, can be approximately true only of systems closely resembling them. The model of the ideal pendulum described so far is still quite far removed from a good many real pendulums. Gravity is generally not the only cause acting on them, nor can they swing only within small angles. Giere thinks we can get a more accurate model of real pendulums by 'deidealising' the model – or, as I say, 'customising' it – to the specific pendulum we wish to treat. On Giere's account this is done primarily by introducing extra elements into the model and, correspondingly, correcting the equation of motion by introducing additional force terms to represent these features. The model and its defining laws will become increasingly complex. Thus, we can allow greater angles in the swing, but then the model does not obey the equation of motion for harmonic oscillators. We can also introduce terms representing frictional forces, such as those operating on pendulums in clocks. But the more realistic the model, the more complex will its characterisation be, and so will the corresponding form of Newton's law of motion and the form of its solutions.[5] In this sense, for Giere, getting more accurate representative models of a situation requires that the laws originally put forward in that situation be corrected. Nevertheless, the required corrections preserve the form of Newton's laws intact even though they modify the specific forms of the force-term in the equations of motion. Thus the more accurate models still satisfy Newton's laws.

Within this view of corrigible models of the theory, controversies arise about the power of the theory to provide corrections. Giere's is a story of ideal models and their gradual 'deidealisation', or customisation, to meet a given case by a series of corrections legitimated by the theory itself given the circumstances of the case. Both Imré Lakatos and Ernan McMullin see the fact that the model improves when we make the changes that a theory itself suggests as a central argument in favour of

---

[5] Although, according to Giere, this complexity would gradually become impossible to accommodate within the limits of our human capacities. See Giere (1988, 77). See also Morrison (this volume) for an account which shows the independence of functions of such 'corrected' models in the context of using the pendulum for measurement.

the theory. For McMullin it argues for the truth of the theory and for Lakatos, for the progressiveness of the research programme in which the theory is embedded. By contrast, I have argued that the corrections needed to turn the models that are provided by theory into models that can fairly accurately represent phenomena in the physical world are seldom, if ever, consistent with theory, let alone suggested by it. One of the central points of *How the Laws of Physics Lie* was that corrections are almost always *ad verum* – they take you away from theory and closer to the truth. And they are generally not made by adding fuller and more literally true descriptions (cast in the language of the theory) of the circumstances of the case at hand.[6] When they do, we have what Morrison describes under the heading 'theoretical models': 'A great many corrections are necessary but the important point is that as a model whose structure is derived from theory (a theoretical model) it is capable of absorbing corrections that provide a highly realistic description of the actual apparatus' (Morrison, this volume, p. 48). When it comes to models that represent situations realistically and also provide accurate predictions, very few in my experience are theoretical models, in Morrison's sense. Indeed, in general the corrections will often yield models that are inconsistent with the theory. So I am at odds with Giere about the power of theory itself to tell us how to correct the models. Mauricio Suárez's chapter in this volume explores the implications of this.

Not surprisingly, we also differ on my more pessimistic claims about the extent to which theory can stretch, which are the central themes of this essay. Giere seems to think that once customising corrections are allowed (bracketing for the moment the question of whether these corrections are or are not consistent with the theory itself) our good theories will be able to stretch very far. Probably all compact masses are Newtonian systems; all pendulums and springs have Newtonian models; so too do all planets and comets and falling bodies. For Giere insists that universality is a common feature of scientific laws. In his account the issue of universality – that is, how widely do laws apply? – is treated by the second component of the scientific theory as he reconstructs it: the hypothesis about what systems in the world have behaviours approximated by the models of the

---

[6] This is particularly apparent in cases where the original model does not provide replicas of the major causal factors in the first place but rather represents their total effect all at once. We may for instance represent the electromagnetic field as a collection of harmonic oscillators or Josephson junctions and SQUIDs as systems of connected pendulums. These are very common kinds of models in both physics and economics, where the predictive successes of theories often depend much more centrally on finding a nice and manageable mathematical form with certain desirable features than on finding a model that replicates the conditions of the problem.

theory. This insistence that the specification of the intended domain is a central part of the formal theory originates with the German Structuralist School of Wolfgang Stegmueller, following the work of Patrick Suppes. Giere states the following: 'The feature most commonly associated with laws is universality. So interpreted, Newton's law is indeed a hypothesis about everything in the universe. Our interpretation is somewhat different. We have interpreted the LAW of Universal Gravitation as being part of the definition of a Newtonian Particle System – a theoretical MODEL' (Giere 1984, 86; emphasis in original). And he adds: 'A SCIENTIFIC THEORY is a GENERAL THEORETICAL HYPOTHESIS asserting that some designated class of natural systems are (approximately) systems of a specified type. The type, of course, is specified by some explicitly defined THEORETICAL MODEL' (ibid., 84; emphasis in original). So he concludes: 'The generalization, "all real pendulums satisfy Galileo's law", is surely false. But the hypothesis that most real pendulums approximately satisfy the law might be true. This is really all that science requires' (ibid., 87).

To the contrary, this hypothesis seems to me extremely unlikely. Only pendulums in really nice environments satisfy the law, even approximately if the approximation is to be very precise. I want to consider real systems as we encounter them. And as we encounter them they are usually subject to all sorts of perturbing influences that do not appear in any way whatsoever to fit the models for perturbing influences available in Newtonian theory. To bring a real compact mass, say a pendulum, into the domain of Newtonian theory, we must be able to provide for it a Newtonian model that will justify assigning to it some particular force function. And the models available for doing so do not, to all appearances, bear a close resemblance to the situations of a great many highly perturbed pendulums. We observe a large number of such systems for which we have never found sufficient similarities to any theoretical model. Giere does not explicitly go into this question. He certainly does not commit himself to the existence of hidden similarities that we might hope to find behind the apparent dissimilarities. So he does not really address the issue of the applicability of, say, classical mechanics, to every observed real system, but sticks rather to questions where approximate truth will answer. But when we do turn to the question of universal validity of the theory, Giere does not draw what appears to me an immediate and important lesson: if the theory[7] can be (even approximately) true only of real systems that resemble its models, the theory will be severely limited in its domain.

[7] On Giere's own terms this means its collection of theoretical hypotheses.

To bring together clearly the main reasons why I am not optimistic about the universality of mechanics, or any other theory we have in physics, or almost have, or are some way along the road to having or could expect to have on the basis of our experiences so far, I shall go step-by-step through what I think is wrong with the customisation story. On this account we begin with a real physical system, say the pendulum in the Museum of Science and Industry that illustrates the rotation of the earth by knocking over one-by-one a circle of pegs centred on the pendulum's axis. And we begin with an idealised model in which the pendulum obeys Galileo's law. Supposing that this model does not give an accurate enough account of the motion of the Museum's pendulum for our purposes, we undertake to customise it. If the corrections required are *ad hoc* or are at odds with the theory – as I have observed to be the usual case in naturally occurring situations like this – a successful treatment, no matter how accurate and precise its predictions, will not speak for the universality of the theory. So we need not consider these kinds of corrections here. Imagine then that we are able to deidealise in the way Giere suggests, until we succeed in producing a model with the kind of accuracy we require. What will we have ended up with? On the assumption that Newton's theory is correct, we will, in the language of *The Dappled World* have managed to produce a blueprint for a nomological machine, a machine that will, when repeatedly set running, generate trajectories satisfying to a high degree of approximation not Galileo's law, but some more complex law; and since, as we are assuming for the sake of argument, all the corrections are dictated by Newtonian theory given the circumstances surrounding the Museum's pendulum, we will *ipso facto* have a blueprint for a machine that generates trajectories satisfying the general Newtonian law, $F = ma$. Indeed, the original ideal model was already a blueprint for a machine generating the $F = ma$ law.

Once we have conceived the idealised and the deidealised models as nomological machines, we can see immediately what is missing from the customisation account. In a nomological machine we need a number of components with fixed capacities arranged appropriately to give rise to regular behaviour. The interpretative models of the theory give the components and their arrangement: the mass point bob, a constraint that keeps it swinging through a small angle along a single axis, the massive earth to exert a gravitational pull plus whatever additional factors must be added (or subtracted) to customise the model. Crucially, nothing must significantly affect the outcome we are trying to derive except factors whose overall contribution can be modelled by the theory. This means both that the factors can be represented by the theory and that they are factors for which the theory provides rules for what the net effect will be when they function in the way they do in the system conjointly with the factors already modelled. This is why I say, in talking of the application of a model to a real situation, that *resemblance is a two-way street*. The situation must resemble the model in that the combination of factors that appear in the model must represent features in the real situation (allowing that we may consider a variety of different views about what it is to 'represent appropriately'). But it must also be true that nothing too relevant occurs in the situation that cannot be put into the model. What is missing from the account so far, then, is something that we know matters enormously to the functioning of real machines that are very finely designed and tuned to yield very precise outputs – the shielding. This has to do with the second aspect of resemblance: the situation must not have extraneous factors that we have not got in the model. Generally, for naturally occurring systems, when a high degree of precision is to be hoped for, this second kind of resemblance is seldom achieved. For the theories we know, their descriptive capacities give out.

Let us lay aside for now any worries about whether corrections need to be made that are unmotivated by the theory or inconsistent with it, in order to focus on the question of how far the theory can stretch. In exact science we aim for theories where the consequences for a system's behaviour can be deduced, given the way we model it. But so far the kinds of concepts we have devised that allow this kind of deductibility are not ones that easily cover the kinds of causes we find naturally at work bringing about the behaviours we are interested in managing. That is why the laws of our exact sciences must all be understood with implicit *ceteris paribus* clauses in front. As I shall argue in the rest of this paper, our best and most powerful deductive sciences seem to support only a very limited kind of closure: so long as the only relevant factors at work are ones that can be appropriately modelled by the theory, the theory can produce exact and precise predictions. This is in itself an amazing and powerful achievement, for it allows us to engineer results that we can depend on. But it is a long distance from hope that all situations lend themselves to exact and precise prediction.

9.4 BRIDGE PRINCIPLES AND INTERPRETATIVE MODELS

I have made a point of mentioning bridge principles, which get little press nowadays, because they are of central importance both practically and philosophically. Practically, bridge principles are a first step in what I emphasise as a sine qua non of good theory – the use of theory to effect changes in the world. They also indicate the limitations we face in using any particular theory, for the bridge principles provide natural boundaries on the domain the theory can command. So they matter crucially to philosophical arguments about the relations between the disciplines and the universal applicability of our favoured 'fundamental' theories. These are arguments I turn to in later sections.

I take the general lack of philosophical investigation of what bridge principles are and how they function in physics to be a reflection of two related attitudes that are common among philosophers of physics. The first is fascination with theory *per se*, with the details of formulation and exact structure of a heavily reconstructed abstract, primarily mathematical, object: theory. I say 'heavily reconstructed' because 'theory' in this sense is far removed from the techniques, assumptions, and various understandings that allow what is at most a shared core of equations, concepts, and stories to be used by different physicists and different engineers in different ways to produce models that are of use in some way or another in manipulating the world. The second attitude is one about the world itself, an attitude that we could call Platonist or Pauline: 'For now we see through a glass darkly, but then face to face. Now I know in part; but then shall I know even as also I am known.'[8] It would be wrong to say, as a first easy description might have it, that these philosophers are not interested in what the world is like. Rather they are interested in a world that is not our world, a world of appearances, but rather a purer, more orderly world, a world which is thought to be represented 'directly' by the theory's fundamental equations.

But that is not the world that contemporary physics gives us reasons to believe in when physics is put to work to manage what happens in that world. One reason for this has to do with bridge principles and the way they attach physics concepts to the world. Many of our most important descriptive terms in successful physics theories do not apply to the world 'directly'; rather, they function like abstract terms. I mean 'abstract' in a very particular sense here: these terms are applied to the world only when certain specific kinds of descriptions using some particular set of more concrete concepts also apply.

The quantum Hamiltonian, the classical force function and the electromagnetic field vectors are all abstract. Whenever we apply them there is

---

[8] King James Bible, I Corinthians 13:12.

always some more concrete description that applies and that constitutes what the abstract concept amounts to in the given case. Being a stationary charge of magnitude q, located at a distance r from a second charge q′ is *what it is* to be subject to a Coulomb force $qq'/4\pi\epsilon_0 r^2$. Mass, charge, acceleration, distance, and the quantum state are not abstract concepts. When a particle accelerates at $32\ \mathrm{ft/sec^2}$ there is nothing further that constitutes the acceleration. Similarly, although it may be complicated to figure out what the quantum state of a given system is, there is nothing more about the system that is *what it is* for that system to have that state.

The bridge principle that assigns the gravitational force to a model is another familiar example. A compact mass, m, can be described as *subject to a force of GMm/r^2* when it is *located a distance r from a second mass, M*. The first is an abstract description; the second is the more concrete description which must also apply whenever the abstract description is true.[9] An example of a more complex situation is the Hall effect. Take the case of a conductor carrying a uniform current density **J** of electric charge nq with velocity v parallel to, say, the y-axis, $J_y = nqv_y$. From Ohm's law,

$$\mathbf{J} = \sigma\mathbf{E}$$

(where the constant 'σ' represents the conductivity of the material), we know that there is an electric field parallel to the current (as in the previous example). Now when the conductor is placed in a magnetic field **B** parallel to the z-axis we have a situation to which we can attach the abstract description, 'an electric field exists across the conductor in the direction of the x-axis'. The magnetic field exerts a force $F = qv_y B_z$ on the moving charge carriers in the current, tending to displace them in the x-direction. But this gives rise to a non-uniform charge distribution, on which piggybacks, in turn, a new electric field in the x-direction. Eventually an equilibrium is reached, and the electric field exerts on the charges a force $F = qE_x$ that balances the magnetic force:

$$qE_x + qv_y B_z = 0.$$

Substituting the expression of **J** above, we can express the new electric field as

$$E_x = -(J_y/nq)B_z$$
$$= -R_H J_y B_z,$$

---

[9] With careful arrangements of other bodies, of course, this very same abstract description (the same size force) can be true of m without this particular concrete description obtaining. The point is that some description or other formed in the right way from the appropriate set of concrete descriptions must be true as well if a description assigning force is to obtain.

where 'R_H' is called the 'Hall coefficient'.[10] The expressions of the electric and magnetic fields and forces provide the abstract description of the phenomenon. These descriptions piggyback on the more concrete description in terms of a material of conductivity σ carrying a density of electric charge nq with velocity v_y in the presence of a magnetic field B_z. In the case of the magnetic field, the term 'B_z' piggybacks, in turn, on some more concrete description – involving its 'source' and material circumstances – that must be available and yet is typically omitted in the descriptions of this effect.

In these cases the assignment of forces and fields is determined by specific interpretative models involving specific mass or charge distributions and their circumstances. The abstract terms 'force' and 'field' require specific kinds of concrete models to apply wherever they do. This is similar to Gottlieb Ephraim Lessing's account of the abstract–concrete relation between morals and fables, about which I will say more in the next section (Cartwright 1993, 55–82). In the case of 'force', for instance, the more concrete descriptions are the ones that use traditional mechanical concepts, such as 'position', 'extension', 'mass' and 'motion'. 'Force', then, is abstract relative to mechanics; and being abstract, it can only exist in particular mechanical models. This is why interpretative models and bridge principles are so important in physics. Abstract terms are fitted out by the concrete descriptions provided by interpretative models.[11] And it is the bridge principles that assign concrete interpretative models to the abstract concepts of physics theories.

I shall argue in the next sections how this feature of central concepts in physics delimits what most successful theories in physics can do. Classical mechanics, for instance, has enormous empirical success; but not a classical mechanics reconstructed without its bridge principles. When I say 'successful' here, I am talking from an empiricist point of view. Whatever else we require in addition, a successful theory in physics must at least account in a suitable way for an appropriate range of phenomena. It must make precise predictions that are borne out in experiment and in the natural world. So when we think of reconstructing some

---

[10] For more examples see Ohanian (1988) and the classic Jackson (1975).

[11] As Jordi Cat has explained, a similar abstract-concrete relation involving electromagnetic descriptions and mechanical models was held by James Clerk Maxwell. According to Maxwell, with the illustration (and explanation) provided by mechanical models of electromagnetic concepts and laws the gulf is bridged between the abstract and the concrete; electromagnetic forces and energy exist and can be understood clearly only in concrete mechanical models. See Cat (1995a and forthcoming).

object we call 'theory'[12] and we ask questions about, e.g., whether some kinds of descriptions are abstract relative to other kinds of descriptions in this theory, the answers must be constrained by considerations about what makes for the empirical success of the theory. Once we call this reconstructed object 'quantum theory' or 'the BCS theory of supercon-ductivity', it will be reasonably assumed that we can attribute to this object all the empirical successes usually acknowledged for these theo-ries. What this requirement amounts to in different cases will get argued out in detail on a case-by-case basis. The point here is about bridge prin-ciples. In the successful uses of classical mechanics, force functions are not applied to situations that satisfy arbitrary descriptions but only to those situations that can be fitted to one of the standard interpretative models by bridge principles of the theory; so too for all of physics' abstract terms that I have seen in producing the predictions that give us confidence in the theory.

Recall the analogue of this issue for the semantic view, again with the most simple-minded version of classical mechanics to illustrate. Does the set of models that constitutes the theory look much as Ronald Giere and I picture it: pendulums, springs, 'planetary' systems, and the like, situ-ated in specific circumstances, each of which *also* has a force acting on it appropriate to the circumstances; that is, do the objects of every model of the theory have properties marked out in the 'interpretative' models and a value for the applied force as well? Or are there models where objects have *simply* masses, forces and accelerations ascribed to them with no other properties in addition? I think there is a tendency to assume that scientific realism demands the second. But that is a mistake, at least in so far as scientific realism asserts that the claims of the theory are true and that its terms refer, in whatever sense we take 'true' and 'refer' for other terms and claims. The term 'geodesic' is abstract, as, I claim, are central terms of theoretical physics, like 'force' or 'electric field vector': it never applies unless some more concrete description applies in some particular geometry, e.g. 'straight line' on a Euclidean plane or 'great circle' on a sphere. But this does not mean that we cannot be realists about geodesics. Or consider questions of explanatory (or pre-dictive or causal) power. The set of models that I focus on, where forces always piggyback on one or another of a particular kind of more con-crete description, will predict accelerations in accordance with the prin-ciple $F = ma$. Even though *force* is an abstract description in this

---

[12] Or, for structuralists, some node in a theory net.

reconstruction of the theory, there is nothing across the models that all objects with identical accelerations have in common except that they are subject to the same force. Although abstraction and supervenience are different characteristics, the issues about scientific realism are similar in the two cases. The exact answer will clearly depend on the exact formulation of the question, but the general point is that for the most usual ways of cashing out 'scientific realism', putting the bridge principles into the theory when we reconstruct it does not conflict with realism. And it does produce for us theories that are warranted by their empirical successes.

9.5 PHILOSOPHICAL ASIDE: MORE ON THE ABSTRACT–CONCRETE RELATION

What I have said so far about abstraction should suffice for understanding my argument about how the use of interpretative models provides natural boundaries for the application of representative models constructed with the aid of a given theory. But it is probably helpful for those involved in other philosophical debates to spell out a few more details. The account of abstraction that I borrow from the Enlightenment playwright Lessing to describe how contemporary physics theories work provides us with two necessary conditions.[13] First, a concept that is abstract relative to another more concrete set of descriptions never applies unless one of the more concrete descriptions also applies. Secondly, satisfying the associated concrete description that applies on a particular occasion is what satisfying the abstract description consists in on that occasion. Writing this paper is what my working right now consists in; being located a distance r from another charge q' is what it consists in for a particle of charge q to be subject to the force $qq'/4\pi\epsilon_0 r^2$ in the usual cases when that force function applies. To say that working *consists in* a specific activity described with the relevant set of more concrete concepts on a given occasion implies at least that no further description using those concepts is required for it to be true that 'working' applies on that occasion, though surely the notion is richer than this.

The abstract–concrete relation has important differences from other nearby notions that philosophers talk about. Consider supervenience. Roughly, to say that one set of concepts supervenes on another is to say that any two situations that have the same description from the second

---

[13] See Cartwright (1993, 55–82).

set will also have the same description using the first set: the basic concepts 'fix' the values of those that supervene on them. This is not the case with the abstract-concrete distinction, as we can see from the example of *work*. Washing dishes may be work when one is paid for it or must do it as a household chore, but may not be when it is viewed as part of the fun of staying home and cooking on the weekend. Whether *work* is an appropriate description depends not only on the more concrete description that might constitute the working but also on other facts involving other equally abstract concepts like 'leisure', 'labour' and 'exploitation'.[14] So the notion of supervenience is in this sense stronger than the abstract–concrete relation described by Lessing. The determinable–determinate relation is also stronger in just the same way[15]. For example, the determinable 'colour' is fixed to hold as soon as any of the determinates that fall under it are fixed.[16]

In general, the abstract concepts that we use in physics do not supervene on the more concrete. Consider the simple case of the bridge principle expressed in Newton's law of gravitational attraction. The force function $F = GmM/r^2$ is not fixed simply by fixing the masses and separation between two compact objects; the gravitational constant, G, is

---

[14] I have noticed that there is a tendency among reductionists of various kinds to try to collapse the distinction between abstraction and supervenience by arguing that in each case the entire abstract vocabulary will supervene on some more concrete description if only we expand the concrete descriptions to cover a broad enough piece of the surrounding circumstances ('global supervenience'). This is of course a metaphysical doctrine of just the kind I am disputing in this paper.

[15] The determinable–determinate relation is also stronger since it requires that the designated determinate descriptions be mutually exclusive.

[16] This notion of supervenience — as well as Lessing's concept of abstraction — is also stronger than the notion of Jordi Cat has shown to be at work in Maxwell's discussions of concrete mechanical models *vis-à-vis* the more abstract descriptions in the energy-based Lagrangian formalism and its associated general principles of energy and work. The generality of the Lagrangian formalism, like that of a more 'abstract' phenomenological representation of electromagnetic phenomena in terms of electric and magnetic forces and energy (for Green, Maxwell and Heaviside), or that of the more 'general' representation of macroscopic media in continuum mechanics (for Stokes), lies in the elliptic assumption of the existence of an unknown underlying molecular structure represented by a mechanical model with hidden mechanisms – in which energy is manifested in motion (kinetic) or stored in elasticity (potential) – together with the realization that an infinite number of more concrete mechanical descriptions can realise (or merely illustrate) the more abstract one. The more abstract one, however, needs independently to satisfy the mechanical principles that regulate and characterize the concepts of energy and force. See Cat (1995a, n.31).

The supervenience relation is also, technically, weaker, for many definitions of supervenience do not formally require the first condition necessary for abstraction: to say that identical descriptions at the base level imply identical conditions at the second level does not imply that no descriptions at the second level apply without some appropriate description from the base concepts, although this is often assumed.

required as well. Similarly, fixing the charges and the separation between two stationary particles does not fix the Coulomb force $(qq'/4\pi\epsilon_0 r^2)$ acting between them; we also need the permitivity ϵ_0.

9.6 REMARKS ON HOW REPRESENTATIVE MODELS REPRESENT

My focus in this paper is on interpretative models. I have little to say about how representative models represent, except to urge a few cautions about thinking of representations too much on the analogy of structural isomorphism. Consider the case of the quantum Hamiltonian, which is the example I will be developing in detail. I think it is important to use some general notion of representation of the kind R. I. G. Hughes remarks on in this volume and not to think of the models linked to Hamiltonians as picturing individually isolatable physical mechanisms, otherwise we will go wrong on a number of fronts. First we can easily confuse my claim that Hamiltonians are abstract descriptions needing the more concrete descriptions provided by interpretative models with a demand that the Hamiltonians be explained by citing some physical mechanisms supposed to give rise to them. The two claims are distinct. Throughout quantum theory we regularly find bridge principles that link Hamiltonians to models that do not describe physical mechanisms in this way. The Bloch Hamiltonian discussed below provides an illustration.

Second, it could dispose one to a mistaken reification of the separate terms which compose the Hamiltonians we use in modelling real systems. Occasionally these Hamiltonians are constructed from terms that represent separately what it might be reasonable to think of as distinct physical mechanisms – for instance, a kinetic energy term plus a term for a Coulomb interaction. But often the break into separable pieces is purely conceptual. Just as with other branches of physics and other mathematical sciences, quantum mechanics makes heavy use of the method of idealisation and deidealisation. The Bloch Hamiltonian for the behaviour of moving electrons in a perfect crystal again provides an illustration. The usual strategy for modelling in condensed matter physics is to divide the problem artificially into two parts: (a) the ideal fictitious perfect crystal in which the potential is purely periodic, for which the Bloch Hamiltonian is often appropriate when we want to study the effects on conduction electrons; and (b) the effects on the properties of a hypothetical perfect crystal of all deviations from perfect periodicity, treated as small perturbations. This kind of artificial breakdown of problems is typical wherever perturbation theory is deployed, but it is not tied to perturbation analyses. As we shall see, the BCS theory relies on earlier treatments by Bardeen and Pines of the screened Coulomb potential that separates the long wavelength and short wavelength terms from the Coulomb interactions because it is useful to think of the effects of the two kinds of terms separately. But this is purely a division of the terms in a mathematical representation and does not match up with a separation of the causes into two distinct mechanisms.

Thirdly, without a broader notion of representation than one based on some simple idea of picturing we should end up faulting some of our most powerful models for being unrealistic. Particularly striking here is the case of second quantisation, from which quantum field theory originates. In this case we model the field as a collection of harmonic oscillators, in order to get Hamiltonians that give the correct structure to the allowed energies. But we are not thus committed to the existence of a set of objects behaving just like springs – though this is not ruled out either, as we can see with the case of the phonon field associated with the crystal lattice described below or the case of the electric dipole oscillator that I describe in 'Where in the World is the Quantum Mechanical Measurement Problem?' (Cartwright 1998).

Last, we make it easy to overlook the fact that when we want to use physics to effect changes in the world we not only need ways to link the abstract descriptions from high theory to the more concrete descriptions of models; we also need ways to link the models to the world. This is a task that begins to fall outside the interests of theorists, to other areas of physics and engineering. Concomitantly it gets little attention by philosophers of science. We tend to try to make do with a loose notion of resemblance. I shall do this too. Models, I say, *resemble* the situations they represent. This at least underlines the fact that it is not enough to count a description as a correct representation of the causes that it predicts the right effects; independent ways of identifying the representation as correct are required. I realise that this is just to point to the problem, or to label it, rather than to say anything in solution to it. But I shall leave it at that in order to focus effort on the separate problem of how we use the interpretative models of our theories to justify the abstract descriptions we apply when we try to represent the world. I choose the quantum Hamiltonian as an example. In the next section we will look in detail at one specific model – the BCS model for superconductivity – to see how Hamiltonians are introduced there.

9.7 HOW FAR DOES QUANTUM THEORY STRETCH?

The Bardeen-Cooper-Schrieffer model of superconductivity is a good case to look at if we want to understand the game rules for introducing quantum Hamiltonians. As Towfic Shomar argues (1998) if we are looking for a quantum description that gives very accurate predictions about super-conducting phenomena we can make do with the 'phenomenological' equations of the Ginsburg-Landau model. These equations are pheno-menological in two senses: first, they are not derived by constructing a model to which a Hamiltonian is assigned, but rather are justified by an *ad hoc* combination of considerations from thermodynamics, electromagne-tism and quantum mechanics itself. Secondly, the model does not give us any representation of the causal mechanisms that might be responsible for superconductivity. The first of these senses is my chief concern here.

The Ginsburg-Landau equations describe facts about the behaviour of the quantum state that according to proper quantum theory must derive from a quantum Hamiltonian. Hence they impose constraints on the class of Hamiltonians that can be used to represent superconduct-ing materials. But this is not the procedure I have described as the correct, principled way for arriving at Hamiltonians in quantum theory, and indeed the equations were widely faulted for being phenomenolog-ical, where it seems both senses of 'phenomenological' were intended at once. The description of the Ginsburg-Landau model in the recent text by Poole, Farachad and Creswick is typical: 'The approach begins by adopting certain simple assumptions that are later justified by their suc-cessful predictions of many properties of superconducting materials.' (Poole, Farachad Jr and Geswick 1995). Indeed it is often claimed that the Ginsburg-Landau model was not treated seriously until after we could see, thanks to the work by G'orkov, how it followed from the more principled treatment of the BCS theory (Buckel 1991)[17].

Before turning to the construction of the BCS Hamiltonian I begin with a review of my overall argument. We are invited to believe in the truth of our favourite explanatory theories because of their precision and their empirical successes. The BCS account of superconductivity must be a paradigmatic case. We build real operating finely-tuned super-conducting devices using the Ginsburg-Landau equations. And, since

---

[17] It remains an interesting question whether this account is really historically accurate or rather reflects a preference in the authors for principled treatments.

the work of G'orkov, we know that the Ginsburg-Landau equations can be derived from quantum mechanics or quantum field theory using the BCS model. So every time a SQUID detects a magnetic fluctuation we have reason to believe in quantum theory.

But what is quantum theory? *Theory*, after all, is a reconstruction. In the usual case it includes 'principles' but not techniques, mathematical relations but little about the real materials from which we must build the superconducting devices that speak so strongly in its favour. *Theory*, as we generally reconstruct it, leaves out most of what we need to produce a genuine empirical prediction. Here I am concerned with the place of bridge principles in our reconstructed theories. The quantum Hamiltonian is abstract in the sense of 'abstract' I have been describing: we apply it to a situation only when that situation is deemed to satisfy certain other more concrete descriptions. These are the descriptions provided by the interpretative models of quantum mechanics.

Albert Messiah's old text *Quantum Mechanics* (1961) provides four basic interpretative models: the central potential, scattering, the Coulomb inter-action and the harmonic oscillator – to which we should add the kinetic energy, which is taken for granted in his text. The quantum bridge prin-ciples give the corresponding Hamiltonians for each of the concrete inter-pretative models available in quantum mechanics. They provide an abstract Hamiltonian description for situations otherwise described more concretely. The point is: this is how Hamiltonians are assigned in a proper theoretical treatment; and in particular it is how they are assigned in just those derivations that we take to be the best cases where predictive success argues for the truth of quantum theory. When the Hamiltonians do not piggyback on the specific concrete features of the model – that is, when there is no bridge principle that licenses their application to the situation described in the model – then their introduction is *ad hoc* and the power of the derived prediction to confirm the theory is much reduced.

The term 'bridge principle' is a familiar one. Like 'correspondence rule', 'bridge principle' has meant different things in different philosoph-ical accounts. C. G. Hempel and Ernst Nagel worried about the meaning of theoretical terms. The internal principles can give them only a partial interpretation; bridge principles are needed to provide a full interpretation in a language whose terms are antecedently under-stood. Here I am not worried about questions of meaning, which beset all theoretical terms equally if they beset any at all. Rather, I am con-cerned about a distinction between theoretical concepts: some are abstract and some are not. Operationalists also use the terms 'bridge principle' and 'correspondence rule', but for them the bridge principles give rules for how to measure the quantity. Again this is not the use of 'bridge principle' I have in mind, for all quantities equally need procedures for how to measure them, whereas bridge principles, as I use the term, are needed only for the abstract terms of physics.

My claim about bridge principles and the limits of quantum physics is straightforward. Some theoretical treatments of empirical phenomena use *ad hoc* Hamiltonians. But these are not the nice cases that give us really good reasons to believe in the truth of the theory. For this we need Hamiltonians assigned in a principled way; and for quantum mechanics as it is practised that means ones that are licensed by principles of the theory – by bridge principles. In quantum theory there are a large number of derivative principles that we learn to use as basic, but the basic bridge principles themselves are few in number. Just as with internal principles, so too with bridge principles: there are just a handful of them, and that is in keeping with the point of abstract theory as it is described by empiricists and rationalists alike.[18] We aim to cover as wide a range as we can with as few principles as possible.

How much then can our theories cover? More specifically, exactly what kinds of situations fall within the domain of quantum theory? The bridge principles will tell us. In so far as we are concerned with theories that are warranted by their empirical successes, the bridge principles of the theory will provide us with an explicit characterisation of its scope. The theory applies exactly as far as its interpretative models can stretch. Only those situations that are appropriately represented by the interpretative models fall within the scope of the theory. Sticking to Messiah's catalogue of interpretative models as an example, that means that quantum theory extends to all and only those situations that can be represented as composed of central potentials, scattering events, Coulomb interactions and harmonic oscillators.

So far I have mentioned four basic bridge principles from Messiah. We may expect more to be added as we move through the theory net from fundamental quantum theory to more specific theories for specific topics. Any good formalisation of the theory as it is practised at some specific time will settle the matter for itself. In the next section I want to back up my claims that this is how quantum theory works by looking at a case in detail. I shall use the Bardeen-Cooper-Schrieffer account of

---

[18] For example, David Lewis, John Earman and Michael Friedman on the empiricist side, and on the rationalist, Abner Shimony.

superconductivity as an example. As Towfic Shomar describes, this account stood for thirty-five years as the basic theory of superconductivity and, despite the fact that the phenomena of type-II superconductors and of high temperature superconductivity have now shown up problems with it, it has not yet been replaced by any other single account.

I chose this example because it was one I knew something about from my study of SQUIDs at Stanford and from our research project on modelling at LSE. It turns out to be a startling confirmation of my point. The important derivations in the BCS paper (Bardeen, Cooper and Schrieffer 1957) are based on a 'reduced' Hamiltonian with just three terms: two for the energies of electrons moving in a distorted periodic potential and one for a very simple scattering interaction. This Hamiltonian is 'reduced' from a longer one that BCS introduce a page earlier. When we look carefully at this longer Hamiltonian, we discover that it too uses only the basic models I have already described plus just one that is new: the kinetic energy of moving particles, the harmonic oscillator, the Coulomb interaction, and scattering between electrons with states of well-defined momentum, and then, in addition, the 'Bloch' Hamiltonian for particles in a periodic potential (itself closely related to the central potential, which is already among the basic models). Superconductivity is a quantum phenomenon precisely because superconducting materials[19] can be represented by the special models that quantum theory supplies. How much of the world altogether can be represented by these models is an open question. Not much, as the world presents itself, looks on the face of it like harmonic oscillators and Coulomb interactions between separated chunks of charge. Superconductivity is a case where, we shall see, a highly successful representation can be constructed from just the models quantum theory has to offer. My point is that with each new case it is an empirical question whether these models, or models from some other theory, or no models from any theory at all will fit. Quantum theory will apply only to phenomena that these models can represent, and nothing in the theory, nor anything else we know about the structure of matter, tells us whether they can be forced to fit in a new case where they do not apparently do so.

9.8 BACKGROUND ON THE BCS THEORY

In the BCS theory, as in earlier accounts of both ordinary conductors and of superconductors, the superconducting material is modelled as a

---

[19] At least low temperature, type-I materials.

periodic lattice of positive ions sitting in a sea of conduction electrons. Earlier theories by Werner Heisenberg, by Hans Koppe and by Max Born and K. C. Cheng had postulated Coulomb interactions between the conduction electrons to be responsible for superconductivity. But the discovery of the *isotope effect* simultaneously by Maxwell at the National Bureau of Standards and by Reynolds, Serin, Wright and Nesbitt at Rutgers, seemed to indicate that lattice vibrations play a central role. Experiments with different isotopes of the same material showed that the critical temperature at which superconductivity sets in and the critical value of a magnetic field that will drive a superconductor into the normal state depend strongly on the isotopic mass. So too do the vibration frequencies of the lattice when the ions move like harmonic oscillators around their equilibrium positions (as they do in standard models where they are not fixed). Hence the hypothesis arises that lattice vibrations matter crucially to superconductivity. A first step towards the BCS theory came in earlier work, by Hans Fröhlich and then by John Bardeen. Bardeen and Fröhlich separately showed that the potential due to electron interactions via lattice vibrations could be attractive, in contrast to the repulsive Coulomb potential between the electrons; and further that when the difference in energy between the initial and final states for the electrons interacting with the lattice were small enough, the overall effect would be attractive.

A second step was the idea of 'Cooper pairs'. These are pairs of electrons with well-defined momenta which repeatedly interact via the lattice vibrations, changing their individual momenta as they do so but always maintaining a total null momentum. The Pauli exclusion principle dictates that no two electrons can occupy the same state. So normally at a temperature of absolute zero the lowest energy for the sea of conduction electrons is achieved when the energy levels are filled from the bottom up till all the electrons are exhausted. This top level is called the 'Fermi level'. So all the electrons' energy will normally be below the Fermi level. The interaction of two electrons under an attractive potential decreases the total potential energy. Raising them above the Fermi level increases their energy, of course. What Cooper showed was that for electrons of opposite momenta interacting through an attractive potential – Cooper pairs – the decrease in potential energy will be greater than the increase in kinetic energy for energies in a small band above the Fermi level. This suggested that there is a state of lower energy at absolute zero than the one in which all the levels in the Fermi sea are filled. This is essentially the superconducting state.

The first job of the BCS paper was to produce a Hamiltonian for which such a state will be the solution of lowest energy and to calculate the state. This is why the paper is such a good example for a study of how Hamiltonians get assigned. The construction of the Hamiltonian takes place in the opening sections of the paper. The bulk of the paper, which follows, is devoted to showing how the allowed solutions to the BCS Hamiltonian can account for the typical features of superconductivity. We will need to look only at the first two sections.

9.9 THE BCS HAMILTONIAN

The BCS Hamiltonian, I have said earlier, is a startling confirmation of my claim that the Hamiltonians we use to treat unfamiliar problems are the stock Hamiltonians associated with familiar models. The BCS Hamiltonian has four very familiar terms. The first two terms represent the energy of a fixed number of non-interacting particles – 'electrons' – with well-defined momenta. I put 'electron' in scare quotes because the particles in the model are not ascribed all the properties and interactions that electrons should have. The right way to think about modelling in these kinds of cases seems to me to be to say that the particles in the model have just the properties used in the model. There is, nevertheless, a good reason for labelling them in certain ways – say, as electrons – because this suggests further features that if appropriately included, should lead to a better representative model, as both Mary Hesse and Imré Lakatos suggest. The third term in the BCS Hamiltonian represents the pairwise Coulomb interaction among the same electrons. The last term represents interactions that occur between pairs of electrons through the exchange of a virtual phonon, as pictured in figure 9.1. This is a standard scattering interaction. The virtual phonons are associated with the lattice vibrations. ('Virtual' here means that energy conservation is briefly violated; phonons are the 'particles' associated with the energy field generated by vibrations of the lattice of ions.)

It may look from what I have said so far as if we just have stock models and Hamiltonians and that's it. Matters are not quite so simple as this initial description suggests, however. For it is not enough to know the form of the Hamiltonian; we have to know in some mathematical representation what the Hamiltonians actually are, and this requires more information about the models. Consider, for example, the first two terms in the BCS Hamiltonian. To specify what the two Hamiltonians are we will have to lay out the allowed values for the momenta of the electrons in the model; and to justify a given choice in these will require a lot more details to be filled in about the model. In fact exactly what structure the model actually has only becomes clear as we see how the third and fourth terms are developed since these latter two very simple terms appear at the cost of complication in the first two. What I shall describe is how each of these terms is justified.

The BCS Hamiltonian is not assigned in as principled a way as I may seem to be suggesting though. For a crucial assumption in the end is a restriction on which states will interact with each other in a significant way. The choice here is motivated by physical ideas that are a generalisation of those involved in Cooper's paper where electron pairs are introduced. BCS assume that scattering interactions are dramatically more significant for pairs of electrons with equal and opposite momenta. As in the earlier work of Fröhlich and Bardeen, electrons with kinetic energies in the range just above the Fermi level can have a lower total energy if the interaction between them is attractive. But there is a limit set on the total number of pairs that will appear in this range because not all pairs can be raised to these states since the Pauli exclusion principle prohibits more than one pair of electrons in a state with specific oppositely directed values of the momentum. So here we see one of the features that quantum theory assigns to electrons that are retained by the electron-like particles in the model.

What is interesting for our topic is that these physical ideas are not built in as explicit features of the model that are then used in a principled way to put further restrictions on the Hamiltonian beyond those already imposed from the model. Rather the assumptions about what states will interact significantly are imposed as an *Ansatz*, motivated but not justified, to be tried out and ultimately judged by the success of the theory at accounting for the peculiar features associated with superconductivity.[20] Thus in the end the BCS Hamiltonian is a rich illustration: it is a Hamiltonian at once both theoretically principled and phenomenological or *ad hoc*. Let us look at each of the terms of the BCS Hamiltonian in turn.

Terms (1) and (2): Bloch electrons. The 'electrons' in our model are not 'free' electrons moving independently of each other, electrons moving unencumbered in space. They are, rather, 'Bloch' electrons and their

---

[20] Because of these *ad hoc* features it makes sense to talk both of the BCS theory and separately of the BCS model since the assumptions made in the theory go beyond what can be justified using acceptable quantum principles from the model that BCS offer to represent superconducting phenomena.

Hamiltonian is the 'Bloch Hamiltonian'. It is composed of the sum of the energies for a collection of non-interacting electrons. Each term in the sum is in turn a standard kinetic energy term for a moving particle – the most basic and well-known interpretative model in quantum theory – plus an externally fixed potential energy: $p^2/2m + V(r)$. In the Bloch Hamiltonian the second term for each electron represents the potential from the positive ions of the lattice, treated as fixed at their equilibrium positions. The crucial assumption is that this potential has the same periodicity as the lattice itself.

A second assumption fixes the allowed values for the energies that we sum over in the Bloch Hamiltonian. The energy is a function of the momentum, which, in turn, recalling wave-particle duality, is proportional to the reciprocal of the wave vector of the associated electron-wave. What facts in the model license a given choice? BCS adopt the Born-Karman boundary conditions on the superconducting material. The lattice is taken to be cubic – although the assumption is typically generalised to any parallelepiped – with the length on each side an integral multiple of the fixed distance between ions in the lattice. Well-established principles from wave mechanics then dictate the allowed values for the wave vectors, hence the energies, in the Hamiltonian.

The German Structuralists have taught us to think not of theories, but of theory-nets. The nodes of the net represent specialisations of the general theory with their own new principles, both internal principles and bridge principles. The introduction of the Bloch Hamiltonian in the quantum theory of conductivity is a good example of the development of a new bridge principle to deal with a specific subject. The Bloch Hamiltonian is an abstract description for situations that are modelled concretely as a certain kind of periodic lattice called a *Bravais lattice*: to be a moving electron in this kind of lattice is *what it is* for these electrons to be subject to a Bloch Hamiltonian. In the Bloch theory this term appears as a phenomenological Hamiltonian. It is not assigned by mapping out the details of the interaction between the (fictitious) non-interacting electrons and the ions that make up the (fictitious) perfect lattice, but rather represents the net effect of these interactions.

This is a good illustration of the difference between the phenomeno-logical-explanatory distinction and the distinction between the principled and the *ad hoc* that I have already mentioned (although the term 'phenomenological' is often used loosely to refer to either or both of these distinctions). The search for explanation moves physicists to look for physical accounts of why certain kinds of situations have certain kinds of effects. Hamiltonians that pick out the putative physical mechanisms are called 'explanatory' as opposed to the 'phenomenological' Hamiltonians, that merely produce quantum states with the right kinds of features. The principled-*ad hoc* distinction, by contrast, depends on having an established bridge principle that links a given Hamiltonian with a specific model that licenses the use of that Hamiltonian. A Hamiltonian can be admissible under a model – and indeed under a model that gives good predictions – without being explanatory if the model itself does not purport to pick out basic explanatory mechanisms. This is just the case with the Bloch Hamiltonian for electrons moving in a Bravais lattice.

The treatment of Bloch electrons below the Fermi energy follows the same pattern.

Term (3): The screened Coulomb potential. Recall that a central claim of the BCS paper is that the Coulomb interactions between electrons should be less important to their quantum state than the interactions between them mediated by lattice vibrations, even though the Coulomb energies are much greater. Their argument depends on the fact that the Coulomb interactions are screened by the effects of the positive ions of the lattice. Electrons interacting under a Coulomb potential will repel each other and tend to move apart. But as they move apart the area in between becomes positively charged because of the ions which are relatively fixed there. So the electrons will tend to move back towards that area and hence move closer together. Bardeen and Pines (1955) had shown that in their model the Coulomb interactions become effectively short-range, operating only across distances of the size of the inter-particle spacing, since the long-range effects can be represented in terms of high frequency plasma oscillations that are not normally excited.

The screened Coulomb potential is represented in the third term of the BCS Hamiltonian. To see where this term comes from consider a gas of electrons moving freely except for mutual interactions through Coulomb forces. The familiar bridge principle for the Coulomb interaction dictates that the contribution to the Hamiltonian for this gas due to the electron interactions should be

$$\frac{1}{2} \sum_{i=1}^{N} \sum_{j \neq i} (e^2/4\pi\epsilon_0 |\mathbf{r}_i - \mathbf{r}_j|)$$

that is, the usual Coulomb potential for a pair of electrons summed over all the ways of taking pairs. The model BCS want to study, though, is not of a gas of free interacting electrons but rather that of a gas of interacting electrons moving in a Bloch potential attributed to a lattice of fixed ions. As I have just described, for this model the Coulomb forces between electrons will be screened due to the effects of the fixed ions. BCS use the treatment of Bohm and Pines for this model, but they also mention one other established way to handle screening, namely, the Fermi–Thomas approach. This is the one Bardeen adopted to deal with screening in his survey 'Theory of Superconductivity' (Bardeen 1956). For brevity I shall describe here only the Fermi–Thomas treatment.[21]

A usual first step for a number of approaches is to substitute a new model, the 'independent electron model', for the interacting Bloch electrons, a model in which the electrons are not really interacting but instead each electron moves independently in a modified form of the Bloch potential. What potential should this be? There is no bridge principle in basic theory to give us a Hamiltonian for this model. Rather the Hamiltonian for the independent electron model is chosen *ad hoc*. The task is to pick a potential that will give, to a sufficient degree of approximation in the problems of interest, the same results from the independent electron model as from the original model. We can proceed in a more principled way, though, if we are willing to study models that are more restrictive. That is the strategy of the Fermi–Thomas approach.

The Fermi–Thomas approach refines the independent electron model several steps further. First, its electrons are fairly (but not entirely) localised. These will be represented by a wave-packet whose width is of the order of $1/k_F$, where k_F is the wave vector giving the momentum for electrons at the Fermi energy level. One consequence of this is that the Fermi–Thomas approach is consistent only with a choice of total potential that varies slowly in space. The Fermi–Thomas treatment also assumes that the total potential (when expressed in momentum space) is linearly related to the potential of the fixed ions. To these three constraints on the model a fourth can be consistently added, that the energy is modified from that for a free electron model by subtracting the total local potential. As a result of these four assumptions it is possible to back-calculate the form of the total potential for the model using standard techniques and principles. The result is that the usual Coulomb potential is attenuated by a factor $1/e^{k_F |r_i - r_j|}$. The Fermi–Thomas approach is a nice example of how we derive a new bridge principle for a quantum

---

[21] In fact I do not do this just for brevity but because the Bardeen and Pines approach introduces a different kind of model in which the electrons and ions are both treated as clouds of charge rather than the ions as an ordered solid. Tracing out the back and forth between this model and the other models employed which are literally inconsistent with it, is a difficult task and the complications would not add much for the kinds of points I want to make here, although they do have a lot of lessons for thinking about modelling practices in physics.

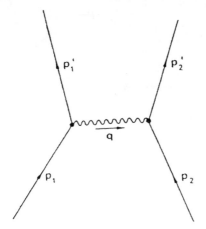

Figure 9.1 Schematic representation of electron–electron interaction
transmitted by a phonon.

theory. The trick here was to find a model with enough constraints in it
that the correct Hamiltonian for the model could be derived from principles and techniques already in the theory. Thus we are able to admit a
new bridge principle linking a model with its appropriate Hamiltonian.
By the time of the BCS paper the Fermi–Thomas model had long been
among the models available from quantum mechanics for representing
conduction phenomena, and it continues to be heavily used today.

Term (4): Ion (or phonon) mediated interactions between electrons. So far we
have considered models where the lattice ions have fixed positions. But,
recall, we are aiming for a model where electrons interact through lattice
vibrations. This kind of interaction was first studied by Fröhlich, following the Bloch theory of thermal vibrations and resistance in metals,
where, Fröhlich observed, interactions between the lattice and the electrons involve the absorption and emission of non-virtual (i.e., energy
conserving) vibrational phonons. For superconductors, he hypothesised,
what matters is the absorption and emission of virtual phonons, as in
figure 9.1. This is a very basic model and appears in almost identical
form across elementary texts in quantum field theory as well as in texts
at a variety of levels on superconductivity.

The first step in the construction of this term in the Hamiltonian is to
represent the ions as a collection of coupled springs. Like the model of
the Coulomb interaction, this is one of the fundamental models of
quantum mechanics and the bridge rule that links it with the well-known

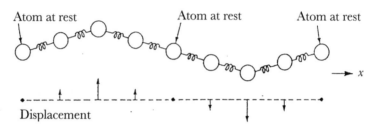

Figure 9.2 The spring model of ions. (G. F. Vidali,
Superconductivity: The Next Revolution, 1993)

harmonic oscillator Hamiltonian is one of the basic (i.e., non-derivative) principles of the theory. The periodic arrangement of the springs in the model fixes the allowed values of the frequency of oscillation. Like the electron–phonon exchange model, the spring model of ions is also pictured regularly in elementary texts in condensed matter physics in an almost identical way. Figure 9.2 shows a typical example.

Although the harmonic oscillator model is basic in quantum mechanics, its use as a representation for the collection of ions in a metal is not simply *post hoc*. We do not use it just because it works but often justify it, instead, as an approximation from a prior model. In our case, the prior model starts, as we have seen, with the picture of a lattice of metal atoms from which the valence electrons get detached and are able to move through the lattice, leaving behind positive ions which remain close to some equilibrium positions that fix the spacing of the lattice.

To get the harmonic model from the prior model, the chain of ions in the lattice are restricted to nearest-neighbour interactions. Thus the energy becomes a function of the total number of electrons and of the distance between any two neighbours. The harmonic model also assumes that ions move only a little about an equilibrium position – the displacements are small in comparison with the distances between ions. That means we can calculate the energy using a Taylor series expansion. Then we keep only the terms that in the expansion correspond to the energy at rest distance and to the second derivative of the energy with respect to the displacements from the distance at rest.[22] All the higher-order terms are the 'anharmonic' terms. In the harmonic approximation, for small values of the displacements, these are neglected.

The harmonic approximation is justified first of all on the grounds of its mathematical convenience. Harmonic equations of motion have exact solutions, whereas even the simplest anharmonic ones do not. Secondly, it allows us to get most of the important results with the simplest of models. Finally, it constitutes the basis for the perturbation method approach to the formalisation of complex cases. On this approach, more accurate results for the original model, which includes interactions between ions at separated sites, can be obtained by systematically adding (anharmonic) corrections to the harmonic term.

The underlying model that BCS work with is one in which there is a sea of loose electrons of well-defined momenta moving through a periodic lattice of positive ions whose natural behaviour, discounting interactions with the electron sea, is represented as a lattice of coupled harmonic oscillators, subject to Born-Karman boundary conditions. I shall call this the 'full underlying model'. As I have already mentioned, a number of important features are indicated about the model by calling the components in it 'electrons' and 'positive ions'. The two central ones are, first, the fact that the particles called 'electrons' are fermions, and thus satisfy the Pauli exclusion principle (hence it makes sense to talk, for instance, of filling each energy level up to the Fermi level since the exclusion principle allows only one electron per energy state), whereas the positive ions are bosons; and secondly, the fact that the particles in the model affect each other through Coulomb interactions. The Coulomb interactions occur pairwise among the electrons, among the ions, and between the electrons and the ions.

It is important to notice that this model is itself just a model. Even though I have called it 'the full underlying model', it is not the real thing,

---

[22] The term corresponding to the linear expansion, i.e., the first derivative, is null, as it expresses the equilibrium condition for the distance at rest.

truncated. It is a *representation* of the structure of some given sample of superconducting material, not a literal *presentation*, as though seen from a distance so that only the dominant features responsible for superconductivity are visible. This more primary model is chosen with a look in two directions: to the theory on the one hand and to the real superconducting material on the other. The model aims to be explanatory. That means both that it should represent basic factors that will be taken as responsible for the superconductivity phenomena derived and that it must bring these factors under the umbrella of the theory, that is, it must use representations which we have a principled way of treating in the theory. It is significant for my central point about the limits of theory that the only kind of Hamiltonian used to describe this underlying model is the one for the Coulomb potential.

As I remarked in the discussion of term (3), it is impossible to solve the equation for the ground state of a Hamiltonian like this. So BCS substitute the new models I have been describing, with their corresponding Hamiltonians. Just as with term (3), if term (4) is to be appropriate, a lot more constraints must be placed on the underlying model before we can expect its results to agree with those from the BCS model. Similarly, the specific features of the Hamiltonians for the BCS model have to be constructed carefully to achieve similar enough results in the calculations of interest to those of the 'underlying' model.

The basis for the further term in the BCS model is the earlier work I mentioned by Hans Fröhlich. Fröhlich's model (1950; 1952) begins with independent Bloch electrons with no interactions among them besides those which will appear as a consequence of the interaction between the electrons and the ions. In this model (as in Bloch's original one) the vibrations of the lattice are broken into longitudinal and transverse, and the electrons interact only with the longitudinal vibrations of the lattice. This, as Bardeen points out (1956), means that the transverse waves are treated by a different model, a model in which the ions vibrate in a fixed negative sea of electrons. The Fröhlich model also assumes that the spatial vibration of the electron waves across the distance between ions in the lattice is approximately the same for all wave vectors. Using perturbation methods in the first instance and later a calculation that does not require so many constraints as perturbation analysis, Fröhlich was able to justify a Hamiltonian of the form of term (4), with an interaction coefficient in it reflecting the screening of the Coulomb force between ions by the motion of the electrons. The model assumes that in the presence of electron interactions the motion of the ions is still harmonic, with the frequencies shifted to take into account the screening of the ion interactions by the electrons. Similarly, in Fröhlich's version of term (4) negative charge density in the model is no longer the density of the original Bloch electrons (which, recall, move in a fixed external potential) but rather that of electrons carrying information about the lattice deformation due to their interaction with it.

Now we can turn to the treatment by Bardeen and Pines (1955). They have a model with plasma oscillations of the electron sea. They deal thus with a longer Hamiltonian involving energies of the individual Bloch electrons, the harmonic oscillations of the lattice, the harmonic oscillations of the plasma, an electron-lattice interaction, a plasma-phonon interaction, a plasma-electron interaction and a term for those wave vectors for which plasma oscillations were not introduced, including the shielded Coulomb electron–electron interaction. They rely on previous arguments of Pines' to show that this model can approximate in the right way a suitably constrained version of the 'full underlying model'. They are then able to show that the electron–lattice interaction can be replaced by a phonon-mediated electron–electron interaction described by a Hamiltonian of the form of term (4). In Bardeen and Pines' version of this term, as with Fröhlich's, the interaction coefficient is adjusted to provide approximately enough for the effects produced by shielding in the underlying model.

We have looked at the BCS theory in detail, but there is nothing peculiar about its use of quantum Hamiltonians. A number of accounts of superconductivity at the time treat the quantum mechanical superconducting state directly without attempting to write down a Hamiltonian for which this state will be a solution. This is the case, for instance, with Born and Cheng (1948), who argue from a discussion of the shape of the Fermi surface that in superconductivity spontaneous currents arise from a group of states of the electron gas for which the free energy is less with currents than without. It is also the case with Wentzel (1951), who treats superconductivity as a magnetic exchange effect in an electron fluid model, as well as Niessen (1950), who develops Heisenberg's account.

Those contemporary accounts that do provide Hamiltonians work in just the way we have seen with BCS: only Hamiltonians of stock forms are introduced and further details that need to be filled in to turn the stock form into a real Hamiltonian are connected in principled ways with features of the model offered to represent superconducting materials. By far the most common Hamiltonians are a combination of kinetic energy in a Bloch potential plus a Coulomb term. This should be no surprise since theorists were looking at the time for an account of the mechanism of superconductivity and Coulomb interactions are the most obvious omission from Bloch's theory, which gives a good representation of ordinary conduction and resistance phenomena. We can see this in Heisenberg's (1947) theory, which postulates that Coulomb interactions cause electrons near the Fermi energy to form low density lattices where, as in Born and Cheng, there will be a lower energy with currents than without, as well as in Schachenmeier (1951), who develops Heisenberg's account and in Bohm (1949), who shows that neither the accounts of Heisenberg nor of Born and Cheng will work. Macke (1950) also uses the kinetic energy plus Coulomb potential Hamiltonians, in a different treatment from that of Heisenberg.

Of course after the work of Bardeen and of Fröhlich studies were also based on the electron scattering Hamiltonian we have seen in term (4) of the BCS treatment, for instance in the work of Singwi (1952) and Salam (1953). A very different kind of model studied both by Wentzel (1951) and by Tomonaga (1950) supposes that electron-ion interactions induce vibrations in the electron gas in the model and the vibrations are assigned the traditional harmonic oscillator Hamiltonian. Tomonaga, though, does not just start with what is essentially a spring model which he then describes with the harmonic oscillator Hamiltonian. Rather, he back-calculates constraints on the Hamiltonian from assumptions made about the density fields of the electron and ion gases in interaction. In these as in all other cases I have looked at in detail the general point I want to make is borne out. We do not keep inventing new Hamiltonians for each new phenomenon, as we might produce new quantum states. Rather the Hamiltonians function as abstract concepts introduced only in conjunction with an appropriate choice of a concrete description from among our set stock of interpretative models.

9.10 CONCLUSION

I have talked at length about *bridge principles* because our philosophical tradition is taken up with scientific knowledge. We focus on what we see as physics' claims about what properties things have and how they relate. Bridge principles have always been recognised as part of the body of precisely articulated knowledge claims. Our discussion reminds us that quantum physics provides more rules than these though for constructing theoretical representations.[23] In the case of the quantum Hamiltonian, bridge principles provide the form for the Hamiltonian from stock features that the model may have. Its detailed content is dictated by less articulated but equally well-established techniques and constraints from other features of the model. The Born-Karman boundary conditions used to fix the allowed values of the momentum are an example. The point of my discussion here is that if we wish to represent a situation within quantum theory – within the very quantum theory that we prize for its empirical success – we must construct our models from the small stock of features that quantum theory can treat in a principled way. And this will fix the extent of our theory.

We are used to thinking of the domain of a theory in terms of a set of objects and a set of properties on those objects that the theory governs, wheresoever those objects and properties are deemed to appear. I have argued instead that the domain is determined by the set of stock models for which the theory provides principled mathematical descriptions. We may have all the confidence in the world in the predictions of our theory about situations to which our models clearly apply – such as the carefully controlled laboratory experiment which we build to fit our models as closely as possible. But that says nothing one way or another about how much of the world our stock models can represent.

REFERENCES

Bardeen, J. (1956). 'Theory of Superconductivity', in *Encyclopedia of Physics*, vol. 15. Berlin: Springer-Verlag, 274–309.
Bardeen, J. and D. Pines (1955). 'Electron-Phonon Interactions in Metals', *Physical Review*, 99:4, 1140–50.
Bardeen, J., N. L. Cooper and J. R. Schrieffer (1957). 'Theory of Superconductivity', *Physical Review*, 108:5, 1175–204.

---

[23] It is well known that relativistic quantum field theory makes central the imposition of symmetry constraints, such as local gauge invariance, for determining the form of Lagrangians and Hamiltonians for situations involving interactions where the force fields are represented by gauge fields. But it cannot be claimed on these grounds alone that the Hamiltonians, say, are built, not from the bottom up, as I have discussed, but rather from the top down. Even in the case of interactions between electrons involving electromagnetic fields the local gauge symmetry requirement that establishes the form of the Lagrangian and Hamiltonian is only formal. It is satisfied by the introduction of a quantity that the symmetry requirement itself leaves uninterpreted. As Jordi Cat has argued, we still need in addition some bridge principles that tell us that the gauge field corresponds to electromagnetic fields and that indeed the Lagrangian or Hamiltonian describe a situation with some interaction with it (see Cat 1993 and 1995b). It is in this sense that it is often argued, mistakenly, that symmetry principles alone can dictate the existence of forces.

Bohm, D. (1949). 'Note on a Theorem of Bloch Concerning Possible Causes of Superconductivity', *Physical Review*, 75:3, 502–4.

Born, M. and K. C. Cheng (1948). 'Theory of Superconductivity', *Nature*, 161:4103, 968–9.

Buckel, W. (1991). *Superconductivity: Foundations and Applications*. Weinheim: VCH.

Cartwright, N. (1983). *How the Laws of Physics Lie*. Oxford: Oxford University Press.

    (1989). *Nature's Capacities and their Measurement*. Oxford: Oxford University Press.

    (1993). 'Fables and Models' *Proc. Aristotelian Society*, 44, 55–82.

    (1998). 'Where in the World is the Quantum Measurement Problem?' *Philosophia Naturalis*.

    (forthcoming). *The Dappled World*. Cambridge: Cambridge University Press.

Cat, J. (1993). 'Philosophical Introduction to Gauge Symmetries', lectures, LSE, October 1993.

    (1995a). *Maxwell's Interpretation of Electric and Magnetic Potentials: The Methods of Illustration, Analogy and Scientific Metaphor*. Ph.D. dissertation, University of California, Davis.

    (1995b). 'Unity of Science at the End of the 20th Century: The Physicists Debate', unpublished pages, Harvard University.

    (forthcoming). 'Understanding Potentials Concretely: Maxwell and the Cognitive Significance of the Gauge Freedom'.

Daston, L. and P. Galison (1992). 'The Image of Objectivity', *Representations*, 40, 81–128.

Fraassen, B. van (1980). *The Scientific Image*. Oxford: Oxford University Press.

Fröhlich, H. (1950). 'Theory of Superconducting State. I. The Ground State at the Absolute Zero of Temperature', *Physical Review*, 79:5, 845–56.

    (1952). 'Interaction of Electrons with Lattice Vibrations', *Proceedings of the Royal Society of London*, Series A: *Mathematical and Physical Science*, 215, 291–8.

Giere, R. (1984). *Understanding Scientific Reasoning*. Minneapolis: Minnesota University Press.

    (1988). *Explaining Science: A Cognitive Approach*. Chicago: University of Chicago Press.

    (1995). 'The Skeptical Perspective: Science without Laws of Nature', in F. Weinert (ed.), *Laws of Nature: Essays on the Philosophical, Scientific and Historical Dimensions*. New York: Walter de Gruyer.

Heisenberg, W. (1947). 'Zur Theorie der Supraleitung', *Zeit. Naturforschung*, 2a, 185–201.

Jackson, J. D. (1975). *Classical Electromagnetism*, Second Edition. New York: Wiley.

Macke, W. (1950). 'Über die Wechselwirkungen im Fermi-Gas', *Zeit. Naturforschung*, 5a, 192–208.

Messiah, A. (1961). *Quantum Mechanics*, Amsterdam: North-Holland.

Mitchell, S. (1997). 'Pragmatic Laws', unpublished paper, UCSD.

Morrison, M. (1997). 'Modelling Nature: Between Physics and the Physical World', *Philosophia Naturalis*, 38, 64–85.

Niessen, K. F. (1950). 'On One of Heisenberg's Hypotheses in the Theory of Specific Heat of Superconductors' *Physica*, 16: 2, 77–83.

Ohanian, H. C. (1988). *Classical Electromagnetism*. Boston: Allyn and Bacon.

Orlando, T. and Delin, K. (1990). *Foundations of Applied Science*. Reading, MA: Addison-Wesley.

Poole, C. P., H. A. Farachad Jr. and R. J. Creswick (1995). *Superconductivity*. San Diego: Academic Press.

Salam, A. (1953). 'The Field Theory of Superconductivity', *Progress of Theoretical Physics*, 9: 5, 550–4.

Singwi, K. S. (1952). 'Electron-Lattice Interaction and Superconductivity', *Physical Review*, 87: 6, 1044–7.

Schachenmeier, R. (1951). 'Zur Quantentheorie der Supraleitung', *Zeit. Physik*, 129, 1–26.

Shomar, T. (1998). *Phenomenological Realism, Superconductivity and Quantum Mechanics*. Ph.D. dissertation, University of London.

Suppe, F. (1977). *The Structure of Scientific Theories*. Urbana: University of Illinois Press.

Tomonaga, S. (1950). 'Remarks on Bloch's Method of Sound Waves Applied to Many-Fermion Problems', *Progress of Theoretical Physics*, 5: 4, 544–69.

Wentzel, G. (1951). 'The Interaction of Lattice Vibrations with Electrons in a Metal', *Physical Review Letters*, 84, 168–9.

CHAPTER 10

Past measurement and future prediction

Adrienne van den Bogaard

10.1 INTRODUCTION

This paper studies the emergence of the first macroeconometric model and the national accounts from a social studies of science perspective. In the twentieth century, mathematics have increasingly been used to describe (parts of) the economy. Mathematisation of the social sciences (following the footsteps of physics) may look like an autonomous, universal and deterministic development. However, it has been the role of the social studies of science to show that these developments may seem autonomous if you look at what has become real and what our current society defines as necessary, but if you look in more detail to the process of establishing it, a much less deterministic, less autonomous, less universal and unavoidable picture arises.

The first macroeconometric model was developed by Jan Tinbergen in the 1930s and set the pattern for such models in economic science in the Western block. His type of modelling became the dominant methodology of the Dutch Central Planning Bureau (CPB) from the early 1950s. Macroeconometric modelling became part of university economics programmes as a consequence of which bureaucrats, Members of Parliaments and ministers increasingly conceived of the economy in those modelling terms. Jan Tinbergen's model has thus had an enormous influence on the international economics discipline as well as on Dutch policy making.

The fact that Dutch economists have all been taught the CPB's macroeconometric models as 'standard knowledge' has resulted in a view that the 'model' constitutes an objective piece of knowledge and consequently Dutch economic policy proposals are routinely 'tested out' on the model. However, economists outside the Netherlands are surprised about this objective view: mostly, countries have competing models which results in a much more moderate view of economic models and their powers.

The first aim of the chapter is to give an interpretation of the emergence of the model. It shows that the emergence of the practice of modelling involved a series of translations of representations of the economy. Instead of portraying the model as a 'revolution' as has often been done, this chapter describes the development of the model in terms of 'continuity'. The 'model' will not be described as something which was suddenly invented but as a result of a gradual process of shifts in the quantitative representations of the economy within the Central Bureau of Statistics (CBS). Each translation meant a new 'answer' to the question: What aspects of economic life can be subject to mathematisation? And which aspects are delegated to other sorts of knowledge, like verbal theory or policy expertise?

This paper also aims to give a broader account of what the 'model' is. Arguments about the model as a theory, or a method, or a distortion of reality, all focus on the model as a scientific object and how it functions in science. Without denying this dimension of the model at all, this paper wants to broaden the perspective by claiming that the model is also a social and political device. The model will be understood as a practice connecting data, index numbers, national accounts, equations, institutes, trained personnel, laws, and policy-making.

My case material focuses on the story of national econometric models and national accounts in the Central Bureau of Statistics in the Netherlands in the period 1925–1955. In general, economists take for granted an analytical distinction between models and data in their daily work, whether they work in academia or in policy-oriented domains. For example, the Dutch textbook for undergraduates *The Dutch Economy. Description, Prediction and Management* is an example of how Dutch economists are taught the essence of national accounts and models:

The national accounts neither deal with *explanation* of economic events nor with *prediction* about what is going to happen, but the national accounts merely deal with the registration of past economic activity. Knowledge of macroeconomic data of the past are not only indispensable for economic theory and the construction of empirical models that follow from economic theory, as for example in the case of the Central Planning Bureau models, but also for the formulation of *concrete* policy advice. (Compaijen and van Til 1978, 1; emphasis in original)

This introductory paragraph tells us what I would call the standard-view: national accounts describe economic activity and deliver the data on which models are based. Establishing national accounts, models and policies are described as three distinct activities. In the Netherlands this is reflected in the fact that these activities nowadays take place in three different institutes, respectively the Central Bureau for Statistics, the Central Planning Bureau and the Ministry of Economic Affairs of the government. The word 'registration' suggests that the economy presents its phenomena unambiguously to the economist who only needs pen and paper to write down the facts. A recent book on the Dutch National Accounts described the national accounts as 'the mirror of the modern national economy' (CBS 1993, 9). Assuming that the mirror works properly in a technical sense, it is a neutral device, which means that this registration activity is considered neutral and objective. When the economy has been registered without distortion, economists can start to explain and predict future phenomena on the basis of theory and models. Besides, national accounts provide the necessary information to make good policies. Therefore, data are input for both models and policy.

By examining empirically the history of the model-data distinction it will be shown that this analytical distinction is more problematic than is usually assumed. Until 1925 the production of (economic) data mainly consisted of gathering figures and constructing means, ratios and index numbers to measure all sorts of economic phenomena. Section 10.2 will show that even from the current perspective most simple constructs like index numbers already involve series of choices on how to gather and manipulate data. In the United States, the wish not only to measure past events but also to be able to predict in some way economic booms or crises had already emerged just after the First World War. This work concentrated on developing an economic barometer that would tell you when the economy would rise or decline. Section 10.3 will describe how such a barometer was constructed in the Netherlands and show that the barometer was in fact another step in manipulating data in order to fulfil a new purpose, namely prediction.

The emergence of the model was a new step designed not only to tell you when the economy would rise or fall but also to tell you which element in the economy was responsible for this change. The development of the model will be told in section 10.4. Section 10.5 will describe the next step in quantifying the national economy: the development of the national accounts. The national accounts are not only a set of data but also represent the structure of the economy in terms of flows from one part of the economy to the other. Finally, section 10.6 focuses on social and institutional aspects of the development of the model and the national accounts.

10.2 MEANS, RATIOS AND INDEX NUMBERS

10.2.1 The case of unemployment

Unemployment figures have been gathered since 1906 within the CBS (den Breems 1949, 10). When the first Statistical Yearbook was published in 1924, different kinds of figures of unemployment were given under the heading of 'the economic and social situation of the population' (CBS Statistical Yearbook 1924, 14–16). What sorts of figures were given to fulfil this purpose?

One table consisted of absolute figures: numbers of persons who had asked for a job, numbers of employers who had asked for a person seeking employment and the amount of people who had actually found a job. Although this seems easy and clear, it is more complicated. Of course, the CBS could not go to every single individual to ask if he or she was unemployed. Therefore, the CBS table consisted of figures that were given to them by employment offices and 'some unofficial job centres included as far as we know them'. This means that these employment offices had to give these figures in an appropriate, and probably consistent way. This meant 'work' to do by these offices. The figures represent numbers of people that actually went to the official employment offices, but people who found their way differently, via the unofficial job centres or another way via family or friends, were much more difficult to count. The point of this little story is that even the most simple-looking figures, i.e. absolute numbers, reflect an institutional structure, embody work of people (filling out questionnaires) as a consequence of which the actual figures represent a specific group of unemployed, namely those who found their job via the official employment offices.

The question then was the precise relation between the measurement system (the employment offices) and the actual entity (unemployment). This problem was later taken up by the economist Kaan in 1939 (Kaan 1939, 233). He discussed the 'distorted' relation between the measurements and what might be the 'real' value of unemployment by doing comparative research between the Central Bureau for Statistics and the municipal Bureaux for Statistics of Amsterdam, Rotterdam and The Hague. He showed that although the four bureaux used the same measurement system they published different kinds of unemployment figures. He gave for example, a table of 'domestic job seekers' (i.e. housekeepers etc.) (table 10.1).

Table 10.1. *Personnel working in domestic services*

Year	Amsterdam		Rotterdam	
	Registered as looking for job	The number included in the unemployment statistics	Registered as looking for job	The number included in the unemployment statistics
1930	1695	109 or 6.4%	1377	827 or 60.0%
1937	1999	316 or 15.8%	1674	1291 or 77.1%

Source: Kaan (1939, 233)

These figures differed because Amsterdam did not register people looking for a 'domestic' job as unemployed while Rotterdam did. This is a clear example of the choices between definitions of unemployment and the contribution of the unemployed to the measured value of the actual entity.

A second table consisted of unemployment index numbers. Yearly figures were given for the period 1915–1922, and monthly figures were given for the year 1923. This index number was defined in the following way:

The ratio – in percentages – between the real number of days unemployment per week and the total number of days people, whose unemployment has been checked, could have been unemployed in that week, which is six times the number of persons. (CBS 1924, 16)

First of all, the fact that unemployment was counted in days reflects a characteristic of the structure of the labour market in that period. People apparently could have a job for two days, be unemployed for two days and then get a new job at another company. The unit of time chosen, the day, to count unemployment would not make much sense today. It would be too detailed.

Also the ratio itself embodies choices. This ratio is a 'past-figure': it tells you the number of days in the past that people did not work in relation to the total number of days they could have worked. It does not tell you the current state of the labour market: if one read this table in 1924 to get an idea about the state of the labour market, the percentage unemployment figure for example, one would not be able to find it out. These are past-measurements because you only know the actual unemployment-days afterwards. Other ratios could also have been used, for example the ratio between the number of days unemployment in some week and the number of days unemployment in the previous week, which would tell you the changes in unemployment through time per week.

These figures look very descriptive. Nevertheless, their use reflected theories about unemployment available at the time: cyclical, structural and frictional unemployment. The cyclical unemployment theory explained unemployment as a sort of 'epiphenomenon' of the business-cycle itself. Frictional unemployment means that it takes some time to find a job although there is a job for you. Structural unemployment is the result of a structural change in the economy which means that some sort of labour becomes superfluous while shortages on a different form of labour emerge.

The third figure in the CBS *Yearbook* (CBS 1924, 16) is a sort of graph representing the yearly index numbers. (The horizontal axis represents time, the vertical axis the index numbers.) The reason to show something in a graph might be to see the 'dynamics' in the figures more immediately. The graph shows a development through time. It shows the wish to compare different years: this is precisely one of the aims of a standardisation like an index number. This graph might reflect the idea of cyclical unemployment. Unemployment is not a constant thing but a dynamic entity.

If we recall the heading of these two tables and one graph, 'the economic and social situation of the population', these tables tell us about absolute numbers of people asking for jobs, and in more abstract sense the development of some sort of ratio of unemployment. They do not allow you to draw any conclusions about the relation between unemployment and other aspects of the economy. Neither do they allow you to draw any conclusions about the current state of the economy – they are all past-measurements.

10.2.2 Data as embodiments and quantifying separate phenomena

The example of unemployment was used to show that data are embodiments of choices – choices related to the measurement system and choices of possible relations between the measurements and the theoretical notion of unemployment. Index numbers require furthermore the choice of a definition.

In this historical period the objectification of the economy consisted of the definition of separate phenomena (unemployment) and their specific measurements. This was complex enough in itself. Although there were theories on unemployment available these were not necessary to the CBS work in the sense that they did not start from theories of unemployment when they set out to count the unemployed. There were people looking for a job and they were observable. The mathematisation aspect consists in the problem of how to measure unemployment if you can't observe all of them at the same time.

If the quantification of unemployment succeeded, it made new inferences possible about the specific phenomenon under study. The number of unemployed in 1924 compared to 1910 is a quantitative question: comparison through time became possible by using index numbers or graphs.

In the next section I will describe how this number-work evolved in such a way that new things, apart from comparisons through time of the same entity, became possible: such as relating different sorts of entities with each other and being able to draw some conclusions about the future, i.e. prediction.

10.3 THE ECONOMIC BAROMETER

Since the early twenties, time-series data have been developed for a variety of economic quantities. These time series contained yearly or monthly measured values of such quantities as 'number of labourers in the mines', 'ships being constructed', 'number of unemployed' and 'export of products'. The measurements were represented by the graphical method, i.e. by drawing a line between the points, which mostly resulted in irregular curves. The question then arose if any relation could be discovered between the lines, i.e. the separate phenomena. For example, it was reasonable to think that there was a chronological relation between a line representing imports of a good and a line representing commodities which were made of those imported goods. The problem was the irregularities of the curves drawn through what I would call the raw data, although we know from the previous section that even these raw data were embodiments. How could these curves be made comparable in order to say something about the economy?

The development of the Dutch barometer, and of other European barometers as well, was heavily influenced by the work done by the Committee on Economic Research which had been appointed by Harvard University in 1917. As many aspects of the method used and the accompanying arguments seem to be almost literally copied from the work of W. M. Persons, who worked at this Committee, I will start in this section with a description of his work before I turn to the Dutch case.

10.3.1 The Harvard Barometer

In 1919, the journal *Review of Economics and Statistics* was established by the Committee on Economic Research. The *Review*'s first issue stated in it's Prefatory Statement that it wanted to 'provide a more accurate record of economic phenomena'. It wanted to work on the development of methods of measurement of economic phenomena in order to be able to make judgements on 'fundamental business conditions' as well as to contribute to the 'general progress of economic science'.

Its first article was entitled 'Indices of Business Conditions' and was written by W. M. Persons. In this article he criticised the way economic data had generally been analysed: 'The usual method is to compare the figure for the current week or month with that for the corresponding week or month of the preceding year, or that of the preceding week or month of the same year. Sometimes these comparisons are improved by giving them in percentage figures' (Persons 1919, 5). He did not criticise the act of comparing as such but the way it had been done. The aim of his article was to 'contrive a method of handling business statistics which will make it easy to determine the significance or lack of significance of each item in indicating current business conditions and possibly those of the immediate future' (7). Comparing data was a step towards providing a measure that could indicate the state of the economy as a whole.

Comparing time series was not an easy task. All sorts of 'fluctuations' appeared in the time series due to a variety of causes. As Persons formulated it: 'Those fluctuations are thus a confused conglomerate growing out of numerous causes which overlie and obscure one another' (8). Therefore the question was 'how to isolate and measure the influence of different sets of causes or forces.' Persons distinguished four types of fluctuations:

1. A long-time tendency or secular trend, the growth-element (for example due to the growth of population);
2. A seasonal movement (for example due to the fact that less laborers are needed in winter than in other seasons);
3. A wave-like or cyclical movement;
4. Residual variations (for example due to wars or national catastrophes).

(Persons 1919, 8)

Persons said about the third type that 'the assumption that there is a cyclical movement in series of fundamental statistics few will controvert' (33). However, concerning the first two types of fluctuations he explicitly argued that 'the existence and measurement were to be established, not on *a priori* grounds, but by examination of the statistics themselves'.[1] Despite Persons' empirical attitude, a 'backbone of theory' underlies the classifications of these types as Schumpeter wrote:

they [the Harvard approach] used what may be termed the Marshallian theory of evolution . . . they assumed that the structure of the economy evolves in a steady or smooth fashion that may be represented (except for occasional changes in gradient, 'breaks') by the linear trends and that cycles are upward or downward deviations from such trends and constitute a separate and separable phenomenon . . . this view constitutes a theory, or a backbone of one. (Cited in Morgan 1990, 2–3)

When the irregular curves through the raw time-series data had been transformed into rather smooth lines representing the cyclical movement of the economy (by removing fluctuations of types 1, 2 and 4 out of the data) Persons started to compare the lines. The idea was to find packages of lines that lagged one quarter of a period. The statistical work was inspired by the notion of a sine-function, i.e. that the cycle might be a periodical phenomenon. However, the cyclical movements were to a large extent irregular which meant in this experimental research that they did not use this sine-function to describe reality. Nevertheless, they did use the interpretation of a time-lag in terms of phases by quarters of periods. As Persons formulated it:

There had been the idea of the economy as behaving like waves, rhythmic movements, but that did not necessarily mean the periodicity i.e. the idea of equal periods: In the present knowledge therefore we are justified in conceiving cyclical fluctuations to be rhythmic but we are not justified in postulating periodicity. (Persons 1919, 35)

With the aid of correlation techniques and the graphical method, described above and more fully explained in Morgan (1990) he ended up with a set of three packages of lines, the famous A-B-C curves called speculation, business and money which can be seen as reproduced for a Dutch audience in figure 10.1.

10.3.2 Developing a Dutch equivalent?

The CBS could not just apply the American Barometer to the Dutch economy. For example, they considered the difference between an open (NL) and a closed economy (US) of major importance. But there were also doubts about the task at hand. The Dutch strived for

---

[1] Persons (1919); about the trend, p. 8, about the seasonal movement, p. 18.

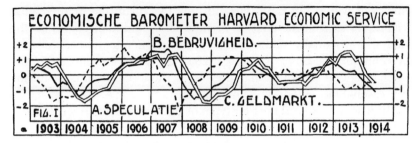

Figure 10.1 The Harvard economic barometer
(*De Nederlandsche Conjunctuur* 1929 (4), 12).

an economic barometer to consist of a small number of time series or curves, which all represent a category of economic life in the best possible way: on the one hand because of transparency, on the other hand to meet the League of Nations' Financial Committee's wish to take the Harvard barometer as an example. Whether this is possible for the Netherlands can be called into question.[2]

Considerable doubt was expressed in 1925 when the major differences between the American and national European barometers emerged at a meeting of League of Nations' experts, met to advise about the 'scientific and technical aspects . . . of economic barometers . . .'.[3] As the United States had refused to sign the Versailles Peace Treaty as a consequence of which they were not a member of the League of Nations, the Americans did not contribute to this committee. Considerable doubt as to the philosophy underlying the barometer was expressed in the Dutch notes of this meeting concerning 'The desirability to get rid of the idea of representing the national economy as a unity, as if its development could be described by a unique set of lines'.[4] Apparently people felt unsure about the barometer's philosophy from the start. Nevertheless, a Dutch barometer was developed consisting of four instead of three lines. A figure of the Dutch barometer is given in figure 10.2.

Within the Dutch Central Bureau of Statistics, the engineer M. J. de Bosch Kemper directed the research programme on the development of the Dutch barometer. In a series of articles in the CBS journal '*The Dutch Business Cycle*' (*De Nederlandsche Conjunctuur*) he presented the method

[2] H. W. Methorst, *Report on Business-cycle Research* [*Nota betreffende conjunctuuronderzoek*] 19 October 1925 CBS ARA number 12.330.1 file 381.

[3] M. J. de Bosch Kemper *Report on Business-cycle Research* [*Nota betreffende conjunctuuronderzoek*] CBS Notes of the League of Nations' expert meeting [*verslag van de vergadering van commissie van deskundigen van de Volkenbond*] 12 March 1927 CBS ARA number 12.330.1 file 381.

[4] See previous note.

employed within the CBS. His aim was to explain the mathematical techniques used in the analysis of time series and their curves as 'this analysis is essentially mathematical, in its design/intention as well as in its technique' (de Bosch Kemper 1929a, 92). His series consisted of five parts, a general introduction to the curves, the secular trend, the means, the seasonal movements and correlation.

His work introduced the four types of fluctuations described above as 'the time series data consisting of four components due to different causes'. What he called the cyclical movement was precisely the raw-data time-series purified of the trend and seasonal movement. According to de Bosch Kemper, the cyclicity was due to an internal economic law-like regularity [*interne wetmatigheid*] that made a certain state of the economy return to a previous 'state' over and over again. For what this regular law might be, he referred to the business cycle theories that were being developed at the time and presented in an overview by Persons (1926). He himself held the conviction that all these theories could exist side by side: economic life was too complex to think that one theory could always explain the cycle.

In Persons' overview, Albert Aftalion's theory was discussed which would later be used by Jan Tinbergen in his work on the model. Aftalion explained the cycle by the time-lag between the decision to produce a new commodity and the actual production of that commodity. The actual production could not take place until the producer had the industrial equipment available for the production. The construction of this industrial equipment necessarily took a certain amount of time. The forecasts on prices the producer would get for his commodity were based on insights on the economy at a particular moment. By the time the commodity was actually being produced and ready to be sold, the economy had changed and thus the producer received different prices from those he had expected to get. As Aftalion (1927, 166) explained it,

Slight fluctuations are inevitable in economic life. But the capitalistic technique of production, by its influence on forecasts and prices, magnifies these fluctuations, transforms them into wave movements of wide amplitude, and causes the fairly long cyclical variations which characterize modern economic development.

So we see that although the CBS did not work themselves on such business-cycle theories their (statistical) work was embedded in a framework of business-cycle theories which legitimated the idea of a cyclical movement. Let us now turn back to the actual steps that needed to be taken to construct the barometer.

The purification of the irregular curves of the seasonal movements and the trend has been described above. To be able to find relations among the cyclical lines, which was fundamental to the idea of a barometer, a further series of manipulations needed to be done to make them comparable. As you can imagine, the absolute numbers differed a lot among the measured quantities. Therefore the CBS' 1927 annual report said 'only the most highly necessary manipulations were executed to make the lines comparable with respect to level and scale'. First, for each series of purified data, the arithmetical mean was calculated. This became the equilibrium-line, normalline or zero-line. Secondly the deviations of each datum from the equilibrium were calculated. The average value of the deviations got the value '1' in the graphical representation. This process meant that lines that consisted of quantities of highly diverging magnitudes, became more smoothed, while lines with little variation became stretched out.

When this had been done, one could finally start comparing the lines, to find if there were any *correlations* between the lines. Apparently, one of the ideas was to develop one line representing the state of production. The moment of choosing the right standard was, of course, a moment of negotiation. For example: Which standard could be used for production? 'Imports of raw materials for industry' seemed a reasonably good choice. Then a decision had to be made whether to count value or weight. Both had their pros and cons. Counting weight meant that more 'forecasting potential' (*voorspellingspotentie*) was attached to heavy materials than to light ones like seeds.[5]

Later, it was proposed to change the standard from 'imports of raw materials' into 'expenditure on coals'. However, Mr Fokker, who was a member of the Central Committee of Statistics (the CBS's advisory board), considered it indefensible to think that the consumption of coals could tell you anything about the course of production in agriculture for example. He was not against the idea of gathering data, but drawing conclusions from those like 'coals determining the course of total production' was not justifiable.[6]

Finally, the Dutch barometer (figure 10.2) ended up with no line representing production at all. Instead of the Harvard barometer that consisted of three packages, the Dutch barometer consisted of four,

[5] 19 October 1925 CBS ARA-number 12.330.1 file 381.
[6] Mr E. Fokker to Mr H. W. Methorst (director of the CBS) 7 March 1926, CBS ARA-number 12.330.1 file. 381.

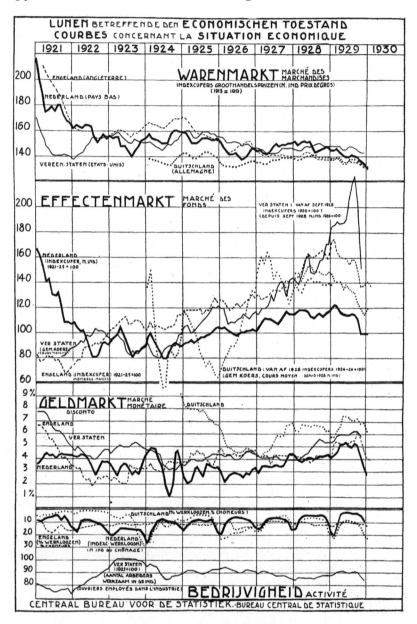

Figure 10.2 The Dutch economic barometer (*De Nederlandsche Conjunctuur* 1930(2), 11).

which were coined the commodity-market, the stock-market, the money-market and economic activity. The Dutch package 'economic activity' mainly consisted of index numbers of unemployment (de Bosch Kemper 1929, 10). This can be seen in the fourth panel of figure 10.2. The commodity-market consisted of index numbers of whole-sale prices. In the Harvard case, the index numbers of retail prices were part of the package called 'business'. So we see that the highly 'experimental' method of constructing the barometer led to quite different national barometers.

A second difference between the American and the Dutch barometer was the graphical representation. While the Harvard lines were represented on one picture, the Dutch lines were represented in four pictures. The Dutch solution seems strange with respect to the aim of the barometer. If we recall the aim of the barometer, it should be able to predict an upward or downward movement of one line when we know the current movement of another line: the movement of one line at a certain moment in time tells us something about the movement of another line in the (near) future. The relation between the lines is much more visible in one picture than in four separate pictures. However, as suggested above, the Dutch had doubts about the philosophy of the barometer and this way of representing it might be considered as their modification of the whole enterprise. Another reason might be the CBS's problem with prediction which will be dealt with in section 10.6.2.

A third difference between the Dutch and American barometer also lies in their representation. The Harvard barometer only portrayed packages of lines representing the US economy. The Dutch also represented comparable lines of the foreign economies of Britain, Germany and the United States. The difference might reflect the self-image of their economies: the United States as a closed economy and the Dutch as an open economy. It might even go further: maybe the CBS held the opinion that one (aggregated) line like 'commodity market' correlated better with foreign equivalents than with some other Dutch line. For example, the number of German unemployed might be a better predictor to the Dutch number of unemployed than the Dutch commodity market. This means – if my hypothesis makes sense – that the Dutch conception of a barometer was slightly different from the American one: the Dutch barometer makes the relation between different phenomena a weaker element, while the relation between equal (foreign) phenomena is emphasised.

10.3.3 The barometer: quantifying relations between phenomena

The continuity between the barometer and the index numbers developed previously can also be shown by the fact that the barometer was sometimes actually called an 'index of the economy' (by Persons for example). The development of the barometer consisted of several steps to transform raw data into some sort of cyclical movement. First, the time series were purified of the trend and the seasonal movement. Secondly, the lines were scaled. Then, with the aid of correlation techniques those time series were put together into one index number representing 'business' or 'money market'. To the element of 'comparing' figures (one of the reasons why index numbers had been developed in the first place), now the element of 'prediction' was added. Whereas index numbers were mainly about separate economic entities like 'unemployment', the barometer aimed at representing in a quantitative manner relations between complex phenomena. In that sense it was a next step in the objectification of the economy.

At first sight, the barometer looks like a descriptive device. The barometer represented relations as correlations which are quantitative relations. The problem of not being able to deal with causation in a quantitative way, not to be able to show exclusive causes, was already felt in the late 1920s. Persons *et al.* believed that the 'three factors shown in our index-chart are in constant interaction; that each of them affects the other two; and that no one is free from the influence of the others.' (Bullock, Persons and Crum 1927, 80). The question was then what these patterns of interaction looked like. As they formulated it:

If we could, by scientific experiment, isolate economic factors, we could deal with the problem of causation in a much more satisfactory manner . . . But even with these disturbing factors removed [trend and seasonal movements] we are far from having conditions similar to those which the experimental scientist can secure in his laboratory, and are therefore seriously hampered in our study of causal relations.

Persons et al. apparently thought it was the economic matter that prevented the study of causal relations in a quantitative way. The economy did not 'allow' you to do laboratory experiments. Persons would have liked to have measured causes: that was precisely what the barometer did not do – it measured 'sequences'.

However descriptive the barometer might seem, the idea of four types of fluctuations was not completely anti-theoretical. Second, the idea of a cyclical movement as such was supported by all sorts of business-cycle theories. These theories dealt with questions of causation of the cycle and most theories had different ideas about them. These theories were not mathematised: although they were supported, as in the case of Aftalion, by statistical material, the argument as such was qualitative (verbal).

The conclusion I want to draw so far is that the barometer was a next step in the series of abstractions of the economy. The measurement and definition of unemployment data and index numbers were a 'first' abstraction of the economy. It meant the creation of a group of people sharing one quality – this was an abstraction from individual unemployment, every individual having their own characteristics. The barometer went one step further. The cyclical movement was a (quantitative) construction based on the first abstraction. The cyclical movement was also a theoretical entity. And this theoretical entity supported the quantitative work.

The aim of the barometer was to be able to make some sort of a prediction. The fact that the barometer failed to predict the crisis of 1929–33 was one reason why research went on to develop more satisfactory predictors of the national economy. Mathematical treatment of phenomena and prediction of the course of these phenomena had become closely related in the twentieth century. Therefore the search for a mathematical formalism to represent the economy went on.

10.4 THE MODEL

10.4.1 Towards a new research programme

Changing a research programme is, apart from an intellectual development, also an economic affair. Usually it goes together with the need for more or new personnel. This was also the case with the CBS's Department II which had worked on the economic barometer and wanted to change the research programme towards the model in the second half of the 1930s. This required legitimation of the enterprise: arguments why the model should be developed at all, and what kind of 'creature' it actually was, had to be made explicitly. Therefore this procedure is discussed in some detail.

In 1935, the CBS proposed to expand Department II in an internal report which was later sent to the Ministry of Economic Affairs.[7] The CBS considered business cycle research to be the following:

---

[7] 11 July 1935 CBS ARA-number 12.330.1 file 381.

A. The diagnosis of the cyclical movement which consists of a *description* of the movements of some of the most important economic phenomena as well as the judgment of the *phase* of the cyclical movement; to know if the economy has more or less reached the equilibrium-state is of particular importance;

B. The verification of the causal relations between the movements of different economic quantities and the figures that tell us how much a movement of the one entity influences another; this must be done to judge the *automatic* movements of the economy as well as to create 'building-stones' for evaluating possible economic policies.[8]

This statement shows first the continuity between the barometer and the business-cycle research. The barometer was a descriptive device and aimed to analyse the phases of the cyclical movement to know if a depression or a boom was going to occur. Part A seems to be an immediate continuation of this kind of work. However, if we read the additional text (unquoted), we see that this part of the programme had become more ambitious. They wanted to develop a '*tableau economique*', which would describe statistically the main *flows* of goods and money. This is the development of the national accounts which will be dealt with in section 10.5.

Part B of the programme shows that the previous intellectual gap (the people who developed the barometers were different from those who developed the business-cycle theories) between the descriptive barometer and the business-cycle theories was going to be solved by trying to quantify the causal relations. The phrase 'automatic' movements is new: the introduction of this idea of an automatic movement of the economy and to stabilise the economy announces cybernetic ideas (cybernetics itself only emerged during World War II). The wish to manage or control the economic system was associated with a view that pictured the economy as a machine, an 'automaton' (Kwa 1989, 9). A machine can be described in terms of causes and effects, and this stimulated the possibility of representing the economy in terms of causal relations. The connection to steering the economy and the wish to evaluate policies are of considerable importance as will be shown in section 10.6.

Additional information on part B explicitly stated that 'the causal relations between separate economic entities' movements must be known quantitatively'. Therefore, the report went on: 'it is better to examine a heavily simplified model of reality in all its details than to design a complex picture of reality which is impossible to handle and which

---

[8] 24 December 1935: Central Committee to the Ministry of Economic Affairs, directed to the Minister himself. CBS ARA-number 12.330.1 file 381; emphasis added.

necessarily contains gaps'.[9] Here we see the term 'model' being introduced. The report included an example of an algebraic model for the US economy written in terms of 'physical entities', 'prices' and 'sums of money'. (This model will be discussed in section 10.4.2 below.) The problem then was, in order to make such a model for the Netherlands, they needed to formulate the equations of the model and gather data before they could calculate the statistical relations and thus compute its parameters. Therefore the expansion of the department was also needed to increase the gathering of data in much more detail than had been done so far. This area of research was to work out details for sectors, and improve statistics, both aiming at the 'possibility to establish policies for individual sectors'.

In an accompanying letter, another motivation was given to expand Department II:[10] 'Barometers did not represent a closed system of variables, i.e. a system of variables which formed a simplified scheme of economic reality as a whole.' A reference was made to Mr J. Tinbergen, who had already suggested such a scheme (model).[11]

According to the procedure, the Ministry asked the Central Committee for Statistics (CCS) for advice. They advised positively. Clearly, the department's aims had become much more focused. They were now stated as 'the diagnosis of the cyclical movement meant to be a description of movements of the most important economic phenomena aiming at providing material for the judgment of future economic policies'.[12]

In a report by Department II sent to the CCS in 1936, the term 'model' was mentioned again.[13] The formulas in the model represented 'as it were the dynamic laws which control (*beheerschen*) economic development'. In a kind of rhetorical way, the report implicitly stated that the model would be the basis of further research. The following was underlined: '<u>In order to have a fruitful discussion, objections to the presented model should refer to a specific equation or indicate which detail of the model is considered necessary.</u>' Objecting to the model as such was not fruitful: critiques should be focusing on choices made in the model but not on the development of the model as a proper knowledge device. The

---

[9] 24 December 1935 (see n. 8), p. 2.   [10] Letter 23 October 1935 CBS ARA-number 12.330.1 file 382.

[11] They referred to a lecture by Tinbergen at the fifth meeting of the Econometric Society held in Namen in 1935.

[12] CCS to Ministry of Economic Affairs 24 October 1935 CBS ARA-number 12.330.1 file 381.

[13] Report Department 2 '*Nota betreffende het algemeene conjunctuuronderzoek*' September 1936 CBS ARA-number 12.330.1 file 381.

model was also described as a 'scientific barometer'. The CCS did not quite understand this new product and asked for explanation. The CBS answered by describing the model as a 'method of refining and systematising experience' (*methode van verfijnde en gesystematiseerde ervaring*), which in fact had been adopted for ten years now (referring implicitly to their barometers). The only new aspect was that 'the analysis was based on a closed system'. Another description of the 'model' was the 'best vision' (*beste visie*):

Although one never elaborates on it explicitly, a 'model' of reality underlies every sound reflection on the economy. However, never before, has such a model been adapted to real figures as well as the represented model. Therefore, deductions from this model are currently the best approximation of the reality-as-a-whole.[14]

So, we see that by the late 1930s the model had become a central area of research within the CBS. This model, however, caused quite a bit of confusion: was it a scientific barometer, a vision, a machine, or what?

10.4.2 The model: a mathematical formulation of the business cycle

Within the CBS, Jan Tinbergen had taken over the lead of Department II from M. J. de Bosch Kemper. Tinbergen had started working at the CBS first as alternative employment for military service in 1928 and he had been fully appointed in 1929.[15] He had finished his thesis on physics and economics in 1929 and then worked on the development of a dynamic quantitative representation of the economy. Tinbergen was not alone, as Morgan (1990, 74) shows: 'At the same time as the statistical genesis of business cycle data was being investigated in the late 1920s and early 1930s, there were parallel developments in the mathematical formulations of dynamic theories of the business cycle.'

By 1935, Tinbergen had done many studies on specific questions like the relation between a price of a good and its demand or supply, of which many were published in the CBS journal '*De Nederlandsche Conjunctuur*' (*The Dutch Business Cycle*) We could describe these studies on specific instances

[14] Report Department 2 '*Nadere toelichting op de nota betreffende het algemeene conjunctuuronderzoek*' 15 October 1936 CBS ARA-number 12.330.1 file 381.
[15] Tinbergen took over the directorship from de Bosch Kemper between 1930 and 1935. M. J. de Bosch Kemper might have been heavily disappointed by the ambitious task of predicting the course of the economy. There is nothing biographical to find about him; the only things he did after his work at the CBS which I managed to find are translations of occult works.

as follows: what is the specific change in one specific entity under the influence of a known change in another entity given a certain time-lag. The guiding question as he formulated it in 1931 was this: 'The basic problem of any theory on endogenous trade cycles may be expressed in the following question: how can an economic system show fluctuations which are not the effect of exogenous, oscillating forces, that is to say fluctuations due to some "inner" cause?' (Tinbergen 1931, 1).

In 1935 he published a critical paper in German called 'Quantitative Questions on Economic Politics/Policies (*Konjunkturpolitik*)' (Tinbergen 1935). (It is probably this 1935 article that was referred to in the previous section.) As this article is one of his first in which he presented a model and actually called it so, I will go into this article in some detail.

Tinbergen started his 1935 article by stating that the economic crisis of the 1930s had brought business cycle policy and its problems to the fore. In Tinbergen's view, these problems of business-cycle policy could not be solved without a proper business-cycle theory. Under the heading 'The Dynamic Structure of Business Cycle Theory' he argued that a mathematical treatment was fundamental:

The theoretical-economic structure of problems of business cycle policies has not been rightly assessed. If one wants to arrive at . . . numerical explanations/judgments [*aussagen*] one needs a quantitative scheme [*Rekenschema*] in the first place which means nothing else than the knowledge of the theoretical-economic structure. . . . Questions of economic policies that have fully and adequately been solved are those which belong to the statistical realm . . . However problems of business-cycle policies are even more complicated. For in that case one asks for a dynamic movement of the economy . . . The strong reason to look at dynamic changes is very often not recognised . . . And if one recognises the dynamic nature of the problem rightly, one rarely fully understands the systematic reactions of an intervention in the economy . . . The only possible way [to understand the relations between economic entities] is a mathematical treatment. If we don't want to work 'ins Blaue hinein' we also need to be informed about the value of the reactions. What we need therefore is a combination of theoretical-mathematical and statistical investigations. (Tinbergen 1935, 366–7)

This summary shows that Tinbergen wanted to represent the structure of the economy in a mathematical way. The statistical part should investigate which mathematical representation corresponded best to economic reality. In that sense 'reality' meant the 'data' of this reality (like the unemployment data in section 10.2).

The structure of the economy consists of pre-determined variables and other variables whose values are determined within the system. Pre-determined variables are ones like the interest-rate for example, and they

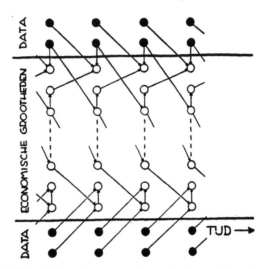

Figure 10.3 The relations between predetermined variables and entities in time
(Polak 1939, 65).

determine the cyclical movement of the economy. Endogenous variables are the ones that need to be explained. The terminology used at the time was rather more confusing: Tinbergen (and his followers) used the term 'data' for these pre-determined variables. (This means that whenever he uses the terms 'data' one needs to be aware what type of data he actually means.) 'Every entity at some moment in time is determined by data (predetermined variables) and entities of some earlier moment in time or the same moment in time.' (Tinbergen 1935, 368) In 1939, Tinbergen's student J. J. Polak visualised this view which is represented by figure 10.3.

In this picture the entities are the white points and the data are the black points. The arrows represent the relations. The picture shows that every entity is 'directly' determined by other entities and data of one previous time-unit or of the same moment in time. The repetition of this 'causal structure' for every time-unit means that every entity is influenced 'indirectly' by entities of many earlier moments in time. The interesting point is that the causal structure itself remains constant over time.

The structure of the 'system' and the values of the constants determine the endogenous movement which he defined as the 'eigen-movement' of the system. Prediction now becomes possible: 'If we know the constants and the beginning values of the system, we can predict the system movements' (Tinbergen 1935, 369). It was of crucial importance to business-cycle policies to know the 'eigen-movements' of the system.

Therefore Tinbergen continued by defining a 'simplified model and its equations':

> In order to know the eigen-movements we have to construct a model which only consists of regular relations . . . If we want to be able to do something with it, reality must be represented in a simplified manner. Business-cycle descriptions have been doing this already as the introduction of different sorts of index numbers has shown. Instead of investigating the development of every single price, we only use one price-index. This way of looking was coined 'macro-dynamics' by R. Frisch. (Tinbergen 1935, 371)

We get a clearer idea about a model now: a model consists of dynamic relations between aggregate entities and pre-determined variables which together define the structure of the economy and its endogenous movement. The next step was to calculate the values of the parameters of the model. As appropriate data to calculate his 'formal' model were not at Tinbergen's disposal, he could not do this. In his article he only showed some difficulties with establishing such fixed relationships.

The question then was how the model could deliver the right policies:

> The mathematical meaning of intervention in the economy is the change of the values of some key parameters . . . The movement of the model boils down to the value of the key parameters. . . . The main question is then: how do we have to establish or change the key parameters in such a manner that the development of the economy will react in the best possible way to outside shocks. Mathematical considerations lead to the insight that damping the cycle is the best solution to this problem. (Tinbergen 1935, 380–1)

The mathematical 'considerations' were, as Boumans (1992) has described, the reduction of the model to the 'characteristic' equation and the solution of this equation in terms of a combination of sine-functions with different periods. Basicly, this means that Tinbergen represented the economy in terms of a sine-function. Whereas in the case of the barometer the sine-function was a 'source of inspiration' to find regular movements in the data, the sine-function was now used to describe economic life.

It was only in his 1936 model of the Dutch economy that Tinbergen succeeded in testing his model and proposing the best business-cycle policy: devaluation. This is why one usually assumes that the first model appeared in 1936. The CBS programme, described in the previous section, referred to the 1935 model described above. The word model was clearly used already in that year 1935. But this was not a completed, calculated model.

As Boumans shows in his work, 'economic reality does not prescribe any particular kind of formalism' (1995, 2). Therefore, economists like Jan Tinbergen had to make a choice about what kind of formalism would explain the business cycle (see also Marcel Boumans' chapter, this volume). As I wrote in the previous section, all sorts of business-cycle theories had become available that explained the business cycle in various ways. Tinbergen was not totally naive about what such a cycle formalism could be, but on the contrary very well informed.

What he was looking for was a theory that explained the cycle in an endogenous manner. As Tinbergen phrased it, 'The aim of business cycle theory was to explain endogenous movements, and only a dynamic theory could explain these' (cited in Boumans 1995, 6). And this he found in the above-described theory of Aftalion, who had explained the business cycle by the time lag caused by the time needed to construct industrial techniques. So, Tinbergen used Aftalion to explain the cycle in terms of a time lag, and this time lag was already available as a descriptive element in the barometer.

Tinbergen wanted to describe the economy as a 'mechanism'. He described this mechanism in a system of equations. This system had to fulfil at least one important condition, namely that the number of variables equalled the number of equations. These equations formed a network of causal relations. The choice of the equations themselves was also based on theories available (Morgan 1990, 102). Whereas in the barometer research causality was mostly delegated to the qualitative theoretical domain, Tinbergen wanted to quantify causality.

The macroeconometric model Tinbergen developed was therefore an integration of mathematical techniques (the formalism), bits of cycle theories (Aftalion), and statistical techniques and data (to calculate it) (on integration, see Boumans, this volume). If we see this piecemeal work of Tinbergen, the confusion of the CBS and the CCS becomes apparent. In a way, the model could be understood as a 'scientific barometer' in the sense that the periodicity which was first *a priori* assumed and phrased as rhythmicity had now become mathematically expressed, i.e. scientifically validated. Tinbergen himself used the description 'examples of theoretically founded barometers' in his 1934 article (Boumans 1992, 75). Many of the aspects of the barometer had been taken over, the time lag, the rhythmic movement, the trend. In the model these got 'theoretically integrated'. In his 1936 preadvice, Tinbergen called the model 'a mathematical machine'. If the structure, constants, parameters and beginning values are known, mathematical techniques calculate (that is what the machine does) the outcomes of the entities. The machine itself does not change the economic meaning – that is why Tinbergen called it a mathematical machine. The word 'mechanism' makes the term 'machine' understandable. The model was also a vision because it was based on theoretical ideas on the business cycle.

10.4.3 Continuity

The continuity between barometer and model was again visible in the terminology: model = barometer, as was the case with the barometer: barometer = index number. We have seen that many aspects of the barometer were taken over in the model research: the notions of periodicity of the cycle, the level at which economic entities are measured, the time lag, the time series, the wish to predict, the notion of relations between entities and also the theoretical notions about causal explanations of the cycle.

The barometer aimed at predicting when a boom or crisis of the national economy was likely to occur. The model added to this a more specific purpose – it should tell the policy maker which entity or entities in the economy was (were) responsible for the changes in the economy.

In the previous sections 10.2.2. and 10.3.3 two stages in the 'objectification of economic life' have been described: first the data representing distinct phenomena abstracting from individual/specific qualities, and second the barometer as an attempt to represent the economy as relations between temporal, rhythmic and aggregate sequences. The model was the third step in the objectification of the economy. Relations between aggregate phenomena were now mathematised in terms of aggregate entities and their mutual relations – the relations had been given a mathematical form. In this sense the model translated the former qualitative business-cycle theories into a mathematical representation.

One of the big problems for Tinbergen in the development of his 1936 model of the Dutch economy was the lack of data. As he wrote himself, 'Measurement of the variables was only possible in a limited way. Considering the current state of the statistics it was hardly possible to get something better out of it and considerable lacunas are there in the figures that are there at all' (Tinbergen 1936, 69). If the model was to be successful, better data needed to be developed. Remember that the aim of the model was to establish consistent business-cycle policies. This means that the formal structure must get empirical meaning and validation. This resulted in the development of the national accounts, which is the topic of next section.

10.5 THE NATIONAL ACCOUNTS

10.5.1 Measuring the economy

One line of research within the CBS from the barometer to the model was described above. The second line of research was the development of national accounts. National accounts have come to be seen, as was already said in the introduction, in economists' terms, 'a neutral registration of the economy'.[16] The research on national accounts was first announced in 1935 (as noted above, section 10.4.1). Full-fledged articles and monographs in this area of research by CBS people started to appear only in 1939.[17]

The development of national accounts was a very complicated problem in which three issues emerged but none of which could be dealt with independently. First, it required a definition of national income. This definition was related to the economic categories to be measured. Secondly, not everything one wants to measure is measurable. The question of what could in fact be measured was therefore important. And thirdly, what was required from the model, and what could be gained from the model in terms of both definitions and measurements.

The development of the accounts depended on the definition of national income which remained controversial until the beginning of the 1940s.[18] The economic entity 'national income' itself has, however, a much longer history (see, for example, Studenski 1958). O. Bakker, who worked at the CBS until 1946, proposed the following definition of national income in 1939: 'A social product, i.e. the total added utility embodied by goods and services, reduced by that part that is equivalent to the replacement of capital, produced in a closed economy.' The aim was to measure this social product. Even if it were possible in a very small economy to sum up the production (in terms of separate commodities and services), in a large economy that was impossible. This meant that all economic goods and services adding up to the social product should be lumped together. The solution was to express these goods and services by the value we attach to them in terms of money. This meant that goods and services which could not be translatable in terms of

---

[16] For example B. Compaijen, R. H. van Til (1978).

[17] Two dissertations had already appeared in a very short time 1936 and 1937 (by Groot and Kruisheer).

[18] C. Oomens (1985 24) and C. van Eijk in CBS (1993); both started their histories of the national accounts in the 1940s.

money did not contribute to the national income. 'The main entry in this respect is the value of labour of those who work the hardest, have the longest working-days and who usually get hardly any appreciation for their work, the *housewives*' (Bakker 1939, 232).

In a 1940 article on the issue of measuring national income, Derksen (the later head of Department 4, previously Department II) proposed more or less the same definition, but added a time element and the income from abroad to it. He also spent some time reflecting on the choice of goods and services that should count as economic goods and services contributing to the national income. His answer was to include everything which was visible in exchange (*ruilverkeer*) (Derksen 1940).

His choice was a methodological one. The starting point was that national accounts should be based on counting exchanges between two actors. This created the possibility of controlling measurements: a payment from actor A to actor B was visible in B's accounts as a receipt. This meant that services of housewives, and 'natural raw materials' could not be included in the national income. Domestic labour was even put on a par with services delivered by cars – and of course, these could not considered as increasing national income![19] We can conclude from this that the exclusion of domestic labour became unproblematic during this period.[20]

The second major problem besides definition was that the national income must be measurable which means that a methodology was needed to measure it. Nearly all the CBS literature referred to the method adopted by W. A. Bonger who had done research on measuring the national income and capital (*inkomen en vermogen*) (Bonger 1910; 1915; 1923). He introduced the 'subjective' and 'objective' method. The first meant 'counting incomes/earnings/wages', the second meant 'counting net production/expenses'. The results of both measurements should be equal. The subjective method made use of income-tax statistics (to which he added 10% as a result of fraud!). The CBS used these methods in their research on national accounts. The two methods did not produce the same results, but the differences in measurement were due to incomplete statistics and were not thought to reflect problems in the methods.[21] In a later publication, only one figure for national income

---

[19] No. 2 in the series Monografieen van de Nederlandsche Conjunctuur *Enkele berekeningen over het Nationaal Inkomen van Nederland* CBS 1939: 6/31.

[20] Only recently has the way models and national income deal with unpaid labour become of central concern to feminist economics (and the exclusion of raw materials has become an issue as well). Unpaid labour should be considered as complementary to paid labour. For example M. Bruyn-Hundt (FENN-seminar 8-12-96), J. Plantenga. *Rekenen met onbetaalde arbeid* Advies van de Emancipatieraad April 1995.   [21] For example Monographe nr. 2 (note 19).

was published, and the separate results of both methods were not given anymore.[22] C. J. van Eijk, one of the model makers who later worked at the CPB, said in an interview in January 1993 that 'of course the two methods never produced the same results. Corrections were often needed.' Nevertheless, the methods as such were not contested.

Treating money as the unit in which every economic transaction could be expressed was never discussed, however, it permanently caused troubles. Inflation, for example, made the value of a product not completely comparable with the value of it the year after.[23] Another example was the need for a fixed exchange standard for the value of florins, gold, and the dollar: the fixed units had to be created with regard to their value.

One of the reasons for developing the national income measurement was to create a number that represented the national economy as a whole. This number made the state of the economy comparable with itself through time. It also made the economies of different countries comparable with each other. The status of the national income as *the* number representing the state of the economy was fully established when the United Nations accepted it as *the* standard for a nation's wealth. Membership contributions came to be based on this figure (Rijken van Olst 1949, 91–3).

Derksen in his 1940 article gave a third reason for the development of national accounts, namely their connections with Tinbergen's model of the Dutch economy:

Problems of business-cycle research on the Netherlands are discussed in terms of a system of equations that have been established by the mathematical-statistical method. It has become clear that some of these equations can be established and examined more accurately by using the recent figures on the national income. Tinbergen's method has therefore gained in exactness and suitability. (Derksen 1940, 594)

Derksen suggested that Tinbergen's method could solve a measurement problem concerning the question: Do consumers pay less taxes than the value of governmental services they get in return? and if so, which bit of the national income paid this gap? This related to the question: Which governmental services should count as production and which as consumption? For example: Does a road serve production or consumption? (Derksen 1940, 577–8.) So we see that Tinbergen's model was used as an

---

[22] Monograph no. 4 (1941) in the series Monografieen van de Nederlandsche Conjunctuur, *Berekeningen over het Nationale Inkomen van Nederland voor de periode 1900–1920.*

[23] For example: Monograph no. 4 (1941) in Monografieen, p. 21: there they wrote about the corrections that were needed to account for changes in the value of money.

aid in decisions concerning measurement and was itself used as a reason to develop the measurement system.

The CBS published their first 'calculations on the Dutch national income' in 1939.[24] It was meant to be a first exploratory overview. The work was done in department II, the same department which had developed the barometer and the model. It was executed by Derksen under the supervision of Tinbergen who was director of the department. The economy was divided into three 'view-points' from which one could look at the economy: production, consumption and distribution. Industries make profits and these are measurable. These profits *flow* into the economy via wages, salaries and other sorts of income which are also measurable. These incomes *flow* to the industries in return for the consumption of commodities, which are more difficult to count. Studying these flows delivered insights in the structure of the economy. It became visible which industries contributed what amounts to the national income.

The CBS produced two other monographs in 1941 and 1948 on calculations of Dutch national income. They were mainly attempts to calculate the national income for other periods. It was only in 1950 that the system of national accounts was published as an integrated system described by the flows described in the previous paragraph and represented in figure 10.4. The figure shows that the economy was presented as a closed system. There are no open ends. The open economy was 'closed' by introducing an entry 'abroad' (A). The picture depicts a structure according to the classification of consumers (C), industries (B), government (D), and the amounts of money that flow from the one to the other (F, G). F represents profits, wages and salaries, while the other flow G represents the value of goods and services of one year. In that sense it is a static picture: it tells you which flows are going around in the economy but it does not show any development. Therefore the system does not allow you to make any predictions.

The structure as it had been developed did allow one to make comparisons between static pictures as it was now possible to calculate national income of pre-war years (for example 1938). The comparisons based on the adopted measurement structure showed that the economy had grown since then. The structure itself was apparently seen as timeless, as they thought it possible to use this structure for pre-war years as well as post-war years.

---

[24] Monograph no. 2 in the series 'De Nederlandsche Conjunctuur' *Enkele berekeningen over het Nationaal Inkomen van Nederland*: CBS 1939.

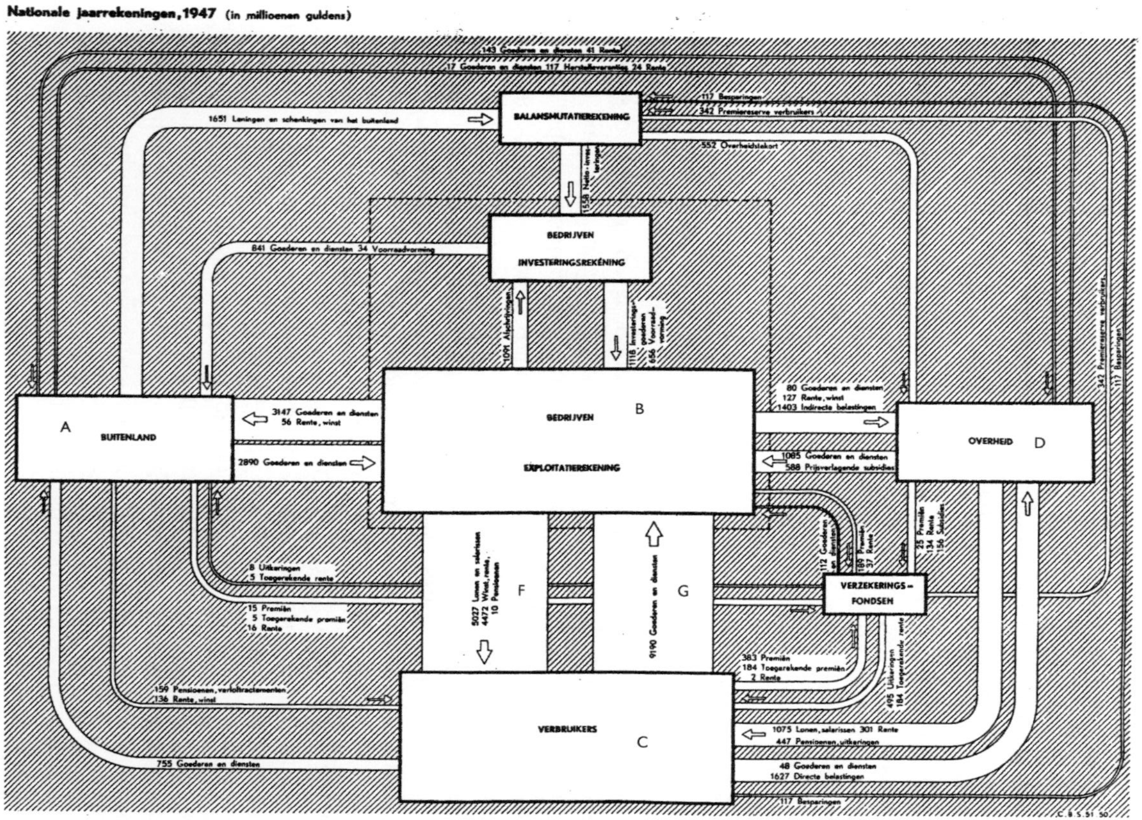

Figure 10.4 The National Accounts (CBS 1950, 105).

10.5.2 National accounts: a timeless data structure

The model, as it was described in section 10.4, aims at representing economic reality. The structure is an abstraction of that reality. But how can we make sure that this abstraction makes sense? In my view, that is precisely what the national accounts do: they are a data structure which is considered an appropriate representation of reality. Results of the model can be matched to a data structure – i.e. reality – to investigate if the abstraction makes sense.

The only way to make this work was to harmonise the national accounts and the model entities. If we compare the Polak picture (figure 10.3, section 10.4.2) with the picture of the national accounts (figure 10.4) the *flows* in the last picture are the *points* in the first. The values of all the flows are a set of data for one period in time. If we make these data-sets for several years it becomes possible to test the model. We calculate the outcomes for a model based on the structure of the 1950 economy and see if its predictions are the figures of the 1951 data.

I described the national accounts as a timeless and static data structure, considered applicable to the economy of every year. In Polak's figure, I also remarked that the structure of entities and their mutual relations repeated itself for every moment in time. Polak himself formulated this as follows: 'The question arises, if the plane can be extended infinitely. In the direction of time we can go back infinitely into the past indeed – and in principle we can go further infinitely into the future' (Polak 1939, 66). This quotation shows that the structure of the model and thus of the economy was considered constant. This assumption is of course fundamental to the whole enterprise. Otherwise it would have been impossible to develop any consistent time series. At the same time, it was the price that had to be paid to make the mathematisation process successful – it is not at all clear *a priori* that the structure of the economy would remain the same over time.

10.6 ORGANISATIONAL AND INSTITUTIONAL ASPECTS

10.6.1 Gathering data

The development sketched above from primary data (reflecting an institutional framework, norms and values) to what came to be called a model and national accounts, took place in a period during which policy intervention in economic affairs was heavily discussed.[25] It was not only discussed: actual policies were set up, the first being protectionist regulations on agriculture in the early thirties, and as a result agrarian statistics had expanded (de Valk 1941, 321–3). After 1935, when the threat of war was increasingly felt, a rationing system was set up and agencies mediating between government and industries were installed. Ideas of planning and increasing political involvement in the economy created a desire for more numbers to sustain and rationalise the political interventions. Idenburgh, the later director of the CBS, formulated it in these terms:

... it is understandable that statistics have received particular interest because 'conscious management' [*bewuste leiding*] of social and economic life has become increasingly important in our time. The development of affairs [*ontwikkeling der dingen*] can be entrusted to the free market [*vrije spel der krachten*] less than ever ... In circles of trade, industry, banking and commerce, desire for methodically gathered and manipulated numbers which could underlie decisions of authorities was felt too. (Idenburgh 1940)

A few months later, it was even stated as a fact that the 'government could only intervene in economic life successfully if they knew the factual circumstances and if they were able to calculate the effects of given policy measures' (van der Zanden 1941). Rationing led to increasing budget-statistics for the authorities wanted insight in people's incomes and their spending-patterns. Stock statistics had been set up in the anticipation of shortages (de Valk 1941).

Not everybody was happy to supply data to the government. The first law established to oblige a specific group of people to give statistical information to the CBS was the Law of 1 December 1917. The First World War had made it necessary to collect data on the production and consumption of raw materials by industries. This law enforced industries to provide their data and proved to be very useful to the task of the CBS, 'especially in small and medium-sized businesses many objections had been raised to fill out the questionnaires despite meetings with manufacturers. Without coercion we would not have achieved much.' (den Breems 1949, 16.)

One of the effects of the different lines of research described above (models and national accounts) and the increasing demand for numbers in the 1930s, was the expansion of those economic statistics collected. To

[25] The political background is described in van den Bogaard, 1998. It describes the development of the CPB as the emergence of a 'package of planning'; a package which tied certain (economic) politics, mathematics, institutions together.

enforce industry, banks etc. to give their numerical information to the CBS, a new law had come into force in 1936[26] to solve two problems. First, more organisations: industry, banks, and others, were forced to provide the data the CBS asked for. Secondly, protection against abuse had to be regulated. One of the clauses stated that 'data acquired under law should not be made public in such a way that others could get information from them about the specific providers'. To refuse to give the required data was punishable by three-months imprisonment or a fine. Providing wrong data was punishable by six-months imprisonment or a fine.

This law was a good weapon in the hands of the CBS against those who were unwilling to give up secret information. Cases for which it was unclear if the specific institute was covered by the law were mostly won by the CBS. For example, small banks considered themselves free from this law.[27] However, the minister forced them to abide by the law.[28] The minister was most of the time on the CBS side and so the definition of what counted as 'economic statistics' was broadened in 1938, to cover more organisations.[29]

Before the war, the CBS was obliged to ask permission from the minister whenever they wanted to start collecting new statistics. Soon after the outbreak of the war the CBS asked Hirschfeld, the head of the Ministry of Economic Affairs (*Secretaris Generaal bij afwezigheid van de minister*), to authorise the CBS to start collecting new statistics, which he approved of.[30] This meant more centralised power for the CBS.

Before and during the war new institutes had been established dealing with all sorts of economic affairs: a system of distribution, agencies mediating between government and industries and a corporative organisation (Woltersom-organisatie) had been set up. All these institutes had 'planning' tasks. As a result they had all started to collect their own data. The CBS wanted to supervise the establishment of questionnaires, and oversee definitions, 'to assure uniformity'. In 1942 they were given this monopoly by the Germans.[31] Every policy organ was obliged to ask the CBS for approval of new statistics, new questionnaires, the aim of their research, etc. By this new rule the CBS gained control of definitions of categories, the way questionnaires

---

[26] 28 December 1936 *Staatsblad van het Koninkrijk der Nederlanden* nr. 639DD.
[27] 20 October 1937 CBS' director to the Minister of Economic Affairs CBS ARA-number 01.3 file 80. [28] CBS ARA-number 01.3 file 80.
[29] CBS ARA-number 01.3 file 80 H. W. Methorst to the Minister of Economic Affairs 2 March 1938/The Minister's answer to the CCS 1 April 1938 / CBS to the Minister about the positive CCS' advice 15 November 1938. [30] 30 August 1940 CBS ARA-number 01.3 file 80.
[31] 3 November 1942 *Nederlandsche Staatscourant*.

were drawn up, etc. The other institutions only kept the executive right of gathering data.

After the war, the CBS did not want this rule to be abolished, and proposed to transform this rule in a Royal Decree.[32] To legitimate their claim, they argued that the 1942 rule had improved uniformity of investigations. It had made the work much more efficient. The reason for their proposal was that immediately after the war, the national agencies (*rijksbureaus*) and the industrial organisations had immediately started gathering their own data again without asking the CBS' approval, referring to such reasons as 'the urgency of the situation', and the 'bad infrastructure which made negotiating with the CBS a complex matter'.[33] The minister refused to consent to the Royal Decree, but supported the wishes of the CBS by giving a directive which, in terms of content, was practically the same as the 1942 rule.[34]

The increase in the construction of numbers led to a demand for educated statisticians. Statistics education was however in a very bad state (van Zanten 1939). There was no scientific basis underlying much of the data-gathering work: statisticians hardly talked to each other as a result of which no common terminology existed; and there was no separate statistical journal. Mr van Zanten, who was the chairman of the Statistical Committee of the Society for Tax-Sciences, therefore proposed that statistics be established as a separate discipline at universities (van Zanten 1939). To improve the situation in the meantime, the CBS did not wait for universities' decisions but organised their own course to educate new employees. It clearly met a need because the first year that the course was given the number of enrolments was 150.[35]

10.6.2 Prediction: a CBS task?

When the CBS had been established, they were given the task of gathering and producing data that could relevant both to scientific investigations and to practical purposes. The organisation had been set up in such a way that independence and neutrality could be guaranteed. The 'objectivity' of a statistical institute is usually crucial to its establishment and legitimation. But the goals of gathering data relevant to society and policy on the one hand and remaining neutral on the other, might be in

[32] CBS ARA-number 01.3 file 80 letter CBS' Director to the Minister of Industry, Trade and Agriculture 18 June 1945. [33] CBS ARA-number 01.3 file 80 Letter CBS to CCS 5 July 1946.
[34] CBS ARA-number 01.3 file 80 The Minister of Trade, Industry and Agriculture to the CBS 7 May 1946. [35] CBS ARA-number 02.651 file 213 27 November 1939.

conflict. Given the importance of a neutral and objective image, how did the CBS deal with forecasting?

10.6.2.1 The case of the barometer

The aim of the barometer was to be able to 'forecast' fluctuations in entities which were considered specifically relevant. This created the possibility for policymakers to regulate the economy to some extent.[36] Although there were considerable doubts about the possibility of a barometer, and its underlying philosophy, as we saw in section 10.3, the CBS worked to produce such a barometer. One might think therefore that the CBS involved itself in prediction. This was however not the case. Mr Methorst, who was the CBS director in the period 1906–1939, formulated it like this in 1925:

> Especially in the United States but also elsewhere, much fuss has been made of the Economic Barometer which would allow us to shed a modest light on the future. Also in our country the Barometer attracted attention. Therefore it might not be superfluous to state explicitly that it is not my intention to let the CBS make predictions. Predicting is of course none of the CBS' business. But, considering the fact that various people feel the need for Barometers, I think it is the CBS's task to provide them with the time series they need and to publish those in such a way that they provide a sound basis for that purpose.[37]

Let us assume that this statement is fully in accordance with the CBS's institutional mission. On the one hand the CBS wanted to provide the data people asked for by establishing the time series and on the other hand they withdrew from forecasting because that would harm their objective and neutral image. The way the CBS published the barometer supports this. The question then becomes: Why did they think that forecasts based on the barometer would harm their status?

The barometer was a descriptive product of knowledge based on the inductive method. The graphical method to judge the lines was very important and also the role of the trained eye to make these judgements. No single explanatory theory underlay the path from past data to future data. This means that the future information does not follow mechanically

---

[36] CBS ARA-number 12.330.1 file 381 Report 12 March 1927 M. J. de Bosch Kemper. For those readers who speak Dutch: In de nota wordt geschreven 'Het streven was erop gericht schommelingen te kunnen voorzien zodat het beleid regelend kon ingrijpen'. De term 'regeling' werd vaak gebruikt in de jaren dertig, net als ordening en planning, om het tegengestelde uit te drukken van het vrije-markt-principe.

[37] CBS ARA-number 12.330.1 file 381 15 May 1925 Report signed by the CBS' director H. W. Methorst.

from the past data, but that some sort of judgement or argument is needed. Is the past period chosen meaningful to the future-period? Have any important structural changes occurred? These kinds of problems might have been precisely the CBS' reasons for avoiding prediction, in the sense that they might be seen by others to be (politically) biased. This lack of trust in the forecasting potential of the barometer was of course increased when it turned out that the big crisis of 1929 was not foreseen.

10.6.2.2 The case of the model
By 1938, when a prototype of the model was already developed but they were only just starting to work on the national accounts, Methorst wrote an article in which he concluded that research on the cyclical movement could be done within the CBS – which meant research on the model – but that the CBS could not go into forecasting.[38] The CBS had been attacked by a reader of its quarterly journal *De Nederlandsche Conjunctuur* who argued that the journal had to come out more often for its readers to be able to make predictions. Methorst's answer was that an infallible prediction was impossible to make. As, in fact, the prediction was only a 'projection of the diagnosis of the business cycle onto the future' it should be enough to the reader to be informed about the diagnosis of the economy. The readers could then draw their own conclusions about the future. The phrase 'diagnosis of the business cycle' was explicitly used in the first CBS note announcing the new research programme on the model (as was described in section 10.4.1).

Whatever Methorst meant precisely, it is in any case clear that he did not consider predictions based on the mechanics of the model sure enough to see them as 'objective and neutral'. Although the CBS had started a research programme on the model, this had apparently not changed his views on the dangers of forecasting for the CBS. The barometer had proven that prediction was a very risky thing to do and these views were transfered to the model. Therefore the CBS systematically abstained from forecasting. The CBS journal *De Nederlandsche Conjunctuur* should give overviews of the current state of the economy, and articles somewhere between scientific analyses and market messages, but not foretell the future.

10.6.2.3 The case of the model and national accounts
The separation of the Central Planning Bureau (CPB) from the CBS had taken place by the sixteenth of May, eleven days after liberation.

[38] CBS ARA-number 12.330.1 H. W. Methorst (1938) 'De Conjunctuurbarometer'.

Table 10.2. *Institutional development of the business-cycle research*

1899   1955

1899–May 1945	(Proposal) May 1945:	September 1945: CBS
CBS – Department II: economic statistics, barometer (1924–) model-research (1935–) national accounts research (1938–)	Public Service for Statistics and Research: CBS, BESO (national accounts and model), BURESO	CBS (Department 4: national accounts) and CPB (model) become separate institutes

Idenburgh who was director of the CBS then, had already finished a report making a plea for its reorganisation, designed to be introduced as a Royal Decree (*Koninklijk Besluit*) in Parliament. The CBS should be transformed in a 'Public Service (*Rijksdienst*) for Statistics and Research'. This National Agency should consist of three divisions, a Central Bureau of Statistics, a Bureau for Research on Economics and Statistics (Bureau voor Economisch-Statistisch Onderzoek BESO) and a Bureau for Research on Regional Statistics (BURESO).[39] Only his proposal to create the BESO is of interest here. He gave this reason for the establishment of the division: 'Our society is being transformed from a chaotic [*ongerichtheid*] into a directed state, from amorphism into structural organisations, from planlessness into "planning", from free forces [*vrije spel der maatschappelijke krachten*] into economic order [*ordening*].' (ibid.) Government policy should influence economic activities; therefore, systematic reconstruction of the economic system was of outstanding importance. 'Every form of counter-cyclical policy needed a service/machinery (*apparaat*) which would make statistical observations, examine the diagnosis and which would guide the planning of actions.' (ibid.) This machinery was already there in Department II, which had developed 'insights in quantitative relations of the economy.' (ibid.) The BESO was to be successor to the 'old' Deparment II (see table 10.2).

Work in this 'new' institute (BESO), observing, diagnosing and forecasting, would be 'dependent on the kind of counter-cyclical policy the government adopted.' At the same time work on national acounts should be continued. These should be the basis for the 'past-calculations' and for the analysis of the economic structure and its 'mechanism of the business cycle'.

[39] CCS archive file 8 Ph. J. Idenburg *Rijksdienst voor Statistiek en Onderzoek* May 1945.

Idenburgh tried to combine the different elements of quantitative economics, the national accounts, the economic mechanism (he did not use the word model) and mathematic statistics. He wanted to base the forecasts on the national accounts: by doing so *past and future acquired the same structure.*

The framework Idenburgh and Tinbergen (who agreed with this proposal) pictured was something like this:

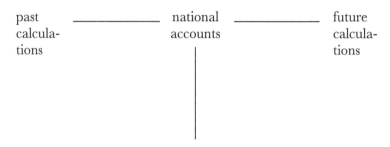

structure of the national economy

The national accounts provided a structure of the national economy on the basis of which past and future calculations could be based. As the national accounts were a static structure, the model was needed to be able to make the predictions. But the structure of the model was itself based on the national accounts, which were considered to be timeless (section 10.5). The boundary between past and future became arbitrary: the boundary is determined by the year one choses as the 'basis year' which gives you the initial conditions.

The model/accounts relationship is comparable with the ballistic trajectory. The mathematical representation of the trajectory is a differential equation which is a movement. If we know at some point in time the place and velocity of the ball we are able to solve the equation. The result delivers the actual path the ball follows. Therefore the initial condition determines its past and its future.

In this vision, the BESO should not enter the domain of policy institutions, but provide the right information for them not to need other research institutions. In a following version of the report, the BESO was called 'Central Planning Bureau'.[40] The idea of a Central Plan was formulated as a tool to coordinate the economy. This CPB would evolve

---

[40] It is not entirely clear who wrote this second version. I could only find the copy criticised by Tinbergen.

from the work done by Department II. They could handle the responsible task of assisting in drawing up a plan. We can conclude from these reports that initial ideas were to keep the CPB within the CBS. On the one hand the CBS probably wanted to keep competent staff within their own institution. On the other hand, *there was no good epistemological reason* to split the responsibilities for *the past and the future shared the same structure within their business-cycle research*. The integration of the national accounts and the model was critical in this respect, for it was only in 1945 that the national accounts had come into existence, and this had apparently changed the situation in such a way that the CBS could start making predictions.

It was because of political reasons that the CPB became a separate institution. W. Drees senior, the Minister of Social Affairs in the 1945 Cabinet, thought that a CPB and CBS merging into each other would unnecessarily complicate contacts with industry and the scientific community (Duynstee and Bosmans 1977, 493). Whatever the real reason, in September 1945, Tinbergen rewrote the report discussed above and reformulated the BESO as a division of the CBS constructing national accounts. They were also allowed to investigate relations between economic entities and keep in close contact with the CPB who were going to make the model.[41] Although the Royal Decree had not been introduced in Parliament, the CPB started their work under Tinbergen's leadership in September 1945 in the CBS-building (and continued under a different law passed only in 1947). In February 1946 they moved, taking many CBS personnel with them (CBS Annual Report 1945).

The model-data distinction created confusion between the CPB and the CBS. The CBS (still lead by Idenburgh) was not very happy with the separation of the CPB as illustrated by the following paragraph from their annual report of 1945:

Drawing up a 'Plan' could be considered as 'applied national accounts'. The report of Department II, written in August 1944, called 'National accounts, aims, problems, and results' containing tables and a diagram 'Money-flows in the Dutch Economy', formed the basis of the construction of the Plan for 1946.

Basically they argued that the CPB got the honour for work that had been done in the CBS.

---

[41] CCS archive file 8 J. Tinbergen *Rijksdienst voor Statistisch Onderzoek* September 1945.

Table 10.3. *The finally established solution*

CENTRAL BUREAU OF STATISTICS	CENTRAL PLANNING BUREAU
PAST	FUTURE
NATIONAL ACCOUNTS	MODEL

Only after the CPB's removal was the allocation of tasks discussed.[42] The CPB's main task would be to produce future calculations and the CBS would produce the past calculations (as shown in table 10.3). The institutions should not fight each other's territory. From now on, the CPB was obliged to request the CBS for new (or changes in) statistics if they needed these. The communication between the two institutions was formalised.

Nevertheless, much remained to be negotiated. For example, in a report made for Marshall Aid, CPB's numbers were used. The CBS contested this by claiming that these forecasts would be denied by reality. CBS's numbers would have been much more scientific.[43] The CPB answered by denying that their numbers were forecasts: the numbers published referred to a moment in time so closely after the time-coordinate of the numbers of the past they were based on, that one could not speak of their numbers as 'forecasts'.[44]

The argument continued in 1953, when Department 4 of the CBS, which was the successor of Department II, had constructed their own model and wanted to publish it. Idenburgh thought it better to discuss the model with the CPB before publication and asked the relevant staff of the CPB to discuss the allocation of such tasks. 'It might be the case that this work [the model] could preferably be done by the CPB'.[45] Tinbergen answered that according to him and others within the CPB the CBS model did not meet the CPB's standards. He proposed instead to start a common research project among the CPB, CBS, National Economic Intitute (NEI) and the Mathematical Centre dealing with economics, statistical method and data, and calculations.[46] Department 4 in the CBS did not want this. They believed that the CBS should develop

---

[42] CBS archive 01.72-D1188 file 1811 Notes of the meeting between CBS' Board and Department's directors and the CPB i.o. 22 March 1946 (meeting held at 18 March 1946).

[43] CBS ARA-number 12.330.1 file 383 CBS Department 4 *Driemaandelijks verslag van de Administrator van het European Recovery Programme* 9 June 1948.

[44] CBS ARA-number 12.330.1 file 382 CPB to CBS betreffende drie-maandelijks verslag over de EHP-hulp (Economisch Herstel Programma) 23 June 1948.

[45] CBS archive 01.72-D1188 file 1811 18 May 1953 Idenburg to Tinbergen.

[46] CBS archive 1.72-D1188 file 1811 J. Tinbergen to Ph. Idenburg 18 July 1953.

its own model to analyse economic life. 'As the national accounts are constructed by the CBS, the CBS needs a model which has to be accepted by the CPB.'[47] Working together more closely was not deemed necessary. C. Oomens, one of the constructors of the CBS model, answered Tinbergen's proposal negatively: 'I think, we should *not* take any notice of their proposals. We have been trying to negotiate intensively with them, but they have hardly [the word 'never' was changed into hardly!] shown any interest in that. The NEI and the MC have never achieved anything on this kind of work'.[48] Oomens proposed to organise the work done at Department 4 around the model, and to make their department's director, Mr Kuilen, director of the whole CBS! The latter did not happen, of course.

In 1956 the allocation of tasks was again under pressure. The CPB, under the direction of Professor F. Polak, thought that the CBS had crossed the boundary. Idenburgh answered as follows:

You are right: there was a line of demarcation between your bureau and mine, simplistically viewed as the boundary between past and future. I know that I acted against this rule . . . Although, a simple divide between past and future can never be an adequate standard for the allocation of tasks. The CPB cannot abstain from studies referring to the past. On the other hand, all the work done in the CBS is directed to the future. We are not a historical society . . . For us, too, it holds that 'the future is the past [*de toekomst is verleden tijd*]'.[49]

This last sentence is a reference to an often-quoted idea of the model as 'the history of the future'.

The contested boundary between the institutes partly resulted from the fact that the CBS considered 'the model' to be a scientific product of their own, while Tinbergen considered it as a personal achievement which he could bring with him to the CPB. This can be illustrated by a report on work done in CBS's Department 4, which made a reference to their development of a model in the prewar-years without mentioning Tinbergen's name. They considered it as a collective effort.[50] This view makes sense when we recall that the model stemmed from their research on the barometer.

Another reason for the CBS to have difficulties with the past–future boundary was that the numbers published by the CBS were always

---

[47] CBS archive 1.72-D1188 file 1811 Department 4 to Ministry of Economic Affairs 13 August 1953.
[48] Previous note/appendix 2 C. Oomens to Kuiler (head of department 4 of the CBS).
[49] CBS archive 1.72-D1188 file 1811 CBS to CPB 1 November 1956.
[50] CBS ARA-number 12.330.01 file 383 Tasks and work of Department 4 August 1948.

delayed compared to the CPB's numbers. While the CPB could publish numbers about the 1947–economic affairs in 1946, the CBS could only do that in 1948. Rijken van Olst, working at the CBS, made excuses for the late publication of 1946 national accounts compared to the CPB. 'The CBS, dealing with reality, (!) could only start computing them when the 1946 figures were known' (Rijken van Olst 27 April 1949).

10.7 CONCLUSION AND DISCUSSION

Sections 10.2–10.5 have shown the fusion between two different histories. First, we had the process of objectification of the economy: the development of a mathematical formalism representing an abstract structure of the economy. Secondly, we had a history of measuring the economy: the development of national income and national accounts. The construction of the model and the national accounts supported each other. The national accounts, unlike the barometer, were a data-structure which could not be used to predict. Similarly, the model could not explain without the data-structure, but nor could it predict on its own. Data and model were mutually developed, mutually dependent and mutually used. The model-data distinction was therefore not unambiguous at all as was shown in section 10.5.2: the flows in the national accounts are the entities in the model. The paradox described in section 10.6 consisted in the fact that precisely when these two elements became fully united they become institutionalised separately. Nor could the past–future distinction in tasks be maintained in the newly separated institutions because the model-data distinction could not be made clearly. For the CBS, the data tasks were described as 'the future is the past', while for the CPB the model was 'the history of the future'.

The emergence of the model can be understood in terms of a network that connects all sorts of heterogeneous aspects. We have already seen that the model could not work without the national accounts that delivered a structure to analyse the economy. We have also seen that the model consisted in the integration of data, bits of theory and mathematical and statistical techniques (see Boumans, this volume). But the emergence of the model needed even more. The short history given of the laws on economic statistics shows that the model could not have been established without laws that enforced industries, banks and other organisations to deliver their local information. Trained personnel were needed, and were delivered by the statistical education started by the CBS. And let us not forget the institutionalisation of the model into an autonomous institute which gave the model a special status.

The Central Planning Bureau was established with specific policy purposes: the model needed a *demand* side so to speak. It needed politicians and policy-makers who were going to use the model results in their regulations. Economic measurement was not simply a defence against outside pressure, to follow T. Porter (1995), but the process of quantifying the economy was part of a socially and politically supported process towards economic intervention and planning. Or to phrase it differently, if Tinbergen and his colleagues in department II had been the only people interested in intervening in the economy, the model would not have got its dominant place in economic thought and policy making.

The model and national accounts meant a new economics and a new way of doing economic politics. The model – national accounts 'ensemble' can be seen as a 'technology' aimed to deal with a huge problem of economic crises and one which at the same time involved redefining economics. J. J. Polak whose visual representation of the model was given in figure 10.3, section 10.4.2, explicitly noted this change:

From the system of equations we can – after some mathematical manipulations – draw conclusions on economic questions. It is beyond all doubt that the processing of economic relations in a proper mathematical form has extraordinary advantages compared to the difficulties that arise in literary economics. (Polak 1939, 72)

The technology was some sort of 'liberation' both from the uncertainties caused by the whimsical nature of the economy and the woolly theories of the economists. At the same time the technology defined both new tasks: how to make and maintain the model and how to use it; and new roles: roles for economists, roles for politicians, roles for statisticians and roles for universities.[51]

REFERENCES

Archives CBS.
Archives CCS.
Annual reports CBS.
Monographies CBS Series *Speciale onderzoekingen van de Nederlandsche Conjunctuur* nr. 1,2,4,7,8 1938, 1939, 1941, 1948, 1950.

[51] I thank the editors for all their constructive criticisms on this chapter.

Aftalion, A. (1927). 'The Theory of Economic Cycles Based on Capitalistic Technique of Production', *Review of Economic Statistics*, 9, 165–70.

Bakker, O. (1939). 'Het nationale inkomen en de berekening ervan', *De economist*, 219–34.

Barten, (1991). 'The History of Dutch Macroeconometric Modelling 1936–1986', pp. 153–94 in Bodkin, Klein and Marwah (eds.), *A History of Macroeconometric Model-Building*. Aldershot: Edward Elgar.

Bogaard, A. van den (1998). *Configuring the Economy: The Emergence of a Modelling Practice in the Netherlands 1920–1955*. Amsterdam: Thelathesis.

Bonger, W, A. (1910). 'Vermogen en inkomen van Nederland', *De Nieuwe Tijd*, 236–45.

———. (1915). 'Vermogen en inkomen van nederland', *De Nieuwe Tijd*, 226–49, 269–91.

———. (1923). 'Vermogen en inkomen van Nederland in oorlogstijd', *De Socialistische Gids*, 158–92, 342–57.

Bosch Kemper, M. J. de (1929a). 'Korte inleiding tot de analyse der conjunctuurlijnen I en II', *Economische en Sociale Kroniek*, Volume 2, 92–101.

———. (1929b). 'Korte inleiding tot de analyse der conjunctuurlijnen III en IV', *Economische en Sociale Kroniek*, Volume 3, 75–84.

———. (1929c). 'Korte inleiding tot de analyse der conjunctuurlijnen V', *De Nederlandsche Conjunctuur*, Volume 4, 10–17.

Boumans, M. (1992). *A Case of Limited Physics Transfer: Jan Tinbergen's Resources for Reshaping Economics*. Amsterdam: Thesis Publishers/Tinbergen Institute, Tinbergen Institute Research Series, 38.

———. (1995). 'The Change from Endogenous to Exogenous in Business-Cycle Model Building', *LSE CPNSS Discussion Paper Series*, DP 13/95.

Breems, M. den (1949). *Gedenkboek Centraal Bureau voor de Statistiek*, CBS.

Bullock, C. J., W. M. Persons, and W. L. Crum (1927). 'The Construction and Interpretation of the Harvard Index of Business Conditions', *The Review of Economic Statistics*, 9, 74–92.

Central Bureau of Statistics (1924). *Statistical Yearbook*. CBS.

———. (1993). *The Value Added of National Accounting* Voorburg/Heerlen.

Compaijen, B. and R. H. van Til (1978). *De Nederlandse Economie. Beschrijving, voorspelling en besturing.* Groningen: Wolters-Noordhoff.

Derksen, J. B. D. (1940). 'Het onderzoek van het nationale inkomen', *De economist*, 571–94.

Duynstee, F. J. and J. Bosmans (1977). *Het kabinet Schermerhorn-Drees 24 juni 1944 – 3 juli 1946.* Assen: van Gorcum.

Idenburgh, Ph. J. (1940). 'De Nederlandse Stichting voor de Statistiek', *Economisch Statistische Berichten*, 27 November, 708–10.

Kaan, A. (1939). 'De werkloosheidcijfers der arbeidsbeurzen als maatstaf voor den druk der werkloosheid', in *Bedrijfseconomische opstellen*. Groningen: P. Noordhoff NV, 227–46.

Kwa, C. (1989). *Mimicking Nature*. Amsterdam thesis, UvA.

Morgan, M. (1990). *The History of Econometric Ideas*. Cambridge: Cambridge University Press.

Oomens, C. A. (1985). 'De ontwikkeling van Nationale Rekeningen in Nederland', in *Voor praktijk of wetenschap*. CBS: Voorburg/Heerlen, pp. 23–46.

Persons, W. M. (1919). 'Indices of Business Conditions', *The Review of Economic Statistics*, 1, 5–100.

(1926). 'Theories of Business Fluctuations', *Quarterly Journal of Ecnomics*, 41, 94–128.

Polak, J. J. (1939). *Bedrijfseconomische opstellen*. Groningen: P. Noordhoff.

Porter, T. (1995). *Trust in Numbers*. Princeton: Princeton University Press.

Rijken van Olst, H. (1949). 'De berekening en betekenis van het nationaal inkomen van Nederland,' in *Economisch Statistische Berichten*, 2 Feb., 91–3.

Roll, E. (1989). *History of Economic Thought*, rev. edn London: Faber & Faber.

Studenski, P. (1958). *The Income of Nations*. Part I, *History*; Part II, *Theory and Methodology*. New York.

Tinbergen, J. (1959, reprint of 1931). 'A Shipbuilding Cycle?', in *Jan Tinbergen, Selected Papers*, ed. L. H. Klaassen, L. M. Koyck, H. J. Witteveen. Amsterdam: North Holland Publishing Company.

(1934). 'Der Einfluss der Kaufkraftregulierung auf den Konjunkturverlauf', *Zeitschrift für Nationalökonomie*, 289–319.

(1935). 'Quantitative Fragen der Konjunkturpolitik', *Weltwirtschaftliches Archiv*, 42, 366–99.

(1936). 'Kan hier te lande . . .', *Praeadvies*, voor de vereniging voor staathuishoudkunde, 62–108.

Valk, V. de (1941). 'De beteekenis van de economische statistiek in verband met de geleide economie', *Economisch Statistische Berichten*, 21 May, 321–3.

Zanden, P. van der (1941). 'Economische Statistiek in Duitsland', *Economisch Statistische Berichten*, 19 March, 202–4.

Zanten, J. H. van (1939). 'Statistiek en onderwijs', *De economist*, 345–60.

CHAPTER II

Models and stories in hadron physics

Stephan Hartmann

11.1 INTRODUCTION

Working in various physics departments for a couple of years, I had the chance to attend several PhD examinations. Usually, after the candidate derived a wanted result formally on the blackboard, one of the members of the committee would stand up and ask: 'But what does it mean? How can we understand that x is so large, that y does not contribute, or that z happens at all?' Students who are not able to tell a 'handwaving' *story* in this situation are not considered to be good physicists.

Judging from my experience, this situation is typical not only of the more phenomenological branches of physics (such as nuclear physics) but also for the highly abstract segments of mathematical physics (such as conformal field theory), though the expected story may be quite different. In this paper, I want to show that stories of this kind are not only important when it comes to finding out if some examination candidate 'really understands' what he calculated. Telling a plausible story is also an often used strategy to legitimate a proposed model.

If I am right about this, empirical adequacy and logical consistency are not the only criteria of model acceptance. A model may also be provisionally entertained (to use a now popular term) when the story that goes with it is a good one. But what criteria do we have to assess the quality of a story? How do scientists use the method of storytelling to convince their fellow scientists of the goodness of their model? Is the story equally important for all models or do some models need a stronger story than others? These are some of the questions that I address in this chapter. In doing so, I draw on material from a case-study in hadron physics.

It is a pleasure to thank S. Hardt, N. Huggett, F. Rohrlich, A. Rueger and in particular the editors for their very useful comments on the first draft of this paper. Helpful discussions with D. Bailer-Jones, N. Cartwright, J. Ehlers, I.-O. Stamatescu, M. Suárez and M. Stöcker are also gratefully acknowledged.

The rest of this contribution is organised as follows. Section 11.2 clarifies the terminology and discusses various functions of models in the practice of science. In section 11.3, two models of hadron structure are introduced. I present the story that goes with each model and analyse what role it plays for legitimation of the model. Section 11.4 concludes with some more general meta-theoretical implications of the case study.

11.2 PHENOMENOLOGICAL MODELS: A 'FUNCTIONAL' ACCOUNT

Phenomenological models are of utmost importance in almost all branches of actual scientific practice. In order to understand what 'real' physicists do, their role and function need to be analysed carefully. In doing so and to avoid confusion, I will first specify the terminology. I suggest the following (minimal) definition: a phenomenological model is a set of assumptions about some object or system. Some of these assumptions may be *inspired* by a theory, others may even contradict the relevant theory (if there is one).[1]

The relation of a phenomenological model to some possibly existing underlying theory is somewhat bizarre. I cannot give a general account of this relation here, as it seems to depend on the concrete example in question. Very often there is an underlying fundamental theory such as quantum electrodynamics (QED) for condensed matter physics or quantum chromodynamics (QCD) for nuclear physics that is not frequently used in actual calculations for various reasons. Instead, phenomenological models are constructed; they mimic many of the features of the theory but are much easier to handle.

Some of these models can even be exactly derived from the theory in a well-controlled limiting process. In other cases, the explicit deduction is considerably more delicate, involving, for example, many different (and sometimes dubious) approximation schemes – if it is possible to perform the calculation at all. In any instant, there is usually not much to be learned from these deductions, so that many phenomenological models can be considered autonomous. Besides, deduction from theory is not the usual way to obtain a phenomenological model.[2] Theory, as N. Cartwright and her collaborators have recently pointed out, serves only as *one* tool for model construction.[3]

---

[1] For a similar characterisation see Redhead (1980).
[2] For more on this, see Cartwright (1983).   [3] See Cartwright *et al.* (1995).

Our discussion of phenomenological models has focused only on formalism so far. A model has been characterised by a set of assumptions, the relation of a model to a theory has been sought in a reduction of the formalism of the theory to the model's formalism etc. There is, however, more to models than formalism.

In order to appreciate the role that models play in contemporary physics, it can be instructive to point out the various *functions* of models in physics. Without pretending to give a complete account (or something close to it), here are some of them:

a) *Apply a theory.* General theories cannot be applied without specifying assumptions about a concrete system. In Newtonian mechanics, for example, a force function has to be given in order to facilitate detailed calculations.[4]

If the concrete system under study is purely fictional ('toy model'), students of the theory can learn something about the features of the general theory by applying it. Furthermore, they get used to the mathematical structure of the oftentimes very complicated theory by analysing tractable examples.

b) *Test a theory.* If the system under study is real, constructing and exploring the consequences of a phenomenological model may serve to test the underlying theory.[5]

c) *Develop a theory.* Sometimes models even serve as a tool for the construction of new theories. One example is the development of QCD which is (so far) the end point of a hierarchy of models of hadron structure.[6] I suppose that models also play a role in the construction of general theories (such as Newtonian mechanics).

d) *Replace a theory.* Many functions of models can be traced back to the observation that models are easy to work with, or at least easier to apply than the corresponding theory (if there is one). Then the model is used as a substitute for the theory.[7] Since high-powered computers are widely available nowadays, one could argue that this function of models will get less important in the future. It should be noted, however, that the complexity of our current theories will considerably increase as well.

Be it as it may, even if it were possible to solve all theories 'exactly'

---

[4] This view is elaborated in Bunge (1973).
[5] More on this can be found, for example, in Laymon (1985) and in Redhead (1980).
[6] I have demonstrated this in some detail in Hartmann (1995a, b) and (1996).
[7] More on this can be found in Hartmann (1995a) and (1995b).

with the help of computers, models would still be necessary for at least two reasons which I will give below.

e) Explore the features of a theory. Firstly, models help scientists to explore the features of a theory in question. By exploring the consequences of one isolated feature in a numerical experiment, physicists learn something about the consequences of this feature. Is it possible to reproduce the mass spectrum of low-lying hadrons by solely modelling confinement? This is a typical question asked in this context.[8]

f) Gain understanding. The second reason is closely related to the first one. By studying various models that mimic isolated features of a complicated theory, scientists gain (partial) understanding of the system under investigation.[9] Physicists are not satisfied with a numerical treatment of a given theory. They reach for simple intuitive pictures that capture the essential elements of a process.[10]

Here is an example. It is generally believed that QCD is the fundamental theory of hadron (and nuclear) structure. Almost exact results of this theory can, however, be obtained only by applying a sophisticated numerical technique called Lattice-QCD.[11] This method essentially works like a black-box: asked for the numerical value of a certain observable (such as a hadron mass), the computer will start to numerically evaluate *all* possible contributions (however small they are). Having this tool is important because it is the only way to rigorously test QCD at present.

Despite this advantage, Lattice-QCD faces several serious problems. Some are technical and I will not discuss them here; others are more conceptual. Physicist T. Cohen emphasises the following one:

[W]hile high-quality numerical simulations may allow us to test whether QCD can explain low-energy hadronic phenomena, they will not, by themselves, give much insight into how QCD works in the low-energy regime. Simple intuitive pictures are essential to obtain insight, and models provide such pictures. (Cohen 1996, 599)

---

[8] For more on this important function I refer the reader to Hartmann (1998).

[9] M. Fisher (1988) points out the importance of understanding from the point of view of a practising physicist. See also Cushing (1994).

[10] It is interesting to note that all but this function of models relate to 'theory' in one way or the other.

[11] In Lattice-QCD the full action of QCD is truncated on a discretised space-time lattice. The results obtained will, therefore, generally depend on the lattice spacing a. By extrapolating $a \to 0$ from computer simulations for finite a, it is hoped to approach the exact results of QCD. It should be noted, however, that Lattice-QCD is extremely (computer-) time consuming. See Rothe (1992) for a textbook presentation of Lattice-QCD.

Cohen explains:

Condensed-matter physics provides a useful analogy: even if one were able to solve the electron-ion many-body Schrodinger equation by brute force on a computer and directly predict observables, to have any real understanding of what is happening, one needs to understand the effective degrees of freedom which dominate the physics, such as photons, Cooper pairs, quasiparticles, and so forth. To gain intuition about these effective degrees of freedom, modeling is required. In much the same way, models of the hadrons are essential in developing intuition into how QCD functions in the low-energy domain. (Cohen 1996, 599f)

According to Cohen, the task of the physicist is not completed when numerical solutions of a theory are obtained. This is because a numerical solution does not suffice to provide any insight or understanding of the processes inside a hadron. Models, Cohen claims, give us this desired insight. There is, of course, an obvious problem here. Models are necessarily provisional and incomplete. Furthermore, they often do not describe empirical data well. How, then, can they provide insight and understanding? The answer to this question is, I maintain, that good models provide a plausible story that makes us believe in the model.

In order to understand all this better, we must now have a closer look at the models Cohen has in mind.

11.3 CASE-STUDY: HADRON PHYSICS

Before discussing the two models in detail, I will first set the frame of the case study and start with an episodic sketch of QCD that points out the main features of this theory. It turns out that the models we discuss pick out only one of these features.

Hadrons are strongly interacting sub-nuclear particles such as protons, neutrons, and pions. *Hadron physics* is the branch of particle physics that investigates their structure and (strong) interactions. Here is a short sketch of the history of hadron physics.[12]

In 1932, Cavendish physicist J. Chadwick produced in a series of experiments electrically neutral particles with almost the same mass as the positively charged hydrogen nucleus (later called 'proton'). This astonishing observation marked the beginning of hadron physics. It soon turned out that atomic nuclei could be understood as composed systems of protons and neutrons. W. Heisenberg (1935) took advantage of the

---

[12] For details, see Pais (1986).

similarities between protons and neutrons (now called nucleons) by introducing the isospin concept in analogy to the spin concept familiar in atomic physics, and Japanese physicist H. Yukawa (1934) proposed a dynamical model for the short-ranged interaction of nucleons. Subsequently, these theoretical works were extended, but a real research boost did not occur until a wealth of new particles ('resonances', 'hadron zoo') were directly produced after World War II. Besides, the analysis of cosmic rays supplied physicists with plenty of new data.

These findings inspired the development of a variety of models that attempted to organise and systematise these data. I will here only mention the (more theoretical) investigations in the context of current algebra and, of course, the famous (more phenomenological) quark model, suggested independently by M. Gell-Mann and G. Zweig in 1964.[13]

Relying on analogies to QED and with the now somewhat dubious requirement of renormalisability[14] in mind, quarks proved to be an essential part (besides gluons) of the ontology of the then-developed non-abelian gauge quantum field theory of strong interactions, QCD, in 1971. This theory is currently assumed to be the fundamental theory of strong interactions.

QCD has three characteristic features that are isolated and investigated in detail in the models that I now discuss. These features are asymptotic freedom, quark confinement and chiral symmetry.[15] According to *asymptotic freedom*, quarks move quasi-free at very high energies compared to the rest mass of the proton. This theoretically well-established consequence of QCD has also been demonstrated (though somewhat indirectly) in accelerator experiments at facilities such as CERN near Geneva and Fermilab near Chicago. More specifically, the interaction between quarks inside a hadron, characterised by an effective ('running') coupling constant $\alpha_s(q^2)$ ($-q^2$ is the 4-momentum transfer), monotonically approaches zero. Therefore, perturbation theoretical tools, well-known from QED, can be successfully applied in this regime.

At low energies, on the other hand, the opposite effect occurs. For decreasing momentum transfer $\alpha_s(q^2)$ increases and soon exceeds 1, making a perturbation theoretical treatment dubious and practically impossible.

[13] References to the original publications (or preprints in the case of Zweig) are given in Pais (1986); see also Pickering (1984). [14] See, for example, Cao and Schweber (1993).
[15] Technical details can be found in Weinberg (1996, 152f).

This is the energy-regime where confinement and chiral symmetry dominate the scene. *Quark confinement* ('confinement' for short) was originally proposed to account for the fact that so far no single free quark has been observed in experiments. Quarks always seem to be clumped together in triplets (baryons) or pairs (mesons).[16] The confinement hypothesis has, however, not yet directly been derived from QCD. It even does not seem to be clear what confinement exactly is. We will come back to this below.

Chiral symmetry and its dynamical breaking is the second typical low-energy feature of QCD. But unlike confinement, we are better able to understand what it means.

Chirality is a well-known property of many physical, chemical and biological systems. Some sugars, for instance, only show up in a right-handed version. If there were a left-handed version with the same frequency too, the system would be chirally symmetrical. Then the interaction would not distinguish between the left- and the right-handed version.

There are also left- and right-handed states in quantum field theory. It can be demonstrated theoretically that a quantum field theory with explicit mass terms in its Lagrangian density cannot be chirally symmetrical. It turns out that chiral symmetry is (almost) realised in the low-energy domain of QCD because the current quark masses of the relevant quarks in this regime are small (about 10 MeV) compared to the rest mass of the proton (about 1000 MeV). Therefore, every eigenstate of the interaction should have a chiral partner with the same mass but opposite parity. Experimental data, however, do not support this conclusion: Chiral partners with the same mass but opposite parity simply do not exist.

A way out of this unpleasant situation is to assume that the interaction itself breaks the symmetry dynamically. As a result of this supposed mechanism, an effective quark mass is generated.[17] There is much empirical evidence that chiral symmetry is really broken in QCD. For example, there are plenty of relations between hadron masses that can be derived from the assumption of

---

[16] According to the 'naïve' quark model baryons 'are made' of three (valence) quarks and mesons of one quark and one antiquark.

[17] In the formalism, an explicit mass term will then show up in the corresponding effective Lagrangian density. This mass is, by the way, easily identified with the (dressed) constituent quark mass used in non-relativistic Constituent Quark Models ($m_{CQM} \approx 300$ MeV). Here, a field theoretical mechanism provides a deeper motivation and theoretical legitimation of the CQM that has been used for a long time without such a legitimation.

chiral symmetry only. These relations are well-confirmed experimentally.[18]

It is theoretically well-established that confinement and dynamical chiral symmetry breaking cannot be obtained in a simple perturbation-theoretical analysis of QCD. Both are low-energy phenomena and perturbation theory breaks down in this regime. Technically speaking, an infinite number of Feynman diagrams must be added up to obtain these phenomena. Yet this is not an easy task. Whereas chiral symmetry breaking has been demonstrated in the Lattice-QCD approach, it is not really clear what confinement exactly is. There are at least four different proposals as to what 'confinement' means. All of them are *inspired* by the (negative) observation that there are no free quarks. Modelling is a way to explore these proposals.

1. *Spatial confinement:* Spatial confinement means that quarks cannot leave a certain region in space. This explication, of course, accounts for the fact that no free quark has been detected so far. But where does the confining region come from? Is it dynamically generated? There are models that explore the consequences of spatial confinement, such as the MIT-Bag Model. Others even attempt to understand the mechanisms that produce this kind of confinement, such as the Chromodielectric Soliton Model.[19]

2. *String confinement:* String confinement is a variant of spatial confinement, crafted especially for the application to scattering processes in which mesons are produced. Motivated by features of the meson spectrum, it is postulated that the attractive (colour-singlet) quark–antiquark force increases linearly with the distance of the quarks. Quark and antiquark are tied together by something like a rubber band that expands with increasing energy. Therefore, free quarks never show up. However, when the separation becomes large enough and the corresponding laboratory energy exceeds a certain critical value (the so-called pion production threshold), the string eventually breaks and new particles (mostly pions) are created. Models like this string-fragmentation model are fairly successful phenomenologically.

---

[18] A famous example is the Gell-Mann–Oakes–Renner relation (Bhaduri 1988):

$$m_\pi^2 f_\pi^2 = -\frac{m_u + m_d}{2} <\bar{q}q> \qquad (11.1)$$

This relation accounts for the finite pion mass due to an (additional) explicit breaking of chiral symmetry. Here, f_π is the pion decay constant, m_π, m_u and m_d are the masses of the pion and of the u- and d-quark, respectively and $<\bar{q}q> \approx (-250 \text{ MeV})^3$ is the quark condensate.

[19] See Wilets *et al.* (1997).

3. *Colour confinement:* Colour confinement means that only colour-
    singlet states are physically realised asymptotic states. This type of
    confinement was first introduced by M. Gell-Mann to get rid of a
    problem within his original quark model.[20] It is not clear at the
    moment how this variant of confinement exactly relates to spatial
    confinement.

4. *The quark propagator has no poles:* This is a field-theoretical hypothe-
    sis. If the full quark-propagator has no poles, asymptotic quark
    states cannot show up. Therefore, free quarks do not exist. In this
    account, confinement is a constraint on the unknown quark prop-
    agator.

These are only some of the most attractive suggestions currently
explored by physicists. In the rest of this section, I will investigate in some
detail how confinement (choosing one of the above mentioned hypotheses) and dynamical chiral symmetry breaking are modelled. For that
purpose, I focus on two models, the MIT-Bag Model (section 11.3.1) and
the Nambu–Jona-Lasinio Model (section 11.3.2). I will describe how
each model relates to QCD and how it is legitimised. As I hinted already,
a qualitative story turns out to be important in this respect.

11.3.1 The MIT-Bag Model

The MIT-Bag Model is a conceptually very simple phenomenological
model. Developed in 1974 at the Massachusetts Institute of Technology
in Cambridge (USA) shortly after the formulation of QCD, it soon
became a major tool for hadron theorists. From the above-mentioned features of QCD, the MIT-Bag Model models spatial confinement only.[21]

According to the model, quarks are forced by a fixed external pressure
to move only inside a given spatial region (see figure 11.1). Within this
region (the 'bag'), quarks occupy single-particle orbitals similar to nucleons in the nuclear shell model.[22] And just like in nuclear physics the
shape of the bag is spherical if all quarks are in the ground state. This is
the simplest variant of the model. When considering higher excitations,
non-spherical shapes must also be considered. However, this raises additional technical problems.

---

[20] According to Gell-Mann's quark-model, the nucleon and the Δ-particle should have the same
rest mass. Empirically, their mass difference is, however, about $1/3$ of the nucleon mass.

[21] For a comprehensive exposition of the MIT-Bag Model see Bhaduri (1988) and DeTar and
Donoghue (1983).

[22] This is no surprise. The MIT-Bag Model was developed by nuclear physicists.

We will now turn to the mathematical formulation of the model. In order to determine the wave function of a single quark, three cases have to be distinguished: (1) In the *inside*, quarks are allowed to move quasi-free. The corresponding wave function can, therefore, be obtained by solving the Dirac equation for free massive fermions.[23] (2) An appropriate boundary condition at the *bag surface* guarantees that no quark can leave the bag. This is modelled by requiring that the quark-flux through the bag surface vanishes. (3) This implies that there are no quarks *outside* the bag. In this region, the wave function is zero.

The boundary condition yields, as usual in quantum mechanics, discrete energy eigenvalues. For massless quarks (and with units such that $\hbar = c = 1$) these energies are given by

$$\epsilon_n = \frac{x_n}{R}, \tag{11.2}$$

where R is the yet undetermined radius of the bag. The dimensionless eigenvalues x_n (n denotes the set of quantum numbers of a single particle state) are easily obtained by solving a transcendental equation. The lowest value is $x_1 \approx 2.04$.

So far, we have only considered a single quark in a bag. Real hadrons are, however, collections of $\mathcal{N}_q$ valence quarks ($\mathcal{N}_q = 3$ for baryons and $\mathcal{N}_q = 2$ for mesons). If one neglects the interaction between the quarks for the time being, the total kinetic energy of the bag is given by

$$E_{kin}(R) = \mathcal{N}_q \frac{x_n}{R}. \tag{11.3}$$

Here we have assumed for simplicity that all quarks occupy the same orbital state.

The necessary stabilising potential energy results from the external pressure that guarantees the fulfilment of the boundary condition. It is given by

$$E_{pot}(R) = \frac{4}{3} \pi R^3 B, \tag{11.4}$$

where B is the so-called bag-constant that reflects the bag pressure.

The total bag energy is finally the sum of both contributions:

$$E(R) = E_{kin}(R) + E_{pot}(R) \tag{11.5}$$

---

[23] It is sometimes claimed that this simulates asymptotic freedom, see DeTar and Donoghue (1983).

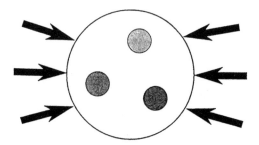

Figure 11.1 A baryon in the MIT-Bag Model.

Minimising $E(R)$ with respect to R yields the equilibrium radius of the system and, after some simple algebra, the total energy of the bag:

$$R_n = \left(\frac{N_q x_n}{4\pi B}\right)^{1/4}, \quad E_n = \frac{4}{3}(4\pi B N_q^3 x_n^3)^{1/4} \qquad (11.6)$$

These are consequences of the simplest version of the MIT-Bag Model. Only the model parameter B is adjusted in order to get a best fit of hadronic observables (masses, charge radii etc.). Since the model is, at this stage, rather crude, the agreement of the predictions with empirical data is only fairly modest. When fixing the nucleon mass to its empirical value ($m_N = 938$ MeV), for instance, the nucleon radius can be calculated to be $R_N = 1.7$ fm (compared to the empirical value of roughly 1 fm) and the pion mass comes out to be $m_\pi = 692$ MeV (compared to 138 MeV).

Despite the shortcomings, the MIT-Bag Model has been entertained by many physicists. This immediately raises the question: why? There is an obvious pragmatic reason for this: The model is (as we have seen already) very easy to apply. It allows it, furthermore, (as we will see below) to incorporate physically motivated extensions in a relatively straightforward manner.

However, this is not the only reason to entertain the MIT-Bag Model, despite its modest empirical adequacy. Incidentally, there are many other models that share several features with the MIT-Bag Model. They are also easy to apply and to extend and do not fit experimental data particularly well. So why do so many physicists choose the MIT-Bag Model and not one of its rivals as a starting point for their theoretical investigations?

I argue that this is because the *story* that goes with the MIT-Bag Model is a remarkably good one. This statement needs some explication. The purpose of the story (that I will soon tell) is to fit the model in a larger theoretical framework and to address open questions of the model in a qualitative way. The larger framework is, in our example, of course QCD. Here are some of the open questions of the model: (1) How does a bag result physically? (2) Where does the bag pressure come from? (3) What about colour confinement? Is it also realised in the model?

This story of the MIT-Bag Model aims at tackling the first question. It will also shed some light upon the other questions as well. So how does a bag result physically? The story starts with a description of the initial state: In the beginning, there was only a highly complicated ('non-perturbative') vacuum with strongly interacting gluons. This is plausible from QCD. In the next step, a single quark is put in this messy environment. What will happen? Well, the quark will 'dig a hole' (to use the metaphoric language of the physicists) in the vacuum by pushing the gluons away due to the repulsive colour-interaction. It turns out, however, that only colour neutral objects can be stable, as the following argument from T. D. Lee shows.[24]

The non-perturbative vacuum can be described by a colour-dielectric function $\kappa(r)$ that vanishes for $r \to \infty$.[25] Consider now the total energy W_c of the colour electric field E_c of a colour-charge Q_c in the abelian approximation. One obtains:

$$W_c \sim \int \vec{E}_c \cdot \vec{D}_c \, d^3r \sim Q_c^2 \int_0^\infty \frac{dr}{r^2 \kappa(r)} \quad (11.7)$$

It is easy to see that the integral diverges for suitably chosen colour-dielectric functions $\kappa(r)$. Hence, the bag is stable if and only if the total colour-charge vanishes: $Q_c = 0$. This is just colour-confinement: asymptotic physical states are colour-neutral.

The bag and consequently the shape of the $\kappa(r)$-profile is effectively generated because the quarks pushed out all non-perturbative gluons from the inside of the bag. The concrete radius of the bag reflects an equilibrium constellation: The pressure from the outside due to the non-perturbative gluonic interactions balances the pressure from the motion of the quarks in the bag interior. Since there are no real gluons present in the inside of the bag, quarks move (almost) freely in this region.

How does this story relate to the formalism of QCD? It is important to note that the story just told cannot be strictly deduced from QCD.[26] Given our knowledge of this theory, the story appears to be *plausible*.

---

[24] Details can be found in Wilets (1989).   [25] In the MIT-Bag Model one chooses $\kappa(r) = \theta(R - r)$.
[26] Some physicists tried to derive the MIT-Bag Model from QCD. Their results, however, were not really illuminating. References can be found in DeTar and Donoghue (1983).

Something like this will probably happen inside a hadron. That's all. And that is why the story is considered a good story. It relates in a plausible way to mechanisms that are known to be mechanisms of QCD and this theory is, after all, supposed to be the fundamental theory of this domain.

The MIT-Bag Model has been a focus of much interest and extended in many directions. I will now discuss two of these attempts and thereby pay special attention to the motivation of the corrections and additions as well as to their relation to QCD.

1. One-gluon exchange. So far, the model neglects the mutual interaction of quarks completely. There should be at least some residual interaction that is not already effectively contained in the bag-pressure. Here is a phenomenological argument for the importance of quark interactions. Hadrons with quarks in the same single particle state are degenerate in the independent particle model used so far. It does not matter, for example, if three spin-$1/2$ quarks in a baryon couple to a state of total angular momentum $1/2$ (nucleon) or $3/2$ (Δ-particle). Both should have the same energy. Empirically, however, the mass difference of these states is roughly 300 MeV; that is one third of the nucleon mass.

This problem can be dealt with by adding another term to the total energy in the model that takes into account the interaction of quarks mediated by gluons. In a perturbation theoretical treatment ('one-gluon exchange'), one obtains for the corresponding energy contribution:

$$E_X = \frac{\alpha_s \mathcal{M}_q}{R}, \tag{11.8}$$

with a matrix element $\mathcal{M}_q$ that depends on the quantum numbers of the coupled quarks. α_s is the strong coupling constant. It is treated as another adjustable parameter of the model.

While the one-gluon exchange term is quite successful phenomenologically, serious problems remain. Most important is that an argument is needed to explain why perturbation theory is applicable at all. This is dubious because, as we have seen, the effective coupling constant exceeds 1 in the low-energy domain of QCD. And indeed, it turns out that the adjusted value of α_s is larger than 1 in the MIT-Bag Model. Therefore, perturbation theory should not make any sense in this regime.

Why is it used anyway? The argument is again provided by the story of the MIT-Bag Model. Since the complicated ('non-perturbative') gluonic vacuum is squeezed out of the interior of the bag due to the presence of valence quarks, the residual interaction between the quarks is very small. All complicated non-perturbative effects are already contained in the bag constant B. Residual interactions can, therefore, be treated in perturbation theory.

Nevertheless, it should be noted that an exact deduction of the one-gluon exchange term from QCD has not been achieved so far.

2. The Casimir Term. This is an additional contribution to the total energy of the bag of the form

$$E_{Cas} = \frac{Z}{R} \qquad (11.9)$$

with a parameter Z that can, in principle, be deduced theoretically. In practice, however, Z is usually treated as another adjustable parameter – for good reasons, as I will demonstrate soon. Here is the physical motivation of this correction. The Casimir term is supposed to represent the zero-point energy of the quantum vacuum. It is a well-known feature of quantum field theory that such fluctuations are always present. A similar term was first considered in the context of QED. Dutch physicist H. Casimir showed that two parallel conducting plates attract each other due to the presence of the quantum vacuum. The case of a spherical cavity is, for technical reasons, much more complicated.[27]

Having an additional parameter, the Casimir term definitely improves the quality of the empirical consequences of the MIT-Bag Model. But this term is problematic, too. The main problem is that theory suggests that the term is negative, while the best fits are achieved by using a slightly positive value.[28] This seems to indicate that something is wrong with the model in the present form.

Here, no story is known that can make any sense of the situation.

The MIT-Bag Model faces one more difficulty: chiral symmetry is explicitly broken on the bag surface. It has been shown that this is a direct consequence of the static boundary condition. Incidentally, this is the central assumption of the model. Therefore, an argument is needed as to why this unwanted symmetry breaking is not a cause for concern. Unfortunately, there is no really convincing argument. Here is a way to get rid of this problem: one first states that the model with the above-mentioned corrections (and maybe others) describes empirical data rather well. Consequently, chiral symmetry is not essential to determine the spectrum of low-lying hadrons and can be neglected.

---

[27] An excellent discussion of the Casimir effect can be found in Milonni (1994).
[28] See DeTar and Donogue (1983), Plunien, Müller and Greiner (1986) and Wilets (1989).

This is, of course, not a compelling argument. A closer look at the data shows that pionic observables especially cannot be reproduced sufficiently well. This is alarming! Since the pion is intimately related to chiral symmetry, it is the Goldstone boson of this symmetry. In order to rectify this, a chiral extension of the MIT-Bag Model has been suggested. In the *Cloudy-Bag Model*[29] a fundamental pion field that couples to the quarks at the bag surface in a chirally symmetric manner is added. This field covers the bag-surface like a cloud. It turns out that the Cloudy-Bag Model is much harder to treat mathematically. The resulting phenomenology is, however, considerably better than that of the MIT-Bag Model.

Some remarks concerning the relation between the story and the quality of the empirical consequences of the model is in order here. One might suspect that no one worried about the violation of chiral symmetry, provided the pionic data came out right. Though often the story seems to be more important than data, comparison with data sometimes suggests new models and helps us to assess a certain model in question.

This will be different in the Nambu–Jona-Lasinio Model that we consider now in some detail.

11.3.2 The Nambu–Jona-Lasinio Model

Whereas the MIT-Bag Model extracts confinement from the features of QCD and neglects chiral symmetry, the Nambu–Jona-Lasinio (NJL) Model picks chiral symmetry but does not provide a mechanism for confinement. Consequently, the NJL Model needs a plausible explanation for this neglect. Before presenting this argument, I will introduce the model.

The NJL Model[30] is a non-renormalisable quantum field theoretical model for dynamical chiral symmetry breaking. Although its only degrees of freedom are quarks, the NJL Model is currently considered to be a model for the low-energy regime of QCD.[31]

Here is the Lagrangian density of the NJL Model:

$$\mathcal{L}_{N\!J\!L} = \mathcal{L}_0 + \mathcal{L}_{int}, \tag{11.10}$$

---

[29] See Bhaduri (1988) for a survey.

[30] For a recent review of the model and further references to the literature see Klevanski (1992). The application of this model to hadron physics is summarised in Vogl and Weise (1991).

[31] It is historically interesting to note that this model, which was introduced in 1961, had a somewhat different purpose, see Nambu and Jona-Lasinio (1961). It became clear only much later that the NJL Model can also account for hadronic properties.

with the free (= non-interacting) Dirac Lagrangian density

$$\mathcal{L}_0 = \bar{q}(i\gamma_\mu \partial^\mu - m_0)q \tag{11.11}$$

(q is the quark field operator) and the interaction Lagrangian[32]

$$\mathcal{L}_{int} = G[(\bar{q}q)^2 + (\bar{q}i\gamma_5 \vec{\tau} q)^2], \tag{11.12}$$

which is designed to be chirally symmetrical. For vanishing quark-masses ($m_0 = 0$), $\mathcal{L}_0$ is also chirally symmetrical.[33] G is a coupling constant of dimension *length²*.

The most important feature of the NJL Model is that it provides a mechanism for dynamical chiral symmetry breaking. This remarkable feature already shows up in the Mean-Field-Approximation (MFA). In this approximation, the Lagrangian density is

$$\mathcal{L}_{NJL}^{MFA} = \mathcal{L}_0 + \bar{q}Mq. \tag{11.13}$$

The 'effective' mass M is obtained from a self-consistent equation:

$$M = \frac{2GM}{\pi^2} \int_0^\Lambda \frac{p^2 dp}{\sqrt{p^2 + M^2}} \tag{11.14}$$

This equation has a non-vanishing solution if G exceeds a critical coupling strength G_{crit} (see figure 11.2). For these values of the coupling constant, chiral symmetry is dynamically broken.[34] The momentum cutoff Λ is introduced because the model is non-renormalisable.

Since the NJL Model has an explicit mechanism for dynamical chiral symmetry breaking, all empirically well-confirmed consequences of this feature are also consequences of the model. The model has, however, several additional consequences that do not follow from symmetry considerations only. For example, details of the meson spectrum have been worked out in an extended version of the NJL Model.[35] It follows that the model can indeed account for many properties of low-lying mesons.

Despite these advantages, the NJL Model also faces some serious difficulties. Here are some of them: (1) The model is non-renormalisable. (2) There are no explicit gluon degrees of freedom in the model. (3) Quarks are not confined in the model.

---

[32] We here choose the Flavour–$SU(2)$ version for simplicity.

[33] Since current quark masses are small compared to the rest mass of the proton, a non-vanishing but small current quark mass does not change the features of the model much.

[34] Equation 11.14 is called 'gap-equation' because of a close analogy to a similar effect in superconductors.

[35] In these works three quark flavours and additional chirally symmetric interactions are considered. See Vogl and Weise (1991).

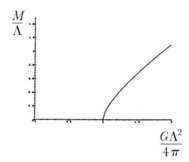

Figure 11.2 Dynamical chiral symmetry breaking in the NJL model.

Problem (1) can be dealt with by pointing out that the NJL Model is an effective field theory that is only defined and applicable at low energies. The corresponding energy scale is set by the cutoff Λ. If the energy in question exceeds Λ, the model does not make sense.

Problem (2) is tackled in a similar way. Consider the quark–quark interaction via gluon exchange at low energies. It turns out that the gluon propagator is then dominated by the squared mass of the exchanged gluon. In this domain, the momentum dependent gluon propagator can, therefore, be replaced by a constant G and the gluons are effectively 'frozen in'. In a more rigorous treatment (Chiral Perturbation Theory), it has been established that the NJL Model is a strict low-energy limit of QCD.

Problem (3) is the most cumbersome. If there is no confining mechanism, it should be possible to 'ionise' a hadron at a finite energy.[36] In order to remove this unwanted effect it has been suggested to consider a position-dependence of the coupling constant. Choosing $G(r)$ appropriately, quark confinement can indeed be achieved, but this proposal has many unwanted consequences. For instance, it breaks Lorentz invariance and destroys the conceptual simplicity of the model and this is probably why it did not attract many physicists.

On the other hand, there is no telling reason motivated by the data as to why confinement should be included as well. In fact, the spectrum of low-lying mesons is reproduced very well with the NJL Model. The situation was different in the MIT-Bag Model. Here, physicists searched for chirally symmetric extensions because pionic observables did not come out very well. There is still another argument why confinement

---

[36] This obvious drawback of the NJL Model is not often mentioned in the relevant literature.

need not be taken into account in the NJL Model. It is because the purpose of the NJL Model is to only explore the consequences of chiral symmetry and its dynamical breaking. Adding other features makes the analysis much more complicated.[37]

To take the NJL Model seriously in a phenomenological sense, it must nevertheless be determined why the lack of confinement does not matter when the model is used to calculate properties of the low-energy spectrum of hadrons. This argument is provided by a qualitative *story*. It goes like this: For low-lying hadrons, most or all of the relevant quarks are in the ground state of an effective potential. At small excitation energies, the probability to excite a quark is very small and the quark will, hence, remain in the ground state most of the time. Then it simply does not matter if the potential has a finite depth or if it is infinitely deep (confinement); the quarks don't 'feel' the difference.

This story is considered to be very plausible and it has several consequences which can be explored theoretically. For example, properties of higher-lying hadron should be more sensitive to the detailed shape of the effective potential. Numerical studies confirm this conjecture.

11.4 LESSONS

I will now draw some more general conclusions from the material presented in the last section. Let me start by comparing the two models introduced in the previous section. Both mimic one isolated feature of QCD. The MIT-Bag Model focuses on (spatial) confinement while the NJL Model chooses chiral symmetry and its dynamical breaking as relevant. It turned out that both models need to motivate why the feature they have chosen is worth considering. After all, QCD has various features, so why should it suffice to model only one of them? In order to legitimise this choice, a qualitative story that goes with the model proves to be important. This story must provide an argument as to why the chosen feature is relevant and why the other features are insignificant.

In the MIT-Bag Model, confinement is relevant, so the story goes, because it results from the complicated interactions that are known to be typical of QCD. Furthermore, extended versions of the model are empirically rather successful. Therefore, possible corrections due to chiral symmetry will presumably only be minimal.

---

[37] For more on this see Hartmann (1998).

In the NJL Model, on the other hand, the motivation for the importance of chiral symmetry is provided by pointing to the mathematical deduction of its Lagrangian density from QCD. Furthermore, the lack of confinement is 'discussed away' by telling the effective-potential-story.

We will now specify more precisely the nature of a 'story'. A story is a narrative told *around* the formalism of the model. It is neither a deductive consequence of the model nor of the underlying theory. It is, however, *inspired* by the underlying theory (if there is one). This is because the story takes advantage of the vocabulary of the theory (such as 'gluon') and refers to some of its features (such as its complicated vacuum structure). Using more general terms, the story fits the model in a larger framework (a 'world picture') in a non-deductive way. A story is, therefore, an integral part of a model; it complements the formalism. To put it in a slogan: *a model is an (interpreted) formalism + a story.*

A disclaimer or two are in order here. The slogan may be reminiscent of the logical empiricist doctrine that a theory is a formalism plus an interpretation of this formalism. Though a story also provides (or supplements) the interpretation of the formalism, it is nevertheless more than just an interpretation. As a consequence of our discussion the minimal (purely syntactic) definition of 'model' given in section 11.2 is incomplete and must be completed by a plausible story.

What criteria do we have to assess the quality of a story? It goes without saying that the story in question should not contradict the relevant theory. This is a necessary formal criterion. All other non-formal criteria are, of course, much weaker. Here are two of them: (1) The story should be plausible. It should naturally fit in the framework that the theory provides. (2) The story should help us gain some understanding of the physical mechanisms. As I have stressed previously, I believe that these are very important aims of modelling and physics.[38] It is, however, very difficult to explicate how a model and its story exactly provide understanding.

Another consequence of our discussion is that the role of the empirical adequacy of a model as a tool to assess the goodness of a model has to be downgraded.[39] It is, however, difficult to make general claims concerning the relative importance of empirical adequacy and a good story. An answer to this question depends on the respective aim of modelling and there is certainly more than one.

There is no good model without a story that goes with it.

---

[38] Cf. Fisher (1988) and Cushing (1994) for a similar view. For a philosophical analysis, see Weber (1996) and Wright (1995).   [39] See also Da Costa and French (1993).

REFERENCES

Bhaduri, R. (1988). *Models of the Nucleon. From Quarks to Soliton.* Redwood City: Addison-Wesley Publishing Company.

Bunge, M. (1973). *Method, Model, and Matter.* Dordrecht: D. Reidel Publishing Company.

Cao, T. and S. Schweber (1993). 'The Conceptual Foundations and the Philosophical Aspects of Renormalization Theory.' *Synthese* 97: 33–108.

Cartwright, N. (1983). *How the Laws of Physics Lie.* Oxford: Clarendon Press.

Cartwright, N., T. Shomar and M. Suárez (1995). 'The Tool-Box of Science', pp. 137–49 in Herfel *et al.* (1995).

Cohen, T. (1996). 'Chiral and Large-N_c Limits of Quantum Chromodynamics and Models of the Baryon.' *Reviews of Modern Physics* 68L, 599–608.

Cushing, J. (1994). *Quantum Mechanics: Historical Contingency and the Copenhagen Hegemony.* Chicago: The University of Chicago Press.

Da Costa, N. and S. French (1993). 'Towards an Acceptable Theory of Acceptance: Partial Structures, Inconsistency and Correspondence,' pp. 137–58 in S. French and H. Kamminga, eds. *Correspondence, Invariance and Heuristics. Essays in Honour of Heinz Post.* Dordrecht: Kluwer Academic Publishers.

DeTar, C. and J. Donoghue (1983). 'Bag Models of Hadrons.' *Annual Review of Nuclear and Particle Science* 33, 235–64.

Fisher, M. (1988). 'Condensed Matter Physics: Does Quantum Mechanics Matter?', pp. 65–115 in H. Feshbach, T. Matsui and A. Oleson, eds., *Niels Bohr: Physics and the World.* London: Harwood Academic Publishers.

Hartmann, S. (1995a). 'Models as a Tool for Theory Construction. Some Strategies of Preliminary Physics,' pp. 49–67 in Herfel *et al.* (1995).

Hartmann, S. (1995b). *Metaphysik und Methode. Strategien der zeitgenössischen Physik in wissenschaftsphilosophischer Perspektive.* Konstanz: Hartung-Gorre Verlag.

Hartmann, S. (1996). 'The World as a Process. Simulations in the Natural and Social Sciences,' pp. 77–100 in R. Hegselmann, U. Mueller and K. Troitzsch, eds., *Modelling and Simulation in the Social Sciences from the Philosophy of Science Point of View.* Dordrecht: Kluwer Academic Publishers.

Hartmann, S. (1998). 'Idealization in Quantum Field Theory,' pp. 99–122, in N. Shanks, ed. *Idealization in Contemporary Physics.* Amsterdam/Atlanta: Rodopi.

Herfel, W. *et al.* eds. (1995). *Theories and Models in Scientific Processes.* Amsterdam/Atlanta: Rodopi.

Klevanski, S. (1992). 'The Nambu–Jona-Lasinio Model of Quantum Chromodynamics.' *Reviews of Modern Physics* 64: 649–708.

Laymon, R. 1985. 'Idealization and the Testing of Theories by Experimentation,' pp. 147–73, in P. Achinstein and O. Hannaway, eds., *Experiment and Observation in Modern Science.* Boston: MIT Press and Bradford Books.

Milonni, P. (1994). *The Quantum Vacuum. An Introduction to Quantum Electrodynamic.* San Diego: Academic Press.

Nambu, Y. and G. Jona-Lasinio (1961). 'Dynamical Model of Elementary Particles Based on an Analogy with Superconductivity. I.' *Physical Review* 122: 345–58.

Pais, A. (1986). *Inward Bound. Of Matter and Forces in the Physical World.* Oxford: Clarendon Press.

Pickering, A. (1984). *Construction Quarks.* Edinburgh: Edinburgh University Press.

Plunien, G., B. Müller and W. Greiner (1986). 'The Casimir Effect.' *Physics Reports* 134: 88–193.

Redhead, M. (1980). 'Models in Physics.' *British Journal for the Philosophy of Science* 31: 145–63.

Rothe, H. (1992). *Lattice Gauge Theories.* Singapore: World Scientific.

Vogl, U. and W. Weise. (1991). 'The Nambu and Jona Lasinio Model: Its Implications for Hadrons and Nuclei.' *Progress in Particle and Nuclear Physics* 27: 195.

Weber, E. (1996). 'Explaining, Understanding and Scientific Theories.' *Erkenninis* 44: 1–23.

Weinberg, S. (1996). *The Quantum Theory of Fields, Vol. 2: Modern Applications.* Cambridge: Cambridge University Press.

Wilets, L. (1989). *Nontopological Solitons.* Singapore: World Scientific.

Wilets, L., S. Hartmann and P. Tang (1997). 'The Chromodielectric Soliton Model. Quark Self-Energy and Hadron Bags.' *Physical Review C.* 55: 2067–77.

Wright, L. (1995). 'Argument and Deliberation: A Plea For Understanding.' *The Journal of Philosophy* 92: 565–85.

CHAPTER 12

Learning from models

Mary S. Morgan

Modern economics is dominated by modelling. For professional economists and students alike, the practice of economics is based around the activity of building and manipulating models of the economy. Even at the earliest stage of study, the use of models forms the basic method of instruction. Economists learn from this modelling method. But what do they learn? What is the process? And what conditions need to be fulfilled for such learning to take place? In order to see what models do, and how they can teach us things, we have to understand the details of their construction and usage. To explore these questions, this essay takes as case study material two models of the monetary system by Irving Fisher (1867–1947), one of the pioneers of scientific model building in economics.

12.1 IRVING FISHER AND HIS CONTEXTS

12.1.1 The contemporary economic view of money

'Money', and in particular the economic laws and government responsibilities which governed its value, formed the most important economic question in America in the 1890s, the time when Irving Fisher took up economics seriously. The institutional arrangements for controlling money were bitterly fought over in the political arena, forming a key

I thank research group colleagues at the University of Amsterdam, at the Centre for the Philosophy of the Natural and Social Sciences, LSE, and at the Wissenschaftskolleg, Berlin, for their pertinent questions and comments when I have talked about the material in this case study. Very particular thanks go to Margaret Morrison, Marcel Boumans and Michael Power, and to Nancy Cartwright, Mars Cramer, Martin Fase, Bill Barber, Pascal Bridel and Perry Mehrling. None bears any responsibility for views expressed and errors remaining. I am also grateful for the comments received when I have presented of this material at various seminar venues. This research, and its presentation at seminars, has been generously supported by the LSE CPNSS, the LSE Department of Economic History, the Tinbergen Institute, the Dutch Royal Academy of Sciences, the Wissenschaftskolleg at Berlin and the British Academy. Finally, I thank Professor George Fisher for permission to reproduce diagrams and text quotes from his grandfather Irving Fisher's work.

element in the presidential and congressional elections of 1892 and more so in 1896, when the farmers rallied behind the political slogan that they were being 'crucified on a cross of gold'. These arguments necessarily entailed discussions of monetary theory and beliefs about economic laws, and many reputable American economists wrote books about the theory, institutional and empirical aspects of money in the period of the 1890s and 1900s.[1] These debates were not just academic, they necessarily involved strong political and commercial interests (as arguments about monetary arrangements always do), so we find that the theory of money was contested not only in the scientific sphere, but also in the economic world. By 1911, when Fisher published his *The Purchasing Power of Money* (hereafter *PPM*), the United States had entered a period of inflation, and though the debate about monetary arrangements continued, it was no longer quite so important an element in electioneering.

The main problem to be treated in *PPM* is the exploration of the factors which determine the purchasing power of money, that is, its value. Since the purchasing power of money is simply the reciprocal of the general level of prices, Fisher presumed that explaining the purchasing power of money is equivalent to explaining the general level of prices. This definition (which provided Jevons with a way of measuring the change in the value of gold) is easily confused with the theoretical claim in monetary theory known as the quantity theory of money. By Fisher's time, this was already a long-standing theory going back to Hume and Locke (and probably earlier – see Humphrey 1974). The theory states that if the quantity of money is doubled, prices will double in the economy. That is, there exists an exactly proportional one-way cause–effect relation going from the quantity of money to the level of prices.

The status of this theory in 1911 was not unproblematic for American economists. Nor were its details fully worked out (by what process did an increase in money quantity create an increase in prices?). For example, Laughlin (1903) held to an alternative tradition: he regarded the quantity theory as a meaningless tautology and claimed, in addition, that the quantity theory did not explain the observed facts (his chapter 8). Instead, Laughlin believed that money was just like any other commodity, its value determined by the supply and demand for it (his chapter 9). Fisher had earlier rejected this position and treated money as a flow, whose use was in transactions, rather than as a stock of a commodity.

---

[1] See Laidler (1991, Chs 2 and 6) for an excellent account of the policy problems, and his book as a whole for a detailed account of the main theoretical positions taken by American and British economists of the period.

Laughlin's 1903 text on money provides a good example of the kind of position, and muddled thinking on the quantity theory, that Fisher was attacking.[2] Yet not all the opponents were so confused, and there was a considerable variety of opinions available amongst American economists. The text (1896) of his respected colleague and Professor of Political Economy (thence President) at Yale, Hadley, shared the same mathematical form of the equation of exchange with Newcomb (1886) and with Fisher (1911), yet Hadley did not believe the quantity theory was adequate. He thought that the theory only provided a good description of the case when the government had intervened to change the quantity of money directly, and this led to an equivalent rise of general prices (his para. 218). But when the volume of money was regulated by a free coinage system, then changes in prices were more likely to be a cause of the changes in money than vice versa (his para. 219). In contrast, Newcomb interpreted the terms of the equation of exchange in a different way to Fisher, and hardly discussed the quantity theory. Yet another economist, Kemmerer (1909), who believed in the quantity theory, had a slightly different equation which he interpreted as a money demand and money supply equation. Finally, there is the position that Laughlin attributed to the English economist Marshall: that the quantity theory of money does hold, but in practice is completely overwhelmed by all the other factors operating on prices.[3]

Irving Fisher was no ivory-towered economist. He certainly believed that the quantity theory of money was a fundamental law of economics, but he did not believe that it provided a simple policy rule. He thought that the quantity theory was being used incorrectly, to justify unsound policy advice, because people did not really understand the theory. When we study his writings on money – both scientific and popular – and when we read about his active lobbying of presidential candidates and business leaders, we find that Fisher's agenda was to convince academic colleagues, people of influence, and the general public, of the importance of maintaining a stable value for the currency.[4] For

---

[2] Laidler (1991), particularly p. 86 note 18, brings clarity to Laughlin's work.

[3] It is not clear that Fisher knew this work (reported in Marshall 1926) before Pigou (1917) made it known to the American audience: see Laidler (1991). The earliest equation of exchange is probably Lubbock, 1840.

[4] Fisher wrote three parallel books for different audiences: *The Purchasing Power of Money* (1911) for the academic and intellectual market, a popular text called *Why is the Dollar Shrinking?* (1915) to cover the same ground, and a policy book, based on chapter 13 of *PPM* called *Stabilizing the Dollar* (1920). Although the *PPM* was published in 1911, it had grown out of earlier papers on various aspects of money and its value (see his 1894 and 1897). It was delayed perhaps by Fisher's long illness with tuberculosis from 1898 to 1903.

Fisher, the important point is that there are a number of economic and institutional laws governing the relationships between money and prices, and the general level of prices might rise and fall for other reasons than those implicit in the quantity theory. His aims in *PPM* were to show under what conditions the quantity theory holds; to explore the process by which the relationship works; and to determine all the other circumstances, economic and institutional, which operate to alter the purchasing power of money. In order to do this, Fisher developed and used a series of models of the monetary system.

12.1.2 Irving Fisher: Model Builder

It is not at all self-evident that a study of money with Fisher's agenda would necessitate any introduction of models for such a method was not part of the standard economics tool-kit of the period. Why did Fisher rely on the use models? If we look to Irving Fisher's personal background for enlightenment, we find the well-known fact that he was a student of the famous American physicist, Willard Gibbs. It is more often forgotten that Fisher was equally an admirer of William Sumner, his economics mentor (one of the foremost economists of the 'old school' and ardent campaigner for free trade), and that Fisher took the chance to change from the mathematics department to join Sumner in the department of political economy at Yale when the opportunity arose in 1895. It is perhaps the case that Gibbs' line was more evident in Fisher's famous thesis of 1891/2 (see his 1925), but for his work on money, we should look first to Sumner, who had written an admired book on the history of American currency (including graphical treatment of the relation between money and prices) (1874). To these two figures, though with reservations, we can add the named dedicatee of his book on money: Simon Newcomb, American astronomer, mathematician and economist of note who had also written on money.[5] These were the intellectual resources, amongst others that Fisher apparently most respected.

But it would be unwise to assume that Fisher himself brought no resources to bear on the problem for it was Fisher who was creative in building and using models, not all these 'influential' characters. Thus,

[5] The reservations are twofold. First, I am indebted to William Barber, editor of Fisher's papers, who has informed me that Fisher did not read Newcomb's book until after he had finished his own! Secondly, as this account proceeds, we shall see how little Fisher took from this resource.

Gibbs was not, in the main, known as a model builder.[6] Sumner was a persuasive writer – with a brilliant way with metaphors – but neither a modeller nor practitioner of mathematical economics (though it was he who suggested that Fisher write a thesis in mathematical economics). Newcomb, whose writings on 'societary circulation' (1886) Fisher found so attractive, used both an illustration and an equation of societary circulation, but their use as models were severely limited compared to the creative work of Fisher in these respects. It was Fisher who broke with the tradition of his teachers and developed constructions which we can now recognise as 'models' in his texts. His son's biography of him (see I. N. Fisher 1956) suggests two points which may be relevant to help us understand Fisher's skills in modelling and some especial features of his models of money.

One point is that Fisher was an inventor. From the age of fourteen at least (after he became the family breadwinner), Fisher expended energy and money on inventing gadgets: a mechanism to improve pianos, a new form of globe,[7] a folding chair, etc. One of his inventions – the visible card file index – was spectacularly successful, making him a millionaire when he sold out to Remington Rand. Though the inventive streak must have been fostered by family necessity, it is also relevant that in the United States it was customary (and had earlier been required by patent law) to provide a working model for any new patent to be granted. Fisher's inventive powers and model-building skills had been made clear in his thesis, where he had designed and built a working hydraulic model of a three-good, three-consumer economy. His use of models in this context was rather self-conscious, perhaps not surprisingly, given how alien the method was at that time in scientific economics (nor was it entirely acceptable in physics).

In addition, as his son reminds us, Fisher was committed to a 'philosophy of accounting' in which economic concepts and terms should be defined and measured in ways which matched those of the business world (of which he was an active participant). We can see this in the very

---

[6] Indeed, the entry by E. B. Wilson on Gibbs in *The Dictionary of American Biography* stresses that Maxwell's response to Gibbs' work was to make a model of part of his work, and to send a copy of the model to Gibbs. On the other hand, Khinchin (1949) characterised Gibbs' approach as depending upon analogical thinking, which, as we shall see, is clearly one of the aspects of Fisher's approach to his material.

[7] His book on globes (1944 with O. M. Miller) in particular shows his ability to think about the many different ways of presenting accurate maps of the world appropriate for different purposes; the importance of variety is evident here – for no single type of globe is best for all purposes. (See also Allen, 1993, for biographical details.)

careful way he set up definitions of all his economic terms at the beginnings of his books[8] and in the way he defines the elements in his models.

Both of these characteristics, the inventive turn of mind and the philosophy of accounting, can be seen at work in the models of money which we find in his book *PPM.* While all economists will recognise Fisher's 'equation of exchange' from this book, the models he built and the ways he used them, are not part of our standard knowledge. This has to be recreated by a careful analysis of Fisher's work. In building a model, a number of choices exist and decisions have to be made about the features of the model. These internal design features are sometimes chosen explicitly with a rational purpose in view (even though, in Fisher's case, it may not be obvious until many chapters later why a particular modelling decision has been made). Sometimes they appear to be chosen almost unconsciously and without any obvious reason, and some features appear to have been chosen arbitrarily. Thus, any historical reconstruction of Fisher's process of modelling based on his texts is necessarily a rational reconstruction. But that doesn't imply that the process was rational for Fisher, for there was no 'model-building algorithm' for him to follow, and modelling is clearly a creative activity as well as a logical one as we shall see.

12.2 THE ACCOUNTING-BALANCE MODEL

Fisher's stated aim in *PPM* (p. 15) is to clarify the quantity theory of money by analysing the equation of exchange. That is, all discussion of the quantity theory has to be related to the equation of exchange. What is the equation of exchange? Fisher presents it as a relation between two flows: of money and goods. (In this, he followed Newcomb's description of societary circulation – but eschewed both his illustration of networked circulation and the associated metaphor of veins and blood flows (Newcomb 1886, Book IV).) Fisher:

> Trade is a flow of transfers. Whether foreign or domestic, it is simply the exchange of a stream of transferred rights in goods for an equivalent stream of transferred money or money substitutes. The second of these two streams is called the 'circulation' of money. The equation between the two is called the

[8] For example, at the very opening of his 1912 book on capital and income he states: 'This book is an attempt to put on a rational foundation the concepts and fundamental theorems of capital and income. It therefore forms a sort of philosophy of economic accounting, and, it is hoped, may supply a link long missing between the ideas and usages underlying practical business transactions and the theories of abstract economics' (1912, vii).

'equation of exchange'; and it is this equation that constitutes the subject matter of the present book. (p. 7)

The equation of exchange is a statement, in mathematical form, of the total transactions effected in a certain period in a given community. It is obtained simply by adding together the equations of exchange for all individual transactions. (p. 15–16)

The equation of exchange is 'illustrated' in three ways in the early chapters of *PPM*.[9] I am going to call these three different illustrations of the accounting identity 'models' of the equation of exchange, because, as I shall show, these are not just 'illustrations'. The reader cannot merely look at them and move on – Fisher puts them to work, and the reader must follow Fisher to understand what is going on. I shall present each of these three model-illustrations in the form Fisher presented them, and then provide an analytical commentary on his process of both building and using the models to see how he learnt from the process.

12.2.1 The arithmetic 'illustration'

The arithmetic version of the model begins with an equation of an individual person with one exchange (p. 16):

70 cents = 10 pounds of sugar multiplied by 7 cents a pound

After a further two pages of text, Fisher presents an aggregate level equation of exchange (p. 18):

$5,000,000 × 20 times a year =
200,000,000 loaves × $.10 a loaf
+ 10,000,000 tons × 5.00 a ton
+ 30,000,000 yards × 1.00 a yard

This example seems so simple that it is easy to miss the many hidden modelling decisions which Fisher made in building up this first model.

One of the main functions of the model is to draw the reader in with a very simple example and thence to move us from that simple level of the individual to the aggregate level. But clearly this is not the whole of the United States with all its goods; rather it is a sort of *halfway house simplification* of three goods and prices, but aggregate-size amounts adding up to

---

[9] In *Why is the $ Shrinking?* (1915), these are overtly labelled: the 'Arithmetical, the Mechanical and the Algebraic Illustrations', in the *PPM* they are the same illustrations, but not labelled as such. Keynes' review (1911) refers to one of the illustrations explicitly as a 'model'.

$100m money value per year. Note that this aggregation apparently results in no differences of kind in three of the elements: prices, quantities of goods or amount of money. These commonsense/folklore elements in the individual level equation are easily recognisable by anyone in terms of their own economic experience and the aggregation creates no changes in the kind of things being added. It is here that we can see his 'philosophy of accounting' (see above) at work, not only in the choice of model-illustration, but also in the careful matching of economic entities to common-usage experience of things economic.

But simplification though it is, the aggregation is not simply a summation of the individual exchange equations as he suggests in his text quoted above (from his p. 15). The fourth element in the equation: '20 times a year' – the velocity of circulation, does not feature at all in the individual exchange equation. It is an additional element which only comes in with the aggregation. The amount of money in the economy is much less than the total value of goods exchanged, each bit of money exchanges many times in the year. At this aggregate level then, we need something else to make the identity hold: we also need the velocity of money's circulation.

This 'hole' in the model, that has to be plugged with a new entity in the 'aggregate' equation, alerts us to the fact that velocity is important and requires interpretation as a separate (independent) factor here. If the accounting identity was a tautology, rather than a law of necessity, why would we need any discussion of it? It could just be found from the rule of three (as pointed out by Humphrey's 1984, p. 13, discussion). The separate character and interpretation given to all four elements has implications for the way the equation is used as we shall see later.

Bringing in a new element necessitates a brief discussion and definition of velocity by Fisher: that each person has their own transactions velocity of circulation which is averaged in the aggregate: 'This velocity of circulation of an entire community is a sort of average of the rates of turnover of money for different persons. Each person has his own rate of turnover [of money] which he can readily calculate by dividing the amount of money he expends per year by the average amount he carries' (p. 17). This keeps the explanation at the level of the individual experience, but an easier way to explain it in the folklore mode (adopted by Newcomb) would be to say something like: 'A exchanges money with B who then spends it to C' – that is, the same cash circulating between different people against different goods. But Fisher did not do this. As we later find out, the funny definition and verbal explanation are needed because he wants us to measure velocity in a particular way. The common concept of velocity as coins changing hands leads us to the wrong method of measurement. Thus the correct concept for the element in the aggregation equation was the average of the individual velocities.

Almost unnoticed in the aggregation is another critical element – the introduction of the time unit.[10] The first simple exchange has no time reference, it does not need one. But velocity cannot be conceptualised properly, let alone measured, without a time unit attached. So the aggregate is also an aggregate for a specific time period – in this case a year. Of course, it is also the case that goods may exchange more than once in the year, but not usually in the same form (e.g. grain, flour, bread, all have different valuations).

Although it is usual to think of models as simplifications of some sort, Fisher makes only one explicit simplification statement at this aggregation level, in which he excludes both other circulation medium (non cash money) and all foreign trade (the illustration includes only 'trade within a hypothetical community' (p. 16)). These elements are to be reincluded later on so that he can proceed 'by a series of approximations through successive hypothetical conditions to the actual conditions which prevail today' (p. 16). Note, these elements are not re-included on this particular model-illustration – the first is reincluded in the mechanical model and the second on the algebraic.[11]

Having built up his model, Fisher then uses his aggregate arithmetic model to work through some examples of what will happen when there are changes in money, velocity, or quantities of goods. He shows that, if the amount of money changes, but velocity and quantities do not change, then *prices must change (on average by the same proportion)*. This is the quantity theory claim. But he also shows, with his manipulated examples, that changes in velocity or quantities would also affect prices (proportionately) provided that the other two elements are unchanged. Since all of these elements could be changing, 'We must distinctly recognize that the quantity of money is only one of three factors, all equally important in determining the price level' (p. 21). It is only when we get to this manipulation of the model that we fully realise why Fisher

---

[10] I am grateful to Mars Cramer for pointing out this element to me.

[11] The extension to foreign trade requires two equations, discussed but not written out on p. 28 and then given and discussed pp. 372–75. Fisher states that since the size of trade flows are a very tiny proportion (1 per cent or less) for the United States, these equations are pronounced 'a needless complication' and dismissed!

has made the aggregation. The quantity theory is an aggregate level claim, it cannot be demonstrated, nor its *ceteris paribus* conditions expressed, unless the model is working at that level.

We can already see several features of Fisher's (and perhaps all) models. The illustration is not just a picture – the work happens in two places. First is the building up of the model. This involves necessary translations from verbal statements, choices of which bits to include and exclude, attention to concepts, definitions and reinterpretations of the parts and an explicit presentation of the structure of any relationships.[12] Secondly, the model is used. The model is worked through to reveal what constraints are entailed, how the interactions work, and what outcomes result from manipulating the relationships in the model: in this case about the equation of exchange and the quantity theory.

12.2.2 The mechanical-balance 'illustration'

This offers a second model of the equation of exchange shown here as figure 12.1 (Fisher's figure 2, p. 21: note that the labels have been added by me).[13]

Here we have an analogical model constructed from the aggregate arithmetic version (with some change of units made for convenience). Some elements maintain commonsense representations (in the form of pictures) some do not. Velocity poses no problem since all four elements can be easily translated from the arithmetical to the analogical model, indeed four elements are necessary to complete the balance. Note, that it is not a fixed equal-arm balance – all four of the elements can, but do not have to, alter in creating the equality.

As before the analogical model is put to work to explain how the changes in the elements are bound by the relationships between the four elements in the system:

An increase in the weights or arms on one side requires, in order to preserve equilibrium, a proportional increase in the weights or arms on the other side. This simple and familiar principle, applied to the symbolism here adopted, means that if, for instance, the velocity of circulation (left arm) remains the same, and if the trade (weights at the right) remains the same, then any increase

---

[12] See Marcel Boumans' account (this volume) for a more general discussion of the integration of bits in a model.
[13] I thank Bill Barber for bringing to my attention Fisher's earlier experiments using a specially constructed mechanical balance to measure nutrition in 1906 and 1907.

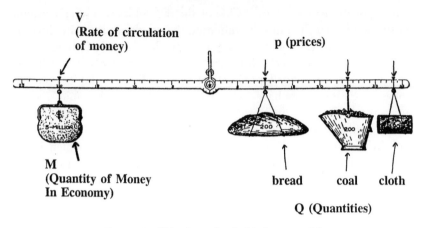

V
(Rate of circulation
of money)

p (prices)

M
(Quantity of Money
In Economy)

bread   coal   cloth

Q (Quantities)

Figure 12.1 Fisher's mechanical balance model.
From Fisher (1911), figure 2, p. 21, with added labels.

of the purse at the left will require a lengthening of one or more of the arms at the right, representing prices. If these prices increase uniformly, they will increase in the same ratio as the increase in money; if they do not increase uniformly, some will increase more and some less than this ratio, maintaining an average. (pp. 22–3)

So any movement in any one of the elements *necessitates* (requires) a movement in one or more of the others to maintain balance. But more than this – it shows much more clearly than the arithmetical illustration the *direction* in which movement must take place because of our familiarity with the working of the balance. As Fisher says 'we all know' how a balance works (p. 22). In thinking about this model-illustration we see that Fisher has focused our attention on the balancing possibilities more effectively than in the arithmetic illustration.

This model-illustration is, like the arithmetic model, at first sight straightforward. But the analogical model is, in two respects, much more complicated than it looks. First this model hooks onto various measurement procedures. Second, the analogical properties of the model cannot be taken for granted; I will return to this point in section 12.3.

The mechanical balance is both an instrument of measurement for merchants in deciding the values of exchange and a scientific instrument for accurate weighing in which the balance is 'in balance' when opposing forces neutralise each other. Mechanical balances act as counting (measuring) machines in the commercial world. Where the goods being weighed are of equal weight units, balances can be calibrated to allow the direct reading of monetary values.[14] One of the difficulties of Fisher's original illustration is that the units are different: the goods are in many different weights and they do not match the units in which money is measured in the left-hand purse. The calibration along the two arms is in different units, one for prices, one for velocity of money. To be useful for counting (measuring), the goods have to be aggregated into similar units and the prices made the same on average in order that the length-×-weight units on both sides of the balance match. To follow up this point we need to look briefly at two important extensions to this model in which Fisher makes the model both more complex and, in another sense, simpler.

The simplification is concerned with converting the goods to one set of units and the prices to an average level to deal with the problem of non-uniform changes noted in the text quoted above. Fisher:

> As we are interested in the average change in prices rather than the prices indi-
> vidually, we may simplify this mechanical representation by hanging all
> the right-hand weights at one average point, so that the arm shall represent the
> average prices. This arm is a 'weighted average' of the three original arms,
> the weights being literally the weights hanging at the right. (p. 23)

The point of averaging prices was mentioned on the arithmetical illustration, but not discussed in any depth. Yet this is a critical point, for the
quantity theory proposes a relationship between the amount of money
and the *general* level of prices. How should one conceptualise and
measure such a general level of prices? This mechanical model is used
to indicate one of the aggregation devices necessary for thinking about
measurement for the price side. The concept of 'weighted average'
explicitly pictured by Fisher on this model (as in the next figure) is a key
concept in price index number theory. Although some aspects of price
index numbers were by now understood, the problem of how to aggregate and average individual prices into one price index number was still
open to discussion, and Fisher's name remains one of most famous in
this literature (see his 1922 book, also *PPM*, appendices 3 and 7 to
chapter 2 and appendix to chapter 10; and Dimand (1998)).[15]

The complication is the addition of other circulating media, bank
deposits subject to check transfer (with their own velocity) onto the
model. This comes as an illustration in his next chapter which again

---

[14] The *Encyclopedia Britannica* (1911 edition) entry on 'Weighing Machines' makes a point of discussing such machines and particularly describes American-made commercial balances of the
period.

[15] Indeed Martin Fase has suggested to me that Fisher's work on price indices was the consequence
of his work on the quantity theory of money. Those of us who have to teach index numbers will
readily agree that this aspect of the model has tremendous heuristic power.

begins with a simple arithmetic illustration of money transferred through the banking system before he introduces such additions onto both the mechanical and algebraic illustrations. This provides for a further two elements on each model: an extra weight in the form of a bank deposit book, with its own velocity, on the left-hand side arm of the mechanical balance (see his figure 4, p. 48 and our next figure 12.2). This allows for the further aggregation to cover all goods and money transfers for the whole economy within the mechanical and algebraic models. This extension works perfectly well within the models, that is, it does not change the way the models work, just effects a higher level of aggregation.

We can now return to our measurement function of the model, with an illustration which comes near the end of the book, when Fisher has measured all six of the elements in the equation of exchange and calibrated those measurements with respect to trusted data at two dates 1896 and 1909 (for which estimates of velocity could be made). He transcribes these measurements of all the elements onto the balance model shown here as figure 12.2 (Fisher's figure 17, opposite p. 306). Though he does not strictly use this statistical version of the model as a measurement instrument, it does serve two measurement purposes for Fisher. First, he uses it as an instrument to display the balanced series of measurements and this enables him to 'show at a glance' how all six elements have changed through recent times.[16] He then uses this statistical model to describe the changes as follows: 'prices have increased by about two thirds between 1896 and 1909, that this has been *in spite of* a doubling in the volume of trade, and *because of* (1) a doubling of money, (2) a tripling of deposits, and (3 and 4) slight increases in the velocities of circulation' (p. 307).

The second purpose is to show at the same time the impossibility of reading off from the model measurements, by simple induction, an empirical proof of the quantity theory since both V and T have changed over the period. That is, the figure allows you to read off the measurements of the equation of exchange, and how the elements in the equation have all changed over time, but not to read off a direct 'proof' or 'disproof' of the quantity theory.[17]

---

[16] The figure reproduced here goes up to 1912 and comes from the 2nd edition of his book. The measurements were afterwards reformulated into a table form of the equation of exchange to enable Fisher to calculate more exactly the effects of changes in the separate elements on the price level (his pp. 308–9).

[17] I agree with Laidler (1991) who claims that Fisher was quite clear that the quantity theory could not be tested from this representation of the figures. But I think that Laidler misses the second point here – Fisher explicitly refers to the empirical work in relation to proof of the *equation of exchange*. This is discussed further below in section 12.3.

Mary S. Morgan

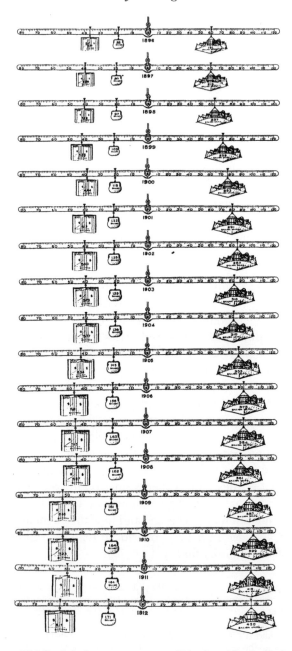

Figure 12.2 Statistical measurements on Fisher's mechanical balance.
From Fisher (1912; 2nd edition of Fisher 1911), figure 17, opposite p. 306.

12.2.3 The algebraic 'illustration'

The final algebraic version of this model defines the elements into symbols (M for money, V for velocity, p and Q for individual prices and quantities) and develops the equation of exchange in algebraic form. This is extended in the next chapter to include both bank deposit money (M' and its velocity V', as we have just seen on the balance illustration) as well as the average price level (P) and all transactions (T) to give the following sequence (pp. 25–6, p. 48):

$$MV = pQ + p'Q' + p''Q'' + \text{etc.} \tag{12.1}$$

$$MV = \Sigma pQ \tag{12.2}$$

$$MV + M'V' = PT \tag{12.3}$$

The same arguments about what else has to change in the equation when one thing is altered are repeated to illustrate yet again the trade-offs within the equation and the circumstances under which the quantity theory of money will hold.

Fisher's stated reason for this algebraic illustration constitutes the oft-made defence of the mathematical method in economics: 'An algebraic statement is usually a good safeguard against loose reasoning; and loose reasoning is chiefly responsible for the suspicion under which economic theories have frequently fallen' (p. 24). A more critical feature for us is that this version moves the model beyond the examples, given in the arithmetic cases, towards the general formula. This is done by denoting symbols for each of the four elements, and a general notation of aggregation (which leads directly to yet another measurement method appendix) and the dropping of any time unit. Fisher treats this as a further simplification: 'This simplification is the algebraic interpretation of the mechanical illustration . . .' (p. 27), but it is more accurately described as a generalisation. With this generalisation, the equation of exchange (and quantity theory) can be treated as applying at all times and for all places.

We might with good reason say that this third illustration is no longer a model (except in so much as a mathematical statement is always a representation or model of the relationships between the real world entities) but rather a general theory or law: the fundamental equation of exchange within which the general quantity theory is embedded. So the generalisation is a double one, a translation into the language of mathematical symbols implies that the model holds generally, and at the same time gives access to an appropriate general method of abstract reasoning.

This final version (equation 12.3) is known as Fisher's equation of exchange (and was derived earlier in his 1897 paper). Three comments put Fisher's algebraic equation of exchange into perspective. First, Fisher's equation will probably strike many readers as completely self-evident. But this is surely part of Fisher's rhetorical strategy – he wants his readers to think his equation is obvious, for it was by no means taken as self-evident by his contemporaries. There is, as is well known, a considerable tradition (see Humphrey 1984) of algebraic formulations of both the equation of exchange and quantity theory. And, amongst these, there are a number of different equations which have been taken as the 'fundamental' equation of exchange, with different definitions of the terms and different causal assumptions within them (see Bordo 1987). Secondly, although a number of mathematical formulations were available to Fisher (see his footnote, p. 25), few of them would count as models. Some of them appear to be simple translations of words to equations, with little or no work done using the mathematical formulations: Newcomb provides one such example. Walras is an exception at the other extreme in providing a full-blown mathematical theorising of deduction from initial assumptions. Neither indulge in the explicit process of model building and usage which characterise Fisher's approach. Thirdly, it might be argued that the equation of exchange hardly deserves the label 'theory' having the status of nothing more than a tautology. This is to misunderstand the equation – it is not a tautology, but an identity, the usefulness of which was a question not only for Fisher, but for many other writers on quantity equations.[18] In his very elegant piece in the *New Palgrave*, Bordo (1987) describes equations of exchange as providing the building blocks both for quantity theories and

---

[18] Fisher discusses the role of identities in scientific research explicitly:

'One of the objectors to the quantity theory attempts to dispose of the equation of exchange as stated by Newcomb, by calling it a mere truism . . . 'Truisms' should never be neglected. The greatest generalizations of physical science, such as that forces are proportional to mass and acceleration, are truisms, but, when duly supplemented by specific data, these truisms are the most fruitful sources of useful mechanical knowledge. To throw away contemptuously the equation of exchange because it is so obviously true is to neglect the chance to formulate for economic science some of the most important and exact laws of which is it capable.' (p. 157)

The question about how to treat and use identities in monetary theory has been a long-running sore, and the arguments and positions are discussed at length in Marget's fine historical study of the question (1938).

for other macroeconomic relations. Humphrey (1974), in his survey of early mathematical quantity theories suggests that it was commonplace to begin with an identity and then to interpret it in functional terms or examine causal relations (such as the quantity theory) within the identity. Fisher's work is an exemplar in both these respects, for we have seen here the fruitful way that Fisher exploited the equation of exchange identity in his model building.

12.3 THE 'WORLD IN THE MODEL'

In the process of modelling the equation of exchange, Fisher has led the reader carefully from the world of his/her individual money-good exchange experience into an aggregate and thence to a general world, in which the laws of economics are denoted in symbols which, while still comprehensible, are divorced from common usage. By starting with a world we know, we are led into believing in the world in the model, because a number of its features still match those of the world we know. Within the world of the model, where the equation of exchange holds, the quantity theory of money can be 'demonstrated' and its *ceteris paribus* conditions (V and T constant) explored. It is in this sense that models 'explain', by demonstrating theoretical claims within the constrained and structured world represented in the model. (In this way, as in others, models can be conceived as having parallels with laboratory experiments where relations or results are demonstrated within a restricted domain.[19])

But there are also new theoretical claims introduced when we enter the world of the analogical mechanical balance model. In the arithmetical model, a simple change in one element necessitates a matching change in one of the other elements to maintain equality, and that is all that is required or implied. In a mechanical balance, a similar change in one element would create an oscillation of the arms of the balance as the mechanism seeks a position of equilibrium. So, the analogical model implies that the economic transition process which follows from a disturbance of any element in the equation of exchange is equivalent to a physical oscillation. There is no necessary prior reason for us to

---

[19] See Wimsatt (1987) for a brief discussion of the similarities between simplifying assumptions in model building and the decisions about what factors have to be controlled in laboratory experiments. Kemmerer's book on money (1909) discusses a 'hypothetical society' in these same terms, and even builds up an equation for the money relations of the society, but hardly uses the relation as a model for analysis.

believe that these analogical model claims fit the economic world, even if we believe the equation of exchange and the working of the quantity theory. The world of the first and last illustrations operate according to the laws of simple arithmetic. Those laws seem to govern our own individual experience of the equation of exchange – so we are willing to believe in them at the aggregate level. The world of the mechanical model is different. It is governed by the laws of mechanics, not accounting, and so introduces new theoretical claims as part of its inherent structure.

This is a good example of Hesse's view of models as analogies, which lead to new theories about the workings of the world. In her work on models and analogies (1966), Hesse proposed that analogies provide models which function at the discovery end of scientific practice. We choose an analogy from another field as a model because of the positive features of the analogy which fit our new field (the notion of equality in the equation of exchange paired with that of balance in the mechanical balance, an analogy supported by the dual use of the balance as both a mechanical and merchant tool). We then look for some other neutral elements (properties or relationships) of the analogy which might suggest new theories or indicate new facts about our economic world: in this case, the oscillation process. The analogy is thus the source of the creative element in theory building.

But there are usually negative elements in the analogy. Though Hesse says little about these, they can not be neglected (see Morgan (1997)). In this case, the negative analogy relates to the accounting identity constraint, which is embedded in the structure of the arithmetic and algebraic equations illustrations, but no longer has to hold in the mechanical balance. There is no constraint built into the mechanical structure which says that the mechanical balance has to rest at equity. Indeed, it is part of the balance model that we can easily imagine increasing the amount of money on the left-side purse, and the mechanism will tilt to the left and come to a position of rest there. In other words, although there is a tendency to an equilibrium, there is nothing to ensure that this point of rest is a point of equal balance! This is a negative analogy with considerable implications for Fisher, for it raises two questions: What makes his money-exchange balance come back into balance and at the point of equity? These are never put explicitly, but a large chunk of the rest of the book is given over to answering them in economic terms. In chapter 4 of the book, he explores the short-run equilibrating mechanisms and in chapters 5–8, he analyses the long-run equilibrium with an exhaustive analysis of the causal laws operating on, and between, the separate economic elements in the balance.

In terms of the modelling problem, one part of the answer to the questions above lies in the way Fisher makes the creative part of the analogy work. The short-run property of the balance to oscillate up and down in an attempt to find a point of rest is the starting point for Fisher's description of the economic transition process whereby a change in one element is transferred into a change into another element. In doing this, he provides a theory of monetary induced cycles in commercial activity. In this short-run adjustment process, he theorises that not only does the quantity theory not hold, but the direction of causality between M and P is reversed (price changes cause changes in money stock) while the level of trade or transactions also alters. Although neither credit-based business-cycle theories nor oscillation metaphors were in themselves new, this is an unusually well-worked out theory for the period. What is new is the explicit integration of these cycles and all disturbance transitions into the same structure as that housing the equation of exchange and the quantity theory relationships.[20] What is important for us is that it is the model that is the device which integrates these two theoretical domains.

The other part of the answer to the modelling problem is that Fisher reinterprets the equation of exchange to neutralise the negative analogical feature. Fisher appeals, in his discussion of the trade-offs, to certain features of the mechanical balance:

We all know that, when a balance is in equilibrium, the tendency to turn in one direction equals the tendency to turn in the other . . . The equality of these opposite tendencies represents the equation of exchange.' (p. 22)

In general, a change in one of the four sets of magnitudes must be accompanied by such a change or changes in one or more of the other three as shall maintain equilibrium. (p. 23)

We note that Fisher replaces 'equality' with 'equilibrium' in discussing this illustration and in doing so, he reinterprets the equation of exchange as an equilibrium system. The mechanical balance model thus, by sleight

---

[20] Laidler (1991) recognises the closeness of the connection when he states that Fisher 'developed the cycles as an extension of the quantity theory of money' (p. 100), but he does not note the structural connection nor that it arises from the mechanical model. Nevertheless, I bow to Laidler's much greater knowledge of the economic theory content of Fisher's work, and note that Laidler treats Fisher's contributions as on a par with the Cambridge School.

of hand, allows Fisher to introduce a completely new interpretation onto the equation of exchange, for in the arithmetic example, equality does not necessarily denote equilibrium, nor is the system a 'tendency' system.

The real answer to the second question raised by the model – what makes the balance come back to a point of balance at equity? – lies outside it. It has to balance because the elements are still governed by the equation of exchange: but now, a reinterpreted equation of exchange consistent with the mechanical balance: the magnitudes of the four 'factors seeking mutual adjustment' 'must always be linked together' by the equation of exchange. But he continues:

> This [equation] represents the mechanism of exchange. But in order to conform to such a relation the displacement of any one part of the mechanism spreads its effects during the transition period over all parts. Since periods of transition are the rule and those of equilibrium the exception, the mechanism of exchange is almost always in a dynamic rather than a static condition. (p. 71)

We should, then, never expect to see the equation of exchange hold exactly.

During the modelling process, the mechanical model, though it allowed Fisher to integrate the theory of economic cycles (transition oscillations) into the same structure as the quantity theory, became temporarily cut off from the original interpretation of the equation of exchange. This had to be re-established as an overriding constraint, but now, reinterpreted as an equilibrium tendency relation. In the process, we can see that Fisher made two important relaxations: first that the quantity theory no longer holds (and even reverses) in the short run; and, second, that in practice, the 'mechanism of exchange' never is exactly in balance. The equation, though reinstated, no longer constrains so tightly as it did before.

The world in the model allowed Fisher not only to extend his theory using the positive analogy, but by taking note of the negative analogy, exploring its implications and finding a way to neutralise it, he effectively used his model to learn about the full theoretical implications of his new world. At some stage, however, if we want to use models to learn about the world, the model needs to map onto the real world. This is exactly what Fisher believed he had achieved when he presented the measurements for the equation of exchange for the United States economy on his balance mechanism (our figure 12.2, shown earlier). Fisher considered his presentation of these equally-balanced results onto the mechanical illustration of the accounting-balance model as an empirical 'proof' of his equation of exchange.[21]

By using this representation, the world in the model had also allowed Fisher to learn some things about the real world. Recall the discussion from section 12.2.2 above, that Fisher first used his measurement-model to establish the real and independent existence of all the individual elements in the equation and he used this statistical version of his model to draw inferences about 'the factors of history', the real changes in the economy. Secondly, he used the model to represent the movements in each of the separate elements in order to make his point about the quantity theory and *ceteris paribus* conditions not as a logical point (as in his earlier model manipulations), but as a matter of empirical reality. Fisher summarised the whole exercise by analogy with Boyle's Law:

Practically, this proposition [the quantity theory] is an exact law of proportion, as exact and as fundamental in economic science as the exact law of proportion between pressure and density of gases in physics, assuming temperature to remain the same. It is, of course, true that, in practice, velocities and trade seldom remain unchanged, just as it seldom happens that temperature remains unchanged. But the *tendency* represented in the quantity theory remains true, whatever happens to the other elements involved, just as the *tendency* represented in the density theory remains true whatever happens to temperature. Only those who fail to grasp the significance of what a scientific law really is can fail to see the significance and importance of the quantitative law of money. A scientific law is not a formulation of statistics or of history. It is a formulation of what holds true under given conditions. (p. 320)

With Fisher, we might judiciously observe that his 'scientific law' of the quantity theory was not a formulation of theory or of statistics or of history, but of his model-building activities.

12.4 THE WORK DONE BY THE ACCOUNTING-BALANCE MODEL

To summarise this rather complex case let me lay out the salient features of the decisions that Fisher's modelling process involved and the results that they entailed.

(a) The modelling sequence moves from the individual to the aggregate to the general level. Why aggregate? Because the quantity

---

[21] The results of his measurements of the individual elements did not establish the equation of exchange *exactly*. Fisher carried out a number of adjustments to the data, of which the first crucial one was a calibration adjustment (such calibrations were also a feature of the measurements for the individual elements); followed by separate small adjustments for errors on both sides of the equation.

theory claim does not hold at the individual level, only at the aggregate. Why generalise? Because the theory is thought to hold at all times and places.

(b) Why does the quantity theory have to be embedded into the equation of exchange modelled as an accounting identity? Because Fisher treated money as a flow, servicing economic transactions (and he wanted to avoid the alternative treatment of money as a commodity for which there was a supply and demand). The equation of exchange accounting model is then used to demonstrate the cause–effect quantity relationship between money and the price level, and to show that this holds only if the other two balancing items remain the same. The model clarifies both the quantity theory of money (treated as a flow), the overall accounting-theoretic framework in which it is operative, and the *ceteris paribus* conditions under which it holds.

(c) The models provide conceptual help with measurement methods; definitions are tied to measurement issues; and one model provides an instrument calibrated to display measurements and is used to describe the historical changes in all the elements over time.

(d) Why use three different model-illustrations when any one could be used to demonstrate the workings of the quantity theory?[22] Rhetorically and functionally, they lead us through various aspects of the problem appropriate to their format, from the simple world of the individual exchange, through the complexity of aggregation, adjustment processes, measurement problems and the like, and thence back to the simple world of the general theory. Not all of these intermediate aspects are fully developed using the models, but the modelling connects with all these problems, which are then dealt with in the rest of the book.

(e) How and what did he learn from his model-building exercise? Even in discussing this apparently simple case, we can see how Fisher learnt to build his models through a series of decisions involving choices of form; analogical reasoning; moves for greater simplification and greater complexity; and for more aggregation and generalisation. In using the model, Fisher was able to (i) explore theoretical frameworks, demonstrate the workings of

[22] This question of 'Why three models?' perhaps relates most closely to the personal aptitude Irving Fisher had for seeing things in lots of different forms for different purposes (e.g. the design of globes depends on whether they are intended for explanation, education or accurate map-reading).

theory, and conditions under which it holds, and create and explore new theoretical claims as an extension to the older ones; (ii) provide various conceptual indications for measurement purposes and add measurements to map the model onto the world; and (iii) use these measurements to learn something about how the quantity theory applied to the world and how the world had been during the historical period. The model played an extraordinary variety of functional roles, each of which provided the basis for learning.

I now turn to another model from *PPM* in which Fisher's modelling was more ingenious, but which entailed a rather different method of learning.

12.5 THE CONNECTING RESERVOIRS MODEL

The accounting-balance model showed the effects of changes in the quantity of money (M), its velocity (V) or the volume of trade (T) on the general level of prices (P) within the equation of exchange, and explored the conditions for the quantity theory of money to hold. Fisher's second model assumes that the quantity theory of money holds and explores the institutional and economic reasons why M changes.

The connecting-reservoirs model was first introduced by Fisher in an 1894 paper dealing with bimetallism. It was reintroduced in chapter 6 of *PPM* and emerges fully fledged in chapter 7. The model grows out of an oft-used metaphor in economics, that of money or goods conceived as flows of water such that the levels of their prices are automatically equalised between two places (as water equalises its level between a lagoon and the ocean). These comparisons rarely move beyond the verbal similes (for example, see Jevons 1875), and no one had formally modelled such a system of money flows according to the hydraulic principles. Irving Fisher had built such a model for his thesis, a clever piece of laboratory-equipment in which flows of water around the system 'demonstrated' a general equilibrium model of an economy – but money was not part of that model. Here (apparently at the suggestion of Edgeworth) he developed a design for an hydraulic model of the monetary system.

In order to understand this second model, we need to focus not just on how it was *designed* and how it can be *manipulated* to give insights into the workings of the system, but we also need to inquire into *the effects of different arrangements of the parts*.

12.5.1 Model design

There are several *general design features* to note. First, this is not the natural system of the verbal simile (lagoons and oceans), but a designed system set up as laboratory equipment. And, although some features are designed to allow intervention and control (as in a laboratory experiment), other features are purposely left open or adjustable – a point to which I shall return.

Secondly, this model does not come from a ready-made analogy – as the balance mechanism did – in which an entire pre-existing structure, with its form and characteristics, is carried over to the new field. Here the model is a specially designed analogical model, in which Fisher carefully chose a set of parts with analogical characteristics and fitted them together. The fact that this is not a straightforward analogical model creates certain difficulties for our analysis. With the mechanical balance model, it was easy to discuss the structural elements of the mechanical system, its negative and neutral features, and to see how they applied or not to economics. Here we have a hybrid model, part hydraulics and part economics; it is not always possible to separate so clearly the operative elements in the model as belonging to one domain or the other – more often they are closely intertwined.[23]

Fisher's model depends upon three main analogical elements: (i) that the economic laws of equal value are analogous to the hydraulic principles governing the behaviour of liquids to equalise their levels; (ii) that the marginalist principles in economics which govern decisions about production and consumption are analogous to the ways in which laboratory equipment can be set up to govern the inflow and outflow of liquids; and (iii) that the government has a level of economic control over the design, arrangement and workings of the monetary system equivalent to the power of the laboratory scientists and technicians over the design, arrangement and workings of laboratory equipment. I will pay particular attention to these analogical features in the following inventory of the bits in the model and what each is designed to do.

Fisher began his account of the money system with a complicated verbal description, but found himself able to deal with the factors only

---

[23] It does not seem to make sense to use terms such as simplification and idealization to describe analogue models. It is difficult to think of Fisher's reservoirs as idealizations or simplifications of anything. Rather, they seem more of a transformation of the quantity theory. A similar case of analogical modelling, involving the adjustment of physical devices into economic models, is discussed by Boumans (1993).

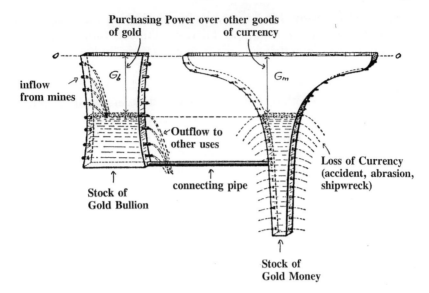

Figure 12.3 Fisher's connecting reservoirs model.
From Fisher (1911), figure 5, p. 105, with added labels.

one at time. In order to show the effects of the individual factors more clearly, and how they interrelate, Fisher introduces his model:

In any complete picture of the forces determining the purchasing power of money we need to keep prominently in view three groups of factors: (1) the production or 'inflow' of gold (i.e. from the mines); (2) the consumption or 'outflow' (into the arts and by destruction and loss); and (3) the 'stock' or reservoir of gold (whether coin or bullion) which receives the inflow and suffers the outflow. The relations among these three sets of magnitudes can be set forth by means of a mechanical illustration given in Figure 5. (p. 104):

I have added labels to Fisher's first version of this 'mechanical illustration' in figure 12.3 (his figure 5) to short-cut the explanations. But as we shall see, the model still requires a considerable amount more description and analysis to understand it, to see how it works and in what domains.[24]

---

[24] One of the striking things about this model of Fisher is the amount of labelling/interpretation and explanation you need (even aside from my analytical comments). It was a new model, used once before by Fisher (1894) but not one which became well known, as Fisher's equation of exchange became. With well-known models, it is easy to forget that they, like theories, need interpretation (or correspondence) statements for the entities and for the domain. In addition, you need a story (or stories) telling you how to make it work and what can be done with it (see Morgan 1999).

The model consists of two reservoirs for gold bullion and gold money (*Gb* and *Gm*) connected by a pipe. There are three sorts of flow; a flow between the two, an inflow into the bullion reservoir and outflows from both.

The stock (liquid level) is depicted to show purchasing power increasing with scarcity in both vessels. This is achieved by defining the purchasing power as the distance from the liquid surface to the top of the vessels (the '*00*' line). So, as the stock of gold rises, its value falls.[25] The effect of this feature is that whenever the liquids are at equal level, they must have equal purchasing power. This is an 'adjustable' design feature, in the sense that there is no measurement scale and there are no structural constraints given by either the economics or hydraulics on the height of the vessels. Fisher could have drawn the vessels taller, thus altering the purchasing power arbitrarily.[26]

The shape of the vessels is specially important for *Gm*, where the shape of the reservoir is designed to represent the quantity theory proportionality relation between changes in the quantity of money and the price level (the reciprocal of the purchasing power of money). That is: a doubling of the money stock should halve prices. This shape is supposed to incorporate the quantity theory so that whatever happens to the level of the money in the *Gm* reservoir, the quantity theory is at work. But, it is not clear that the shape does this at its top (for example, to show hyper-inflation one might expect that the vessel would spread out flat at the very top).

On the other hand, there is neither a theoretical claim nor an empirical observation embedded in the shape of *Gb* so that there is no precise relation between the value of *Gb* and its stock. Fisher does not give any clues here about the shape and size of this vessel, though in his earlier introduction of this model, he proposed that the shape and size of the other reservoirs 'mark the net influence of those causes not explicitly considered – changes in population, volume and character of business transactions, credit, etc.' (1894, p. 330). At this stage then, this design counts as another adjustable feature – there are no constraints, from

---

[25] For the non-economist who might dare to read this paper, it is one of the paradoxes of the quantity theory that (*ceteris paribus*) the more money society has, the less valuable each bit is (more money does not increase wealth); yet the more money the individual has (*ceteris paribus*), the wealthier that individual is.

[26] One could argue that the top is fixed by the minimum cost of extracting gold (the top inflow pipe) but this might not be fixed over time.

either side of the analogy, on the size and shape of *Gb* beyond those required for the pipe flow to function.

The inlet and outlet pipes between the two reservoirs allows for free flow of gold between bullion and coin usage thus ensuring purchasing power equality between bullion and money uses of gold. Fisher interprets this as due to the hydraulic principle: 'The fact that gold has the same value either as bullion or as coin, because of the interflow between them, is *interpreted* [my emphasis] in the diagram by connecting the bullion and coin reservoirs, in consequence of which both will (like water) have the same level' (*PPM*, p. 107). Here we see the importance of considering gold bullion and gold coin as liquid which flows freely. This characteristic enables the first use of fluid mechanics – the principle behind the equalising of levels between the two reservoirs – within the laboratory set up. This is analogous to the economic principle that a commodity cannot have two different values in freely connected markets – there will be a movement of goods to equalise the value. This free flow of gold also represents in part an empirical aspect of the economic side, for when the value of gold as bullion is higher than as coins, people do melt down coins into bullion for other usages.

But more than the hydraulic principles are involved here, for the claim that equalising the level of liquids equalises their economic values depends on the additional feature that Fisher defined purchasing power (value) to be the distance from the top of the liquid to the common '*00*' line on the diagram (see above) and these are, remember, drawn with equal heights *by design*. That is, the operative principle leading to equality is analogous, but the set up has to be designed in a particular way to make the bullion and coin equal at a certain purchasing power value.

The several left-hand side inlets in *Gb* represent the inflows of gold from mines with different costs of production. As mines produce more gold, the liquid rises in *Gb* and the value per unit of all the bullion in the flask falls. The more costly mining operations (those lower down) find it no longer profitable to produce gold and they stop production as the liquid bullion reaches their level, that is, as the value of the product becomes less than their marginal cost of production. This is the economists' marginal principle working for a series of mines with different cost structures to regulate the inflow within the model.

An aside here takes up this point as an example to illustrate the issue of design choices. In this case, there are clearly other possibilities which

Fisher chose not to adopt. Fisher depends on the economic marginalist principle to cut off these flows; the theoretical claim of marginal principles is interpreted in what seems to be a rather realistic way. But, since mines are often observed to continue operating when they make losses, Fisher's earlier version of the model (1894, p. 330), in which he showed mines continuing to produce gold inflow into the Gb stock for some time after they had begun to make losses, might seem to have been even more realistic. The still newish marginal principle was not employed in that earlier version of the model, rather, as Laidler points out, he incorporated the assumption of diminishing returns in extractive industries associated with the older classical school economics. This option might be contrasted with the more idealised perfect competition notion (being developed at that time) modelled, perhaps, as one inflow pipe representing many mines with the same costs. Finally, Fisher could have chosen a purely physical mechanism, consistent with his apparatus, e.g. a valve which cuts off the flow, to model the inflows of gold.

To return to the analysis: the right-hand outflow pipes in Gb embody a similar combination of theoretical claims realistically interpreted. The outflow through the pipes represent gold consumed in other uses (arts, jewelry etc). Once again, the flow is governed or regulated by the new marginal principle which compares the marginal utility of the gold consumed in its artistic and other uses, compared to its value as money for buying other goods: 'Just as production is regulated by marginal cost of what is produced, so is consumption regulated by marginal utility of what is consumed'. (PPM, p. 103). As the purchasing power of bullion falls (liquid rises), the marginal utility of more of the alternative consumptions of gold will be satisfied, shown by the fact that more outflows into the arts below the surface level come into play.

So, the purchasing power of gold bullion (measured vertically from the level of liquid in the flask) is affected by both the production of gold and other demands for it. Because of the interflow between the vessels, the purchasing power of money adjusts to these changes in level (and so purchasing power) in the bullion flask: 'We see then that the consumption of gold [in other uses] is stimulated by a fall in the value (purchasing power) of gold, while the production of gold is decreased. The purchasing power of money, being thus played upon by the opposing forces of production and consumption, is driven up or down as the case may be' (p. 104).

The tiny outflow pipes from Gm are to indicate accidental losses and abrasion, rather than true consumption usage.

12.5.2 The model at work

We now know about the model design and how the individual bits of the model fit together. But, as before, we need to know: How does the model work? Fisher, remember, wants to use the model to understand how the purchasing power of money is determined:

It need scarcely be said that our mechanical diagram is intended merely to give a picture of some of the chief variables involved in the problem under discussion. It does not itself constitute an argument, or add any new element; nor should one pretend that it includes explicitly *all* the factors which need to be considered. But it does enable us to grasp the chief factors involved in determining the purchasing power of money. It enables us to observe and trace the following important variations and their effects . . . . (p. 108)

The variations Fisher discussed were three-fold.

First, an increase in production (due to new mines or new technology) and secondly an increase in consumption (due to changes in fashion), both imagined as an increase in number or size of pipes. These cases can be treated together. Suppose, for example, there is an increased inflow due to the discovery of new mines. Some of this increased bullion stock flows into the currency stock, reducing both their purchasing powers (by the principle of hydraulics, the definition of purchasing power, and in Gm by the quantity theory enshrined in the shape of the reservoir). The increased bullion stock also effects an increase in the consumption of gold in arts (by the joint hydraulic and marginal principle enshrined in the pipes coming into use). At a certain point, production and consumption again become equal and the levels in both reservoirs stabilise.

As we can see, the model is designed so that the level in the Gb reservoir will stabilise after any change in size or flow of input or output pipes. Thus, in the above example, more inflow of gold into the Gb reservoir leads automatically to more outflow, so that the level will eventually settle down at some kind of equilibrium level.[27] Note that this level is not at any one particular or necessary point, since the model determines only which pipes are in use, not the size of flow through them. That is: the level of the liquid in Gb determines *which inflows and outflows* are in use

---

[27] Fisher suggested that 'The exact point of equilibrium may seldom or never be realized, but as in the case of a pendulum swinging back and forth *through* a position of equilibrium, there will always be a tendency to seek it' (p. 108). But unlike the case of the earlier balance model, it is not obvious that this oscillation characteristic is inherent in the model design, nor would it appear to follow necessarily from the way the model was supposed to work either physically or economically. On the other hand, such an oscillation is not ruled out by the model.

according to the marginal principle or according to the hydraulic principle. But the *amounts flowing through the pipes* are not regulated or controlled by either principle in the model. Those amounts are dependent on the discovery of new mines, the thickness of seams, the present fashions etc.

A third type of variation arises from a closure of the interconnecting pipe by a valve (not shown, but discussed). In this case, coin could flow into bullion, but not vice versa, thus partly cutting off the currency reservoir from the bullion reservoir. This enabled Fisher to incorporate certain realistic (i.e. historically accurate) institutional interventions into the model: long experience had shown that you cannot stop people melting down coin into bullion, but the monetary authorities can stop coining money out of bullion. The effect of this is to create a partial independence between the levels in the two flasks, so the purchasing power of bullion might be different from that of currency. Fisher used the recent example of India and their silver currency experience as an illustration for this arrangement. But equally, this closure of the flow could be a policy response to prevent a fall in the value of coins in the face of massive new gold discoveries which would increase the level, and so decrease the value, of bullion.

This third type of variation enabled Fisher to incorporate within the model the possibilities of government regulation of the economy. Everything previously has been dependent on regulation by 'natural' principles: namely by hydraulic principles or the law of equal values, by marginal principles or by the constraints imposed from the quantity theory.

Remember in all this that Fisher's aim is to understand what determines the purchasing power of money. In this model, the quantity theory has to hold, by design, as does the equal purchasing power for equal-level liquids. By design also, Fisher has focused explicitly on the way the value of gold money is related to the value of gold bullion. When we manipulate the model (on paper), we can see how the purchasing power is determined by production and consumption flows (which alter the stocks of liquid in both reservoirs) and by the government's regulatory power over the interconnecting pipe.

As already noted at the design stage, there were choices to be made about each bit of the model. These choices determine the work that the model can do. One way to examine the effects of these choices is to consider, briefly, the (dis)advantages of another choice of model. For example, in similar circumstances, Hadley (1896) used a demand and supply diagram to show the demand and supply of gold in three different uses, and used the same marginal principle to determine the value of gold bullion and thus the purchasing power of money. The alternative that Fisher chose enabled him to include not only the supply of bullion and demand from other uses, but also the entire existing stock (volume) of bullion. In his model, it is not only the inflows (supply of bullion) and outflows (demand), but also the volume of gold bullion and currency and the embedded quantity relation which determine the purchasing power of money. Fisher's choices enabled him to incorporate a wider set of economic principles and, as we shall see, a variety of institutional arrangements which could find no place in Hadley's model.

12.6 REARRANGING THE MODEL

In its simplest form, with just two vessels (as in figure 12.3), the arrangement of the model could be used to interpret beliefs about the workings of the gold standard of the period. This was widely thought, by nineteenth-century economists, to be a system which required no government intervention because it would automatically stabilise the value of coins equal to the value of gold bullion.[28] As we have seen, Fisher's model provided a self-regulating mechanism consistent with those beliefs, for with no (government) controls over the connecting inflow of bullion into the currency reservoir (i.e. free coinage of gold, as in the gold standard) the purchasing power of gold coinage in Gm will automatically adjust to be the same as that of gold bullion in Gb. The way the model worked – the principles of equal value and arrangements of the set up which enabled free flows – represented the 'natural' laws of economics as contemporary economists understood them.

The most innovative thing about Fisher's model is that it does not just cover the natural laws, but also the government imposed monetary system, or as we now say, the institutional arrangements. We have seen some signs of how these might be dealt with in the discussion about introducing a one-way valve into the interconnecting pipe. Fisher easily extended his model to explore how different institutional arrangements would affect the purchasing power of money. He did this by adding an extra element to the model and by arranging the parts in different ways. In the main extension to the model, an additional separate reservoir of

---

[28] Twentieth-century historians however have preferred a different interpretation of the gold standard mechanism, and believed it involved considerable institutional intervention (see, for example, Kenwood and Lougheed (1992) for a textbook account).

silver bullion is added, which is then connected with a pipe to the currency reservoir (as shown in panel (b) in figures 12.4 and 12.5). With this new arrangement of the model, the focus of Fisher's modelling activity changes: 'What we are about to represent is not the relations between mines, bullion, and arts, but the relations between bullion (two kinds) and coins. We may, therefore, disregard for the present all inlets and outlets except the connections between the bullion reservoirs and coin reservoir' (*PPM.* p. 118). These 'disregarded' elements are still in the model, but their effects are being held constant for the moment.

The basic working of the model remains as before: equality in purchasing powers is reached through the levelling of the liquids in all three reservoirs given the definition implied by the '*00*'-line measure. But now it is not just between the two uses of gold as bullion or currency, but between silver and gold bullion and the two metal currencies. When the silver level is above the gold level, the pressure above the mean level in the silver bullion vessel forces the silver into the currency vessel and gold coinage through the pipe into the gold bullion reservoir. The currency reservoir has a movable film so that silver coin cannot permeate into the gold bullion reservoir and vice versa (seen in figure 12.5 (panel (b)). The models' workings are motivated in Fisher's text not so much by an account of the physical working of the system, but by the economic principle that something cannot have two different values in two different but freely connected places: the law of equal value. That is, it is not because the silver will flow into the currency chamber to equalise levels, but because the cheaper currency will flow to where it is worth more. Three examples are illustrated using the model, and many more are discussed.

Let us look in more detail to see how the extended model was manipulated. We begin with Fisher's agenda:

In order to understand fully the influence of any monetary system on the purchasing power of money, we must first understand how the system works. It has been denied that bimetallism ever did work or can be made to work, because the cheaper metal will drive out the dearer. Our first task is to show, quite irrespective of its desirability, that bimetallism can and does 'work' under certain circumstances, but not under others. To make clear when it will work and when it will not work, we shall continue to employ the mechanical illustration. (pp. 115–16)

Bimetallism is the institutional arrangement whereby two metals circulate as legal tender together. The late nineteenth century had seen an international economic and political/commercial debate over the possibilities of bimetallism as an international monetary standard and as a domestic money standard. The debate raged heavily at both the theoretical and historical/empirical level.

Fisher used his extensions to the model to provide answers to these policy and theoretical questions about bimetallism. He also demonstrated the empirical law known as Gresham's Law that: 'bad money drives out good' or as Fisher proposes it 'Cheap money will drive out dear money.' This outcome, observed during the time of the Ancient Greeks (according to Aristophanes) and in medieval Europe (recognised by Copernicus and Oresme), was stated as a principle by Thomas Gresham in the mid-sixteenth century. The law proposes that any two currencies/coinages which are not of equal purchasing power in exchange cannot circulate together – one (the less valuable) will drive out the other. So, while Gresham's Law is a well-attested empirical law, the circumstances under which bimetallism would work, and how and why, were, at least in the 1890s, hotly contested. Anti-bimetallists maintained that government could not fix any ratio to which the market would conform – in other words the government could not overrule the market, so bimetallism could not work. The model built by Fisher implied that bimetallism could work and, within the framework of the model, he was able to outline the constraints on its working. Because Gresham's Law and bimetallism are closely connected outcomes, any model which could claim to answer questions about bimetallism had first to be able to show that Gresham's Law was consistent with the same framework.

Thus, we see Gresham's Law at work in the first case of his extended model arrangements, see figure 12.4 (Fisher's figure 6, p. 116). We can see that the reservoir of silver bullion is rather full and its purchasing power rather low (distance to '*00*' line), while the level in the gold bullion and currency reservoirs is rather low so the purchasing power of gold is high. When the silver vessel is connected up to the system, the 'cheaper' silver pushes the 'dearer' gold completely out of the currency reservoir into the bullion reservoir. (The silver currency cannot circulate with gold bullion, and in the model is held back from entering the gold bullion flask by the movable film.) In this first example, Gresham's law has operated to drive out gold currency (the good/dear money), and the gold bullion still maintains its higher value.

Bimetallism is demonstrated in the second case shown in figure 12.5 (Fisher's figure 7, p. 119). There is less initial difference in levels of the values of gold and silver and so introducing silver into the currency flask by opening the pipe has a different result. In this case the silver effectively pushes enough gold currency into the bullion reservoir to equalise the

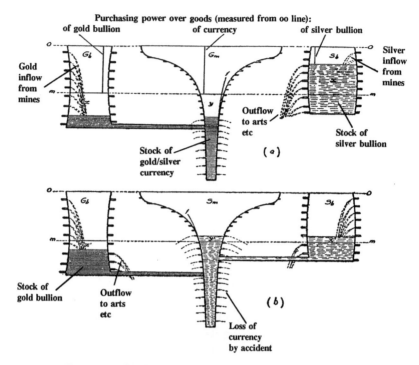

Figure 12.4 Gresham's Law demonstrated with Fisher's model.
From Fisher (1911), figure 6, p. 116, with added labels.

purchasing power value of the liquid in all three vessels while there still remains both silver and gold currencies in the currency reservoir. Gold and silver currencies have the same legal and practical value within the currency reservoir (equalisation to the same level), and the gold currency and bullion have the same value because they have an open pipe. The outcome in this case is that bimetallism works: that is 'governments open their mints to the free coinage of both metals at a fixed ratio [of exchange between the metals]' (p. 117) and both are legal tender and circulate together as currency.[29]

<hr />

[29] The ratio of exchange is realised as the number of grains of each metal contained in each drop of water in the vessels (so a 16:1 ratio means that 16 grains of silver or 1 grain of gold are present in each drop of water: so each unit of water represents a dollar of gold or a dollar of silver).

Fisher has made an unnoted assumption here which is not that important, but perhaps worth mentioning. Equalisation of the fluid levels in hydraulics depends upon the two liquids having equal density. Yet Fisher has defined his liquids as gold and silver grains in water, and quite probably these would have different densities. The effect of this is that physically the liquids' levels would not end up at the same level. In the economic case, where this would mean a different

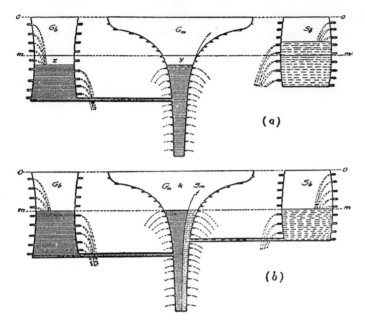

Figure 12.5 Bimetallism demonstrated with Fisher's model.
From Fisher (1911), figure 7, p. 119.

The principle behind these two outcomes is easily seen – they depend on the relative volume of silver bullion in the *Sb* reservoir above the mean level of liquid (in all three reservoirs) compared to the unfilled volume in the currency vessel below the mean level. Whether or not the outcome is Gresham's Law or bimetallism depends on these relative volumes.

Having established both bimetallism and Gresham's Law in the *ceteris paribus* case, as he believes, Fisher then brings back in the complicating factors previously held constant – the flows in and out of the bullion flasks. Of course, these elements are not independent in the model because connecting the silver bullion flask into the system will alter the level of bullion in both silver and gold bullion flasks. This in turn means that production and consumption balances for both silver and gold will be 'upset' (p. 120). Fisher discusses how these inflows and outflows will alter to create a new equilibrium level in all three reservoirs. The arguments resemble those of Fisher's earlier, simpler case of the gold bullion and currency reservoirs.

purchasing power for gold currency than silver currency, Gresham's Law would come into play – cheaper money would drive out dearer until their values were equal. In this case, the economic and physical laws do not work exactly in harmony.

The effect of inclusion of the other factors may be sufficiently great to result in Gresham's law operating after all (as in figure 12.4) for the increased production of silver bullion due to its rising value may be sufficient to push gold currency out of circulation entirely. Thus these elements, originally kept constant to concentrate on the institutional arrangements, are not minor approximating devices, but as important as the institutional arrangements in deciding the outcomes. Bringing them back in enables Fisher to show how the institutional and natural arrangements interact and why the government cannot completely control the situation.

In the context of the debates of the day, Fisher's main point is that the government is free to choose the ratio of exchange between the metals, so that as he notes from his model, 'bimetallism, impossible at one ratio, is always possible at another. There will always be two limiting ratios between which bimetallism is possible' (pp. 123–4).[30] He also discusses, in the context of this model, the various theoretical claims in the bimetallism debate: about whether bimetallism is likely to reduce fluctuations over time in price levels (purchasing power of money), whether it is feasible for an individual nation, etc.

With yet another arrangement of his model, Fisher illustrated the so-called 'limping standard' (in his figure 8 (p. 128), not shown here). Remember that the government can choose to stop up one or both of the pipes between bullion and currency. The limping standard results when bimetallism has once been established and then further currency minting in one of the currencies is stopped, as occurred with silver in the United States and France during part of the nineteenth century. The government can only stop the flow from the bullion reservoir to currency (it cannot prevent people melting down coins) so this creates a situation in which silver in currency is worth more than as bullion, and prevents new discoveries of metals reducing the purchasing power of money.[31]

Fisher uses these examples to attack both bimetallists (p. 136) and monometallists (p. 135) for their failure to understand the mechanisms involved in their standards, which are so clearly shown in his model! He also includes long sections detailing historical episodes in terms of the models. Indeed, these historical accounts form the empirical evidence

---

[30] This is essentially the same finding of Chen (1973), when he dealt with the case of bimetallism in a different form of model.

[31] The case of paper money is discussed in the same context (p. 130), as yet another form of two-currency circulation (paper and gold, for example). But it is difficult to see how paper money can be properly introduced into the model since there would be no 'bullion reservoir' for paper.

that Fisher relied upon. For the accounting balance model, there were statistical 'proofs' – the measurements of history. Such measurements were not possible for this model, so Fisher had to rely on the next best thing – descriptive qualitative historical accounts which could be related to the 'events' shown by manipulating his model under various different institutional arrangements.

These historical elements are important; they are not just a return to an old-style political economy of story telling. On the contrary, they are an essential element in convincing the reader of the credibility of the model. The events described with the model are economic stories, told in terms of the values of metals in the reservoirs, in terms of economic theory and in terms of economic history experience of a variety of countries. Every additional historical experience of currency systems that can be 'explained' within the model, the more that readers might be convinced to take the world in his model seriously and believe it to be a good representation of the world they live in.

12.7 WHAT DOES THE MODEL REPRESENT?

Fisher's connecting reservoirs model handles an impressive array of circumstances: here within the same model he has been able to represent (a) how the 'natural' laws of monetary economics work, (b) how institutional arrangements work, and (c) how they interact. The importance of building the model not by analogy with natural circumstances (i.e. lagoons and oceans) but by analogy with a laboratory set up becomes evident. In order to explore the institutional arrangements, the model has to incorporate some degree of controllability equivalent to that of the monetary authorities. Only with some areas of controllability and some areas determined by the general principles can both sorts of economic behaviour be modelled within one system. For example, the quantity theory is embedded as an assumption of this model, but the working out of institutional arrangements show quite clearly the effect the institutional arrangements have in changing the quantity of money and thus stimulating the quantity theory changes in the purchasing power of money. Using the model, and building onto the model, Fisher helps us to understand both the economic theories about money, and monetary institutions better.

The flexibility of the model depends not only on the mix of law-like and institutional controls that it can be arranged to represent, but also on the adjustable features I mentioned earlier in discussing the original design choices of the model. Indeed, its ability to 'explain' or 'produce' the outcomes of bimetallism and Gresham's Law from the various institutional arrangements within its own analogue world rely crucially on those adjustable features. (By contrast, the mechanical balance model was less flexible, constrained by its structure to pose and answer questions only within that structure.)

This point may be tricky to see, so let me explain. The hydraulics of fluid flows depends in Fisher's model first of all on the silver bullion reservoir being drawn, placed and filled in such a way that the height of the liquid is above the height of the other two vessels and the pipe connections between the vessels being left open. Otherwise there would be no flow of silver into the other vessels. Yet the precise position of the flask and level of liquid in the flask are matters of Fisher's choice – there are no structural constraints provided from the economics side which determine where the flask is drawn and what level of liquid it contains. Similarly, the extent to which silver pushes gold out of circulation depends on the volumes of liquids in the three vessels relative to the mean level. Yet, with the exception of the currency reservoirs, the shapes are arbitrary, and the sizes of all three in relation to each other are also open to choice. The silver bullion reservoir is a different shape to the gold one, but we do not know why. Yet it is on these freely adjustable shapes and sizes that volumes relative to the mean level depend.

Fisher depends on being able to arrange the flasks and adjust shapes, volumes and levels within the flasks, and connect them at will precisely in order to be able to use the model to demonstrate the various different outcomes of gold or silver monometallism or bimetallism. It is this direct sense that the range and flexibility of the model depends on its adjustable design features.

The question arises: what do these adjustable features represent? First, the bullion flasks represent containers in which the stocks of gold and silver bullion available in the world are stored for usage either as currency or in other consumption uses. Yet there is nothing in economic theory nor in the empirical economic world equivalent to the flasks themselves.

Secondly, we have the contents of the bullion flasks: the volume of gold and silver bullion. Despite the many heroic attempts by those interested in bimetallism to measure the annual world production and existing stocks and new flows of gold and silver[32] there was no way to fit these

---

[32] See especially, Laughlin's 1886 *History of Bimetallism*. Fisher also referred to the work of his student; see his *PPM*, p. 324.

measurements onto his model. Unlike those statistical series which could be mapped onto his earlier mechanical-balance model, the statistics of gold and silver could not be transposed onto the reservoirs because there was no calibration which turned weights into liquid volumes. Even if good measurements of weight had been available and weights could be transformed into volumes, there would still be a problem, for the flasks are unconstrained in shape and position so that the relative volumes compared to the mean level remains unknown. Fisher struggled to overcome these difficulties and impose some measurement ideas onto his model (we see the evidence in his various footnotes). But he failed to make them work. The volume of liquid in the flasks remains unconstrained by theory or by the world.

This adjustability and lack of statistical measurement makes it difficult to use this model for the concrete analysis of different monetary institutions in any specific instance. The policy maker can know the existing market exchange ratio between an amount of silver compared to gold, but knowing little about available volumes and nothing of the shapes of the reservoirs, the policy maker must be largely in the dark about what would happen with a new institutional arrangement. In contrast, the currency flask represents the quantity theory of money in its shape, and the quantity of money in the flask can, in principle, be measured and controlled by the monetary authorities.

The possibility of using the model as an instrument to formulate specific policy interventions or to change the monetary institutions is limited because in certain crucial respects the model does not accurately represent, either theory or the world. The model fails on two counts. The bullion flasks do not represent. In addition, as we can now see, the analogy between the laboratory scientist's ability to arrange the apparatus and experiment with their volumes, and the equivalent powers and knowledge of monetary authorities, does not hold. The monetary authorities may have the power equivalent to shutting or opening the connecting tubes, and so they can put into place a gold or silver monometallic standard. But they have no power to alter the shape and position of the bullion flasks, nor do they know their contents. If they want to institute a policy of bimetallism, all they can do is fix the official exchange ratio between the metals, and hope for success. Without quantitative information on the volumes – which they do not have – they will find it difficult to avoid the outcome of Gresham's Law. The model fails to represent accurately that level of controllability which exists in practice, so that although it can represent all the possible policy arrangements, it cannot serve as an instrument to carry out all such policies.

But that does not mean we have not learnt things from this model. We can best think of the model manipulations as offering a counterfactual analysis. They tells us what things the monetary authorities would need to know, and what they need to be able to control, in order to successfully intervene in the monetary arrangements of the economy given the theoretical beliefs of the time. It was precisely this absence of real knowledge on the available world stocks of gold and silver, and thus how various institutional arrangements would work out in practice, which created the debate about monetary systems in the late nineteenth century.

12.8 CONCLUSION

Learning from models happens at two places: in building them and in using them. Learning from building involves finding out what will fit together and will work to represent certain aspects of the theory or the world or both. Modelling requires making certain choices, and it is in making these that the learning process lies. Analogical devices are certainly a help, but do not provide a complete substitute for model-building decisions, as we have seen in these examples. In the first case, the ready-made mechanical analogy had negative features which had to be adapted or neutralised in the process of working with the model. In the second example, the analogical devices guided certain choices, but did not dictate them: a large number of modelling decisions had still to be made.

Learning from using models is dependent on the extent to which we can transfer the things we learn from manipulating our models to either our theory or to the real world. We see, from the analysis of these two models, how the power to represent is intimately connected with the means of learning, but not in any single or straightforward way. In the first case of the mechanical balance, manipulating this representation of the equation of exchange enabled Fisher to learn how to extend monetary theory. At the same time, the statistical version of the same model was an accurate enough representation of the empirical economic world to provide knowledge about certain features of that world. This is perhaps how we might expect the relation of representation and learning to work.

In the second case, however, the rather effective representation of the quantity theory in the connecting reservoirs model taught us nothing. This was because the quantity theory was made a passive assumption in the model, there was nothing about it which depended on other parts of the model or the manipulations of the model: there was no way we could learn anything more about it. By contrast, even though the apparatus pictured in the model failed to provide an adequate representation of the empirical world, by re-arranging the parts and manipulating the system, Fisher was able to represent qualitatively, with a considerable degree of success, a range of empirical phenomena. These manipulations taught him about the interaction of institutional and natural laws in the world of the model, though the extent to which he could transfer this knowledge to intervene in the real world remained limited.

REFERENCES

Allen, R. L. (1993). *Irving Fisher: A Biography*. Cambridge, MA: Blackwell.

Bordo, M. D. (1987). 'The Equation of Exchange', in *New Palgrave Dictionary of Economics*, ed. J. Eatwell, M. Milgate and P. Newman. London: Macmillan.

Boumans, M. (1993). Paul Ehrenfest and Jan Tinbergen, 'A Case of Limited Physics Transfer', in N. De Marchi (ed.), *Non-Natural Social Science: Reflecting on the Enterprise of More Heat than Light*. Durham: Duke University Press.

Chen, C.-N. (1973). 'Bimetallism: Theory and Controversy in Perspective'. *History of Political Economy*, 5, 89–112.

Dimand, R. (1998). 'The Quest for an Ideal Index' in *The Economic Mind in America: Essays in the History of American Economics*. Ed. M. Rutherford, Routledge, New York, 128–44.

Encyclopedia Britannica (1911). 'Weighing Machines' (by W. Airy), 11th Edition. Cambridge: Cambridge University Press.

Fisher, I. (1894). 'The Mechanics of Bimetallism', *Economic Journal*, 4, 527–37.

(1897). 'The Role of Capital in Economic Theory', *Economic Journal*, 7, 511–37.

(1911). *The Purchasing Power of Money*. New York: Macmillan.

(1912). *The Nature of Capital and Income*. New York: Macmillan.

(1915). *Why is the Dollar Shrinking?* New York: Macmillan.

(1920). *Stabilizing the Dollar*. New York: Macmillan.

(1922). *The Making of Index Numbers*. New York: Houghton Mifflin.

(1925, 1892). *Mathematical Investigations in the Theory of Value and Prices*. New Haven, CT: Yale University Press.

Fisher, I. and O. M. Miller (1944). *World Maps and Globes*. New York: Essential Books.

Fisher, I. N. (1956). *My Father Irving Fisher*. New York: Comet Press.

Hadley, A. T. (1896). *Economics*. New York: Putnam.

Hesse, M. (1966). *Models and Analogies in Science*. Notre Dame: University of Notre Dame Press, .

Humphrey, T. M. (1974). 'The Quantity Theory of Money,: Its Historical Evolution and Role in Policy Debates', *Federal Reserve Bank of Richmond Economic Review*, 60 (3), 2–19.

(1984). 'Algebraic Quantity Equations before Fisher and Pigou', *Federal Reserve Bank of Richmond Economic Review*, 70 (5), 13–22.

Jevons, W. S. (1875). *Money and the Mechanism of Exchange*. London: Henry S. King.

Kemmerer, E. W. (1909). *Money and Credit Instruments in their Relation to General Prices*. New York: Holt.

Kenwood, A. G. and A. K. Lougheed (1990). *The Growth of the International Economy 1820–*. London: Routledge.

Keynes, J. M. (1911). 'Review of *The Purchasing Power of Money*' *Economic Journal*, 21, 393–398.

Khinchin, A. I. (1949). *Mathematical Foundations of Statistical Mechanics*. New York: Dover.

Laidler, D. (1991). *The Golden Age of the Quantity Theory*. New York: Philip Allan.

Laughlin, J. L. (1886). *History of Bimetallism in the United States*. New York: Appleton.

(1903). *The Principles of Money*. New York: Charles Scribner's Sons.

Lubbock J. W. (1840). *On Currency*. London: Charles Knight.

Marget, A. (1938). *The Theory of Prices*. New York: Prentice-Hall.

Marshall, A. (1926). *Official Papers of Alfred Marshall*, ed J. M. Keynes. London: Macmillan.

Morgan, M. S. (1997). 'The Technology of Analogical Model Building: Irving Fisher's Monetary Worlds', *Philosophy of Science*, 64, S304–14.

(1999). 'Models, Stories and the Economic World'. Amsterdam: University of Amsterdam Research Memorandum.

Newcomb, S. (1886). *Principles of Political Economy*. New York: Harper.

Pigou, A. C. (1917). 'The Value of Money' *Quarterly Journal of Economics*, 32, 38–65.

Sumner W. G. (1874). *A History of American Currency*. New York: Henry Holt.

Wilson, E. B. (1931/2). 'Josiah Willard Gibbs' entry in *Dictionary of American Biography*. New York: Scribner.

Wimsatt, W. (1987). 'False Models as Means to Truer Theories', in *Neutral Models in Biology*, ed. M. H. Nilecki and A. Hoffman. Oxford: Oxford University Press.

Index

IDEAS IN CONTEXT

Edited by QUENTIN SKINNER (*General Editor*)
LORRAINE DASTON, DOROTHY ROSS,
and JAMES TULLY

Recent titles in the series include:

www.ingramcontent.com/pod-product-compliance
Ingram Content Group UK Ltd.
Pitfield, Milton Keynes, MK11 3LW, UK
UKHW040702180125
453697UK00010B/345